Application, Processing, and Testing of New Progressive Materials

Application, Processing, and Testing of New Progressive Materials

Jozef Dobránsky
Luboš Běhálek

MDPI

Basel • Beijing • Wuhan • Barcelona • Belgrade • Novi Sad • Cluj • Manchester

Editors

Jozef Dobránsky
Faculty of Manufacturing Technologies with a seat in Prešov
Technical University in Košice
Prešov
Slovakia

Luboš Běhálek
Faculty of Mechanical Engineering
Technical University of Liberec
Liberec
Czech Republic

Editorial Office
MDPI AG
Grosspeteranlage 5
4052 Basel, Switzerland

This is a reprint of articles from the Special Issue published online in the open access journal *Materials* (ISSN 1996-1944) (available at: www.mdpi.com/journal/materials/special_issues/progressive_materials).

For citation purposes, cite each article independently as indicated on the article page online and as indicated below:

Lastname, A.A.; Lastname, B.B. Article Title. *Journal Name* **Year**, *Volume Number*, Page Range.

ISBN 978-3-7258-2262-1 (Hbk)
ISBN 978-3-7258-2261-4 (PDF)
doi.org/10.3390/books978-3-7258-2261-4

Contents

Preface . . . . **vii**

María Elena Sánchez Vergara, Emiliano Toledo Dircio, Luis Alberto Cantera Cantera, Lourdes Bazán-Diaz and Roberto Salcedo
Using Recycled Tetrapak and Doped Titanyl/Vanadyl Phthalocyanine to Make Solid-State Devices
Reprinted from: *Materials* **2024**, *17*, 309, doi:10.3390/ma17020309 . . . . **1**

Gerhard Mitaľ, Ivan Gajdoš and Emil Spišák
Research of the Pre-Processing Strategy Influence on the Tribological Properties of PEI Processed by Fused Filament Fabrication Technology
Reprinted from: *Materials* **2023**, *16*, 4527, doi:10.3390/ma16134527 . . . . **19**

Martin Ovsik, Michal Stanek and Martin Bednarik
Evaluation of Cross-Linked Polyamide 6 Micro-Indentation Properties: TAIC Concentration and Electron Radiation Intensity
Reprinted from: *Materials* **2023**, *16*, 2391, doi:10.3390/ma16062391 . . . . **39**

Pavel Solfronk, Jiří Sobotka, Šárka Bukovská and Josef Bradáč
Experimental and Numerical Analysis of the Residual Stresses in Seamed Pipe in Dependence on Welding and Metal Forming
Reprinted from: *Materials* **2023**, *16*, 2256, doi:10.3390/ma16062256 . . . . **55**

Samuel Schlicht and Dietmar Drummer
Eutectic In Situ Modification of Polyamide 12 Processed through Laser-Based Powder Bed Fusion
Reprinted from: *Materials* **2023**, *16*, 2050, doi:10.3390/ma16052050 . . . . **81**

Jan Novák, Luboš Běhálek, Martin Borůvka and Petr Lenfeld
The Physical Properties and Crystallization Kinetics of Biocomposite Films Based on PLLA and Spent Coffee Grounds
Reprinted from: *Materials* **2022**, *15*, 8912, doi:10.3390/ma15248912 . . . . **98**

Jozef Dobránsky, Miroslav Gombár and Tomáš Stejskal
The Influence of the Use of Technological Waste and the Simulation of Material Lifetime on the Unnotched Impact Strength of Two Different Polymer Composites
Reprinted from: *Materials* **2022**, *15*, 8516, doi:10.3390/ma15238516 . . . . **117**

Věra Vodičková, Martin Švec, Pavel Hanus, Šárka Bukovská and Petra Pazourková Prokopčáková
Fe–Al–Si-Type Iron Aluminides: On the Strengthening by Refractory Metals Borides
Reprinted from: *Materials* **2022**, *15*, 7189, doi:10.3390/ma15207189 . . . . **135**

Jana Novotná, Martin Kormunda, Jakub Perner and Blanka Tomková
Comparison of the Influence of Two Types of Plasma Treatment of Short Carbon Fibers on Mechanical Properties of Epoxy Composites Filled with These Treated Fibers
Reprinted from: *Materials* **2022**, *15*, 6290, doi:10.3390/ma15186290 . . . . **153**

Oleksandr Grytsenko, Ludmila Dulebova, Emil Spišák and Bohdan Berezhnyy
New Materials Based on Polyvinylpyrrolidone-Containing Copolymers with Ferromagnetic Fillers
Reprinted from: *Materials* **2022**, *15*, 5183, doi:10.3390/ma15155183 . . . . **167**

Preface

Today, new advanced materials are being developed and applied in all industries, and new materials are still being discovered, capable of replacing other obsolete materials with their properties due to usually being more environmentally friendly and reusable. This has a major impact on the supply chain in all industries. Testing is important for the development and application of new progressive materials in industry, allowing for the specification and verification of their properties; although, at the same time, we need to acquire knowledge for their possible further processing. This Special Issue focuses on the development, testing, processing, and application of new advanced materials for all industries, creating a space for new knowledge, the driving force for the development of other new progressive materials.

Jozef Dobránsky and Luboš Běhálek
Editors

materials

MDPI

Article

Using Recycled Tetrapak and Doped Titanyl/Vanadyl Phthalocyanine to Make Solid-State Devices

María Elena Sánchez Vergara [1,*], Emiliano Toledo Dircio [1], Luis Alberto Cantera Cantera [1,2], Lourdes Bazán-Diaz [3] and Roberto Salcedo [3]

1 Facultad de Ingeniería, Universidad Anáhuac México, Avenida Universidad Anáhuac 46, Col. Lomas Anáhuac, Huixquilucan 52786, Mexico
2 Instituto Politécnico Nacional—ESIME, Unidad Profesional Adolfo López Mateos, Av. Luis Enrique Erro S/N, Gustavo A. Madero, Zacatenco, Ciudad de México 07738, Mexico
3 Instituto de Investigaciones en Materiales, Universidad Nacional Autónoma de México, Circuito Exterior s/n. C.U., Ciudad de México 04510, Mexico
* Correspondence: elena.sanchez@anahuac.mx

Abstract: In this work we studied the semiconductor behavior of titanyl phthalocyanine (TiOPc) and vanadyl phthalocyanine (VOPc), doped with anthraflavic acid and deposited on Tetrapak/graphite as flexible electrodes. The molecular structure was approached using the density functional theory and astonishingly, it was found that the structure and electronic behavior can change depending on the metal in the phthalocyanine. Experimentally, the Root Mean Square was found to be 124 and 151 nm for the VOPc-Anthraflavine and TiOPc-Anthraflavine films, respectively, and the maximum stress was 8.58 MPa for the film with VOPc. The TiOPc-Anthraflavine film presents the smallest fundamental gap of 1.81 eV and 1.98 eV for indirect and direct transitions, respectively. Finally, the solid-state devices were fabricated, and the electrical properties were examined. The tests showed that the current–voltage curves of the devices on Tetrapak and VOPc-Anthraflavine on a rigid substrate exhibit the same current saturation behavior at 10 mA, which is achieved for different voltage values. Since the current–voltage curves of the TiOPc-Anthraflavine on a rigid substrate presents a defined diode model behavior, it was approximated by nonlinear least squares, and it has been determined that the threshold voltage of the sample for the different lighting conditions is between 0.6 and 0.8 volts.

Keywords: Tetrapak electrode; semiconductor film; optical gap; solid-state device; electrical behavior

check for updates

Citation: Sánchez Vergara, M.E.; Toledo Dircio, E.; Cantera Cantera, L.A.; Bazán-Diaz, L.; Salcedo, R. Using Recycled Tetrapak and Doped Titanyl/Vanadyl Phthalocyanine to Make Solid-State Devices. *Materials* **2024**, *17*, 309. https://doi.org/10.3390/ma17020309

Academic Editors: Jozef Dobránsky and Luboš Běhálek

Received: 2 November 2023
Revised: 28 December 2023
Accepted: 5 January 2024
Published: 8 January 2024

1. Introduction

Organic electronic devices have seen rapid development with an emphasis on field-effect transistors [1], optoelectronic [2,3], photodetectors [4], memory devices [5] and gas chemosensors [6]. These devices are made up of organic semiconductors, which owe their behavior to specific properties in their structure, such as electronic delocalization, π–conjugated bonds, and in a significant number of these types of materials, the planarity of their molecular structure. Within these types of molecules, phthalocyanines stand out (Pcs), due to their multifunctional properties [7–10]. Some Pcs behave like semiconductors and exhibit photoelectric sensitivity; in other phthalocyanines, electrical conductivity can be induced by applying an electric field, and some show a photovoltaic effect. These characteristics have made them good candidates for different applications, such as the development of field-effect transistors [1], gas sensors [6], solar cells [11], and for the fabrication of photoswitches [12]. Besides, Pcs have unique physical properties, which allow them to act as dyes since they reflect light in the blue-green region of the visible spectrum. They can also act as photosensitizers, as they exhibit luminescence, fluorescence, and phosphorescence.

Pcs are organometallic compounds consisting of a ring containing four isoindole units linked by aza nitrogen atoms. The two central hydrogen atoms of the structure can be replaced by a wide variety of metal atoms, which can form covalent or ionic bonds with the macrocycle. The alkali and alkaline earth metals form ionic bonds with phthalocyanine. In contrast, other metals form covalent bonds with the two central nitrogen atoms and coordinate bonds with the remaining two nitrogen atoms. The most common metal phthalocyanines (MPcs) are those where the central metal ion has an oxidation state of 2+ (MPc). However, phthalocyanines with metal ions in oxidation states ranging from 3+ to 6+ are also known. According to the oxidation state of the metal, there are atoms or groups of atoms (X) ionically bonded, forming a $(PcM)X_n$ structure. When X is a halogen covalently bonded to the aromatic ring, the structure $(X_nPc)M$ is presented. Additionally, groups directly coordinated to the metal can exist, forming the structure $(PcM)L_n$, where L is a Lewis base. In another case, oxidation of the central metal in MPcs can result in non-planar molecules like the vanadyl (VOPc) and titanyl (TiOPc) phthalocyanines [13]. These oxometal phthalocyanines are prone to two-dimensional π–π stacking, which leads to the presence of convex and concave pairs of these molecules in the crystal structure [14,15]. This results in a larger π–π stacking overlap, which renders promising high-performance organic thin film transistors based on these species [16–19]. However, the utilization of VOPc and TiOPc is rarely reported; these two phthalocyanines have not been extensively studied in optoelectronic or photovoltaic applications.

Besides, an important approach in modern organic electronics is the manufacture of optoelectronic devices, with components that can be recycled [20]. Regarding the above, the VOPc and TiOPc phthalocyanines have not been studied, although the interest in degradable electronics has been growing, with the purpose of reducing electronic waste, usually involving toxic materials that take long to decompose [21–25]. From this point of view, the materials used as a substrate play a major role, since the substrate supports the device and has direct contact with the electrodes. Among the degradable polymers that have been used as substrates are polyethylene terephthalate (PET) [26], polyvinyl alcohol (PVA) [27], poly lactic-co-glycolic acid (PLGA) [28] and naturally derived materials, such as cellulose [29,30]. In this sense, Tetrapak waste could be a good candidate because it is abundant and it is made of 71–75% cellulose, 5% aluminum and 20–24% polymer [20,31]. Besides, according to Platnieks et al. [31], Tetrapak is one of the most extensively used food packaging materials and a large quantity of this packaging ends up in landfills as garbage. Due to the above, the development of applications for Tetrapak waste is in demand, and recycling is a good option for reducing the disposal problem [32].

Therefore, in this work, we have studied the fabrication of flexible devices using recycled Tetrapak-based electrodes. For this purpose, Tetrapak juice containers were cut into pieces and a layer of graphite was deposited on those pieces to make them conductive. After this, we deposited on the graphite, a silver electrode composed of a layer of Tetrapak/carbon/silver (TCS). Finally, simple TCS/MOPc-Anthraflavine/Ag devices were fabricated, using anthraflavic acid-doped TiOPc and VOPc as the active layer. The main novelty of this work lies in the fact that, as far as we know, it is the first time that solid-state devices of this type have been manufactured. Furthermore, Tetrapak is proposed as a non-conventional substrate and electrode; this is a recycled, low-cost, light and flexible material. All these features are very important requirements for organic optoelectronic devices. In other context, the effect of a dopant such as anthraflavic acid on the semiconductor properties of TiOPc and VOPc is also studied. VOPc and TiOPc phthalocyanines have not been widely studied, and finally another novelty in this work is their study as organic semiconductors films.

2. Materials and Methods

2.1. TiOPc and VOPc Doping

Oxytitanium phthalocyanine or titanyl phthalocyanine (TiOPc: $C_{32}H_{16}N_8OTi$), vanadium(IV)phthalocyanine oxide or vanadyl phthalocyanine (VOPc: $C_{32}H_{16}N_8OV$) and

2,6-dihydroxyanthraquinone or anthraflavic acid (anthraflavine; $C_{14}H_8O_4$) (see Figure 1) were obtained from commercial sources (Sigma-Aldrich, Saint Louis, MO, USA) and were used without any further purification. Afterwards, a MOPc-Anthraflavine-doped semiconductor was obtained by dissolution of 200 mg (0.35 mmol) of MOPc and 100 mg (0.42 mmol) of anthraflavine in 6 mL of methanol. Doping was carried out for 30 min at 140 °C in a heated reactor Monowave 50 (Anton Paar México, S.A. de C.V. Hidalgo, México) with a pressure sensor. The reactor was operated with a borosilicate glass vial and manually closed by a cover with an integrated pressure (0–20 bar) and temperature sensor. The system was cooled and brought to atmospheric pressure, the MOPc-Anthraflavine was filtered, washed with methanol, and dried in a vacuum. To verify the main functional groups of the doped semiconductor, a FTIR spectroscopy analysis was performed, using a spectrometer Nicolet iS5-FT (Thermo Fisher Scientific Inc., Waltham, MA, USA) in KBr pellets. SEM micrographs were carried out in a JEOL 7600F Schottky Field Emission Scanning Electron Microscope with secondary electrons at 10 kV. The Energy Dispersive X-ray Spectroscopy (EDS) Oxford Inca-x act was used.

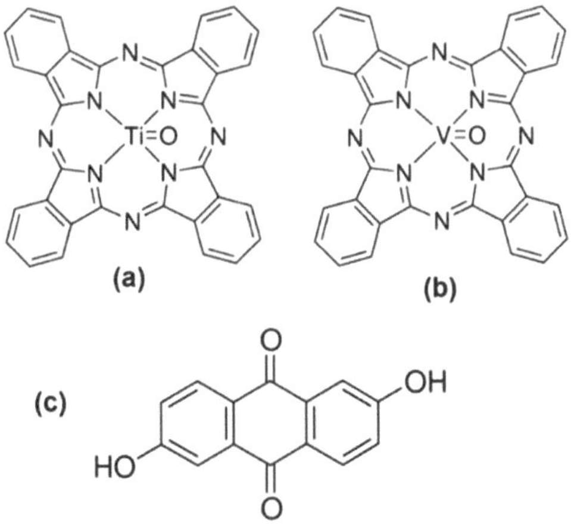

Figure 1. Structure of (**a**) titanyl phthalocyanine, (**b**) vanadyl phthalocyanine and (**c**) anthraflavine.

2.2. Fabrication of the Tetrapak/Carbon/Silver (TCS) Electrodes

The electrode was made with Tetrapak as follows: (i) the Tetrapak obtained from juice containers was washed with water, soap and ethanol several times, and after this, it was cut into pieces of 3 cm × 21 cm, (ii) the Tetrapak pieces were again washed with soap and methanol, and then dried for one hour at 100 °C. Each piece had a side completely covered by an aluminum + polyethylene film, and the other side was a surface of paperboard. (iii) The side covered by aluminum + polyethylene film was covered by a commercial graphitic carbon paint (the carbon paint was made by micro-graphite particles uniformly dispersed in isopropanol, and formulated with a small amount of a special polymer to give extra adhesive characteristics to the paint), with sheet resistance of 1.2 K ohms/sq. (iv) The perfectly painted pieces were left to air-dry at room temperature for one day and once the graphite was completely dried, conductive silver was deposited on the edge of the graphite layer and this would serve as current collectors. (v) Finally, the Tetrapak/carbon/silver (TCS) electrode was cut into pieces of 3 cm × 3 cm. Topographical and mechanical characteristics of the TCS electrodes were investigated with an atomic force microscope NaioAFM EasyScan 2, Naio (Nanosurf, Liestal, Switzerland), using an Ntegra with a measurement accuracy of 0.2 nm. The AFM was operated in static force contact mode, using the Stat0.2LauD silicon tip. The XRD analysis was performed with the θ–2θ technique using Bragg-Brentano Geometry with a diffractometer Bruker D8 Advance (Bruker Nano GmbH, Berlin, Germany) and working with CuK-α (λ = 0.15405 nm) radiation. The samples were measured at 0.6°/min, interval 5–70 and the grazing angle was 1.0°. The samples were also used to determine the optical absorption of the materials with a spectrophotometer Unicam UV300 (Thermo Fisher Scientific Inc., Waltham, MA, USA), in the wavelength range of 200–1100 nm.

2.3. Fabrication and Characterization of the Devices

The TCS electrode was coated by a film of MOPc-Anthraflavine semiconductor. For this, semiconductor films were deposited in a high vacuum deposition system Intercovamex (Intercovamex, S.A. de C.V., Cuernavaca, Morelos, Mexico), using tantalum crucibles, a vacuum pressure of 10^{-5} torr and a rate deposition of 1.4 and 6.3 Å/s for TiOPc-Anthraflavine and VOPc-Anthraflavine, respectively. Next, Ag was deposited on top of the semiconductor layer to act as a cathode, thus the configuration of the device was TCS/MOPc-Anthraflavine/Ag. Another reference device was also fabricated with a fluorine-doped tin-oxide (SnO_2/F)-coated glass slide electrode also called FTO; the configuration in this case was glass/FTO/MOPc-Anthraflavine/Ag. The electrical behavior of the devices was procured through current–voltage (I–V) measurements in an auto-ranging pico-ammeter Keithley 4200-SCS-PK1 (Tektronix Inc., Beaverton, OR, USA). The samples were illuminated with commercial LEDs, using a lighting controller circuit from Next Robotix (Comercializadora KMox, S.A. de C.V., Mexico City, Mexico). I–V curves were obtained under darkness, white, red, orange, yellow, green, blue, and ultraviolet lights at room temperature from −1 to 1 V.

3. Results and Discussion

3.1. Chemical and Structural Characterization

After carrying out the doping of the TiOPc and VOPc phthalocyanines with anthraflavic acid, IR spectroscopy was performed on KBr pellets, to corroborate the presence of the functional groups of both the phthalocyanines and the acid. After the fabrication of the MOPc-Anthraflavine thin films, their spectra were compared with those obtained in KBr pellets. This step was carried out to verify that the films did not suffer decomposition during their deposit. IR spectra of MOPc-Anthraflavine pellets and films are shown in Figure 2; these results indicate that the bands are in fact similar. The values of the representative vibrations of the MOPc-Anthraflavine are shown in the spectra: (i) the band responsible for the pyrrole in-plane stretch vibration in the Pc ring is observed in 1334 ± 3 cm^{-1}, (ii) the bands located in 1288 ± 3, 1163 ± 3 and 1122 cm^{-1} are the result of the interaction between C of the peripheral rings, with the hydrogen atoms [33], (iii) the band located in 753 cm^{-1} is the interaction in the plane of C-H deformation, (iv) the band observed in 1563 ± 2 and 1485 ± 2 cm^{-1} results from a C = C stretching mode [33–36], (v) the signals in 729 ± 2 cm^{-1} and 778 ± 3 cm^{-1} stand for the α-phase and the β-phase, respectively, [34,37–40] and (vi) the signals of the anthraflavine in 3383 ± 3 and 1074 ± 4 cm^{-1} are for C-O and the signal in 1665 ± 1 is for C=O. The IR spectroscopy showed two important results: on one hand, when reviewing the spectra of the pellets, the presence of both phthalocyanine and anthraflavine is observed, which is an indication of a doping of both TiOPc-Anthraflavine and VOPc-Anthraflavine. On the other hand, when comparing the spectra of the pellet with its thin film, the same signals are observed. This is an indication that there was no decomposition or degradation by the MOPc-Anthraflavine semiconductors during the deposition process. These results are important because in the physical vapor deposition technique used to form the films, the MOPc-Anthraflavine is heated above its sublimation temperature creating a vapor, which then condenses on the substrate. According to Cranston et al. [38], it is expected that no vapor-phase chemical reactions occur so that the films are produced strictly through physical means. Conversely, during the chemical doping of the phthalocyanines, the purification of the obtained doped semiconductors was carried out to avoid the presence of impurities. Furthermore, the high vacuum evaporation technique used for the deposition of the material generates high purity films.

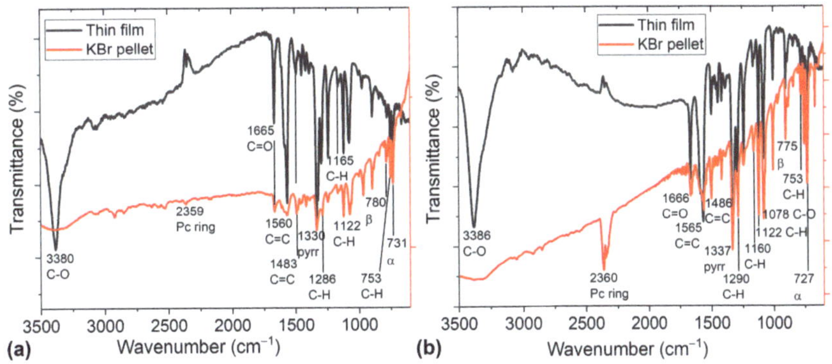

Figure 2. Infrared absorption spectra of (**a**) TiOPc-Anthraflavine and (**b**) VOPc-Anthraflavine in KBr pellets and thin film.

Figure 3 shows the SEM images as well as EDS spectra of both complexes. The morphology analysis of (a) TiOPc-Anthraflavine and (b) VOPc-Anthraflavine, reveals a pronounced high roughness in both cases. Notably, VOPc-Anthraflavine exhibits a larger grain size, representing the highest roughness and displaying a morphology reminiscent of stalactites, whereas TiOPc-Anthraflavine appears predominantly flat. Both films present a homogeneous morphology characteristic of the high vacuum evaporation technique. EDS analysis confirms the purity of the films with specific elements such as N, O, C and Ti or V for TiOPc-Anthraflavine and VOPc-Anthraflavine, respectively. The presence of silicon is due to the substrate.

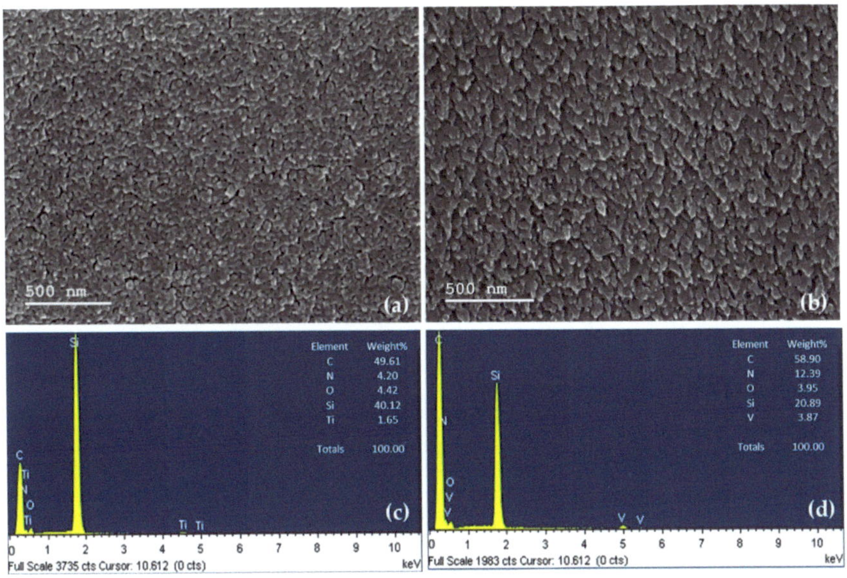

Figure 3. SEM images and EDS spectra of (**a**,**c**) TiOPc-Anthraflavine and (**b**,**d**) VOPc-Anthraflavine.

Solid-state devices require to be integrated by homogeneous films of low roughness, in order to be applied in molecular electronics. The morphology of the MOPc-Anthraflavine films was analyzed by means of AFM. Figure 4 shows the 3D images for the TiOPc-Anthraflavine and VOPc-Anthraflavine films on the n-type silicon substrate and on the TCS electrode. This study was carried out with the purpose of comparing the morphology of films on flexible TCS, and films deposited on a rigid substrate. In phthalocyanines derivatives, two features of the substrate can have an important impact on nucleation and film growth, and they are physical surface roughness and surface chemistry. When comparing the topography of the films on silicon (Figure 4a,b), it is observed that both films are made up of an irregular surface with crests and valleys, formed by preferential directions of growth. These zones are thinner for the TiOPc-Anthraflavine film, whereas the VOPc-Anthraflavine film comprises larger structures, resulting in greater roughness. Table 1 shows the Roughness Average (Ra) and Root Mean Square (RMS) values for each film and shows that the VOPc-Anthraflavine film on silicon is practically twice as rough as the TiOPc-Anthraflavine film. The above is reflected in the different transport of electrical charges exhibited by the two films. Meanwhile, when comparing the films deposited on the TCS electrode (Figure 4c,d), a similar morphology is observed in both films, consisting of coarse grains. In this case, the lower roughness was observed in the VOPc-Anthraflavine film on the TCS electrode, although it should be noted that the films on this electrode exhibit much higher roughness than its titanium counterpart when they are deposited on silicon. This is an expected result because changing the substrate modifies the nucleation and growth process of the thin films, while the silicon used as a substrate has its surface polished, the TCS is made up of cellulose fibers that, as seen in Figure 4e, generate coarse grains and high roughness (Ra = 103.6 nm and RMS = 130.2 nm). Besides, MPcs with a nonplanar structure such as VOPc and TiOPc, are characterized by a change in their film's morphology depending on the substrate material [41]. It was observed that MOPcs-Anthraflavine grew quicker on the silicon substrate which also reveals fewer defects on its surface, forming a greater number of nuclei, but with less growth of them. Furthermore, the high surface roughness in the TCS reduces the barrier for heterogeneous nucleation by decreasing the diffusion distance of molecules but with greater growth.

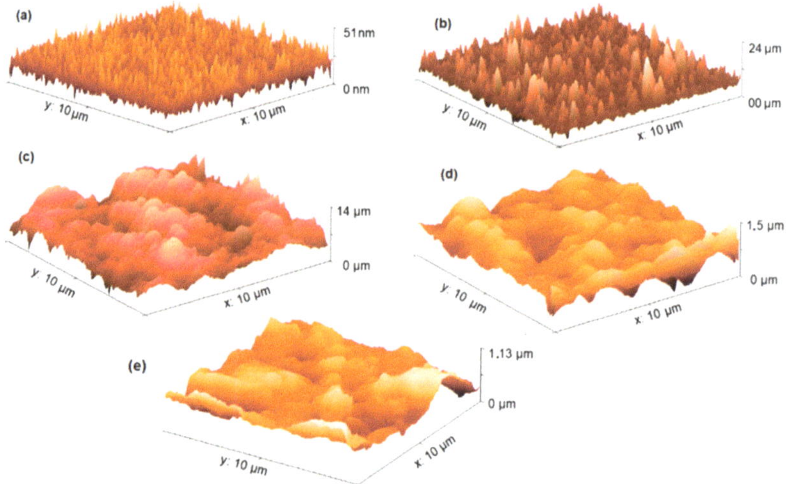

Figure 4. Three-dimensional AFM images of (**a**) TiOPc-Anthraflavine and (**b**) VOPc-Anthraflavine films on the n-silicon substrate and (**c**) TiOPc-Anthraflavine and (**d**) VOPc-Anthraflavine films on the (TCS) electrode. (**e**) TCS electrode.

In order to complement the previous information, X-ray diffraction was carried out on the films on the silicon substrate (Figure 5a) and the TCS electrode (Figure 5b). Physical surface characteristics and the intrinsic surface chemistry of the substrate can have an impact on film nucleation and growth [38]. Thus, a deep analyses of this technique was carried out, searching for answers on these topics. Important differences are observed in the diffractograms of Figure 5. Phthalocyanines have different polymorphs such as α and β [42,43], and these phases are categorized according to the crystalline stacking angle between the molecular plane and b-axis, yielding 65° and 45° for α and β, respectively [42,43]. The MOPc-Anthraflavine films on the silicon substrate exhibits a peak at 9.6°, related to the β-form in phthalocyanine [42] and the same peak is missing for the films on the TCS electrode, indicating that apparently, the structural organization of the β-form is disrupted [43]. However, the peak around 27° for all the substrates in Figure 5, corresponds to the β-form and indicates the growth of small crystallites lying parallel to the substrate surface plane and their stacking axes being inclined to it [44]. Besides, comparing the XRD results obtained for other VOPC-based films [45,46], the diffractograms in Figure 5b present very few peaks; however, in the 2θ region between 20° and 30°, the peaks reported in the literature remain. The differences are related to the used substrate, the deposition mechanism, the type of substituent and the metallic atom in the phthalocyanine. The small peaks in the diffractograms are associated with the low crystallinity of the films, although the crystallinity of the films on TCS electrodes is higher than that on silicon. Finally, and according to the results reported in the bibliography [20,47,48], the XRD patterns of the films deposited on TCS electrodes (see Figure 5b), depicts the band at 38.4°, which are associated with cellulose type I and type III [47,48]. While the peaks at 44.6 and 65.04° are attributed to the graphitic carbon randomly oriented on the Tetrapak surface [32]. It is necessary to highlight that the films on the TCS electrode have greater crystallinity compared to those deposited on silicon as was previously indicated. The TCS surface and its high roughness shows preferential sites for the heterogeneous nucleation and the growth of structures with crystallinity. This greater crystallinity is typically desired for many solid-state applications.

Table 1. Roughness, thickness and mechanical parameters of MOPc-Anthraflavine films.

Sample	Ra * (nm)	RMS * (nm)	Thickness * (μm)	σ (MPa)	HK
TiOPc-Anthraflavine on silicon-n	4.122	5.229	6.5	8.81	4.81
VOPc-Anthraflavine on silicon-n	10.21	16.27	6.1	8.42	5.01
TiOPc-Anthraflavine on TCS	121.1	151.1	6.2	7.91	11.31
VOPc-Anthraflavine on TCS	99.3	124.2	6.7	8.58	11.34

* Measurement accuracy: 0.2 nm.

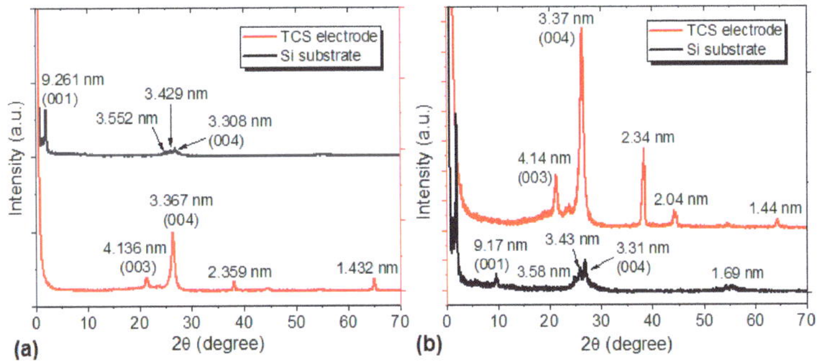

Figure 5. X-ray diffraction patterns on the silicon substrate and the TCS electrode of (**a**) TiOPc-Anthraflavine and (**b**) VOPc-Anthraflavine films.

3.2. Mechanical and Optical Characterization

Regarding the mechanical properties of the films obtained by AFM, Table 1 shows how the maximum stress (σ) does not demonstrate significant variations between the films. These results indicate that the substrate does not exhibit a significant influence on the maximum stress that MOPc-Anthraflavine films can withstand. However, the type of substrate determines Knoop hardness (HK); in this case, the highest values are obtained for the films deposited on the TCS electrode. It is important to note that the HK values obtained are lower than the hardness for polymeric materials, which have an HK of around 22. However, the use of TCS electrodes doubles the hardness compared to rigid silicon substrates, which can be favorable for their use in electronic devices.

In other context, and with the purpose of evaluating the optical behavior of the MOPc-Anthraflavine films, UV-vis spectroscopy was carried out. Figure 6a shows the absorbance spectra of the films. It is important to consider that these spectra are determined by the molecular orbitals within the aromatic π–electrons system and the overlapping orbitals on the central metal in the phthalocyanine [49]. In TiOPc and VOPc, strong absorptions are observed in the visible region between 570 and 720 nm (Q-band) and in the UV region between 370 and 460 nm (B-band) [14,50–54]. The Q-band representing the π–π^* transition [49] originates primarily from electronic transitions from the highest occupied molecular orbital (HOMO) to the lowest unoccupied molecular orbital (LUMO) in the phthalocyanine core [54]. The molecular organization in the films affects the wavelength of the Q-band [50,55], hence the maxima of the Q-band in the spectra of TiOPc-anthraflavine and VOPc-anthraflavine films is located at 578 and 580 nm, respectively. Although the values are very similar, the spectrum of the TiOPc-anthraflavine film presents the second component or the shoulder shifted towards blue, which according to Bonegardt et al. [54], is a sign that in this film, the formation of aggregates with a cofacial arrangement of Pc macrocycles is presented. This shoulder at 509 nm in the TiOPc-Anthraflavine film is associated with the presence of an aggregated species [14,56]. Alternatively, the B-band is assigned to the transition between π–π^* (b2u to eg) orbitals [49] and displays two peaks in the TiOPc-Anthraflavine spectrum and one broad peak as seen in the VOPc-Anthraflavine spectrum [38]. According to Cranston et al. [38], in the region of the B-band, the changes in the absorption spectra between the two films are thought to be originated by orbital overlap of the TiOPc and VOPc rings and the central metal. Additionally, the presence of the peaks in the TiOPc-Anthraflavine spectrum in the B-band region suggests the existence of π–d transitions as a result of the partially occupied d-orbitals of the titanium atom [38]. Finally, in the spectrum of the film with TiOPc, a band appears at 788 nm, referring to the charge transport between phthalocyanine and anthraflavic acid [52].

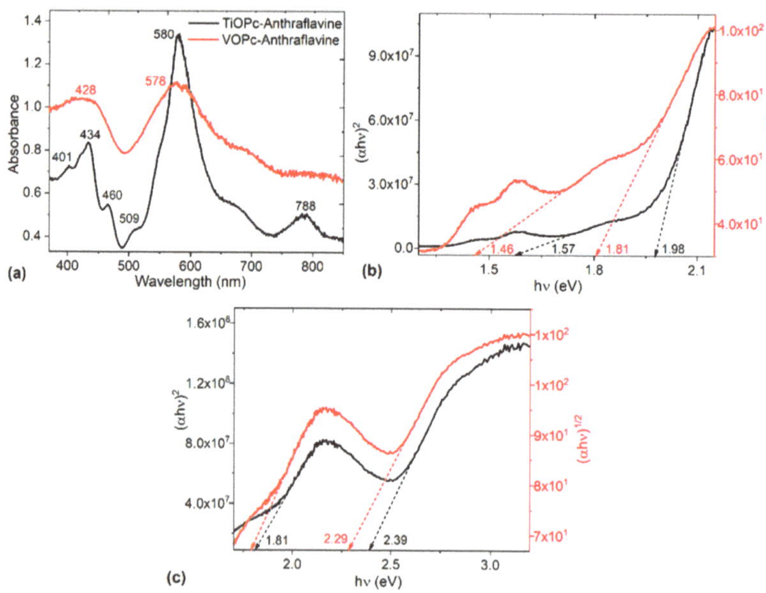

Figure 6. (**a**) Absorbance and Tauc plots for direct transitions $(\alpha h\nu)^2$ and for indirect transitions $(\alpha h\nu)^{1/2}$ of (**b**) TiOPc-Anthraflavine and (**c**) VOPc-Anthraflavine films.

The Tauc method was used to calculate the direct and indirect energy band gap values from the UV-Vis spectra. The energy band gap (Eg) was obtained by analyzing the variation of the absorption coefficient (α) with the photon energy ($h\nu$) in the high absorption region as follows [50,57]:

$$\alpha h\nu = A\left(h\nu - E_g\right)^m \quad (1)$$

where A is a constant and m is an exponent parameter that depends on the type of transition [50,57], m = 1/2 for indirect and m = 2 for direct forbidden electronic transitions. The relation between $(\alpha h\nu)^m$ vs. $h\nu$ was plotted in Figure 6b,c to extract the values of the E_g by extrapolating the straight line portion. These figures and Table 2 present the values of the fundamental and the onset gaps for indirect and direct electronic transitions. The first portion of the Tauc plots can be referred to the energy onset gap and is defined as the formation of a bound electron–hole pair in the semiconductor's materials, or Frenkel exciton [50,57]. The second portion could be attributed to the fundamental energy band gap between the HOMO and the LUMO [50,57]. The obtained values of these band gaps corroborated the calculated theoretical value. The TiOPc-Anthraflavine film is the one with the lowest band gap values, both for direct and indirect transitions. This may be due to the closed shell nature of Ti^{4+}, which reduces the spin–orbital interactions between the Ti^{4+} and the π–electrons system of the Pcs. It is important to consider that the fundamental band gap for indirect transitions obtained in this study is lower than that obtained by some authors [58] for pristine TiOPc films (2.85 eV), and for the film-based planar heterojunction: poly (3,4-ethylenedioxythiophene) polystyrene sulfonate/titanyl phthalocyanine (2.91 eV). These fundamental band gap values for direct transitions are also interesting because they are slightly larger than the one obtained by Bandas et al. [59] of 1.75 eV for hybrid electrodes based on a Ti/TiO2 mesoporous/reduced graphene oxide which can be considered inorganic nature electrodes. Electrodes of organic nature, such as those presented in this work, usually have a much higher band gap than inorganic ones. The above is evidence of the positive effect of anthraflavic acid as a dopant. Additionally, there is no appreciable difference between the fundamental gaps obtained for direct transitions with respect to indirect ones. This reinforces the fact that the doping acid is the one that exerts the greatest influence on charge transport, with respect to the structure of the films and the type of electronic transition (direct for crystalline structure and indirect for amorphous structure).

Table 2. Fundamental and onset gaps for indirect and direct electronic transitions.

Sample	Fundamental Gap for Direct Transitions (eV)	Fundamental Gap for Indirect Transitions (eV)	Onset Gap for Direct Transitions (eV)	Onset Gap for Indirect Transitions (eV)
TiOPc-Anthraflavine	1.98	1.81	1.57	1.46
VOPc-Anthraflavine	2.39	2.29	1.81	1.80

3.3. DFT Analysis

Quantum chemical calculations can help to interpret the last optical parameters obtained. The geometry optimization of MOPc-Anthraflavine was carried out at the level of density functional theory (DFT) using the Gaussian package [60]. The molecular structure of all the involved species was optimized using the B3PW91 method [61,62] included in the Gaussian 16 pack [60], and the calculations were carried out using the basis set 6–31g**. Stable minimum energy configuration of each molecule was approached by calculating their frequencies.

Anthraflavic acid used in this study as a dopant, is a very versatile species [63], which bears several terminal groups with different capabilities of interaction with other molecules, groups or atoms. These groups can establish hydrogen bonds, and both electrostatic and covalent interactions, therefore they can participate in chemical processes of different nature. In particular its terminal aromatic rings resemble that of the phenol molecule, and it is known that phenol has acid behavior because the hydrogen atom joint to the oxygen in the hydroxyl group has a relatively weak bond, due to the aromaticity. It is said the aromatic current in the center of the ring compels the electrons of the oxygen to participate in the delocalization motion, and the consequence is that the hydrogen atom can be separated as a positive ion yielding an acid environment. Therefore, anthraflavic acid shows a similar behavior as phenol; it can lose the hydrogen atoms of its ends as positive ions and manifest the acid behavior. In the present case, the interaction between this compound and the metallic phthalocyanines can take two different paths. First, the ionized species can interact with the central metallic atom of the phthalocyanine group via the ionized terminal oxygen or can follow the hydrogen bond route with the interaction of the oxygen joint to the metal atom with the terminal hydrogen atoms of the anthraflavic acid species forming a classical hydrogen bond. The curious situation is that both possibilities are found in the present case because the first one is followed by the vanadium complex whereas the second one is the case of the titanium complex and in this sense none of the structures has a net charge when the interaction form the definitive complexes. Both cases will be analyzed for their part.

The structure of VOPc-Anthraflavine is shown in Figure 7a and two phthalocyanine molecules interact with a central anthraflavic unit. The bond V-O is 1.95 Å at both ends whereas the Wiberg index for this same bond is 0.72 on average. This value confirms the presence of a moderate coordination covalent bond, the phthalocyanine moiety experiments a little deformation, when the planar square adopts a little ruffle-like shape with a deviation angle of 15.5°. The HOMO and LUMO of these species are shown in Figure 7b and as can be seen, the HOMO is located in the anthraflavic acid molecule and the LUMO is located in the phthalocyanine. From the above it is deduced that in thin films of this semiconductor, applied to electronic devices, the electrons would flow from the acid towards the VOPc.

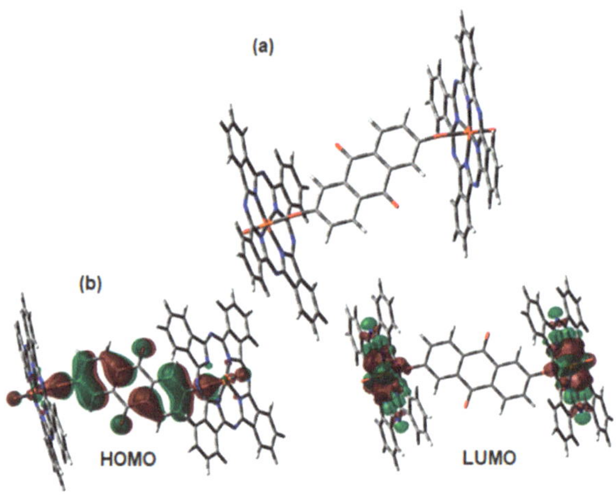

Figure 7. (**a**) Optimized structure of VOPc-Anthraflavine. (**b**) Frontier molecular orbitals of the VOPc-Anthraflavine.

After the successful analysis of the vanadium complex, the analogue of titanium was similarly studied at the beginning. However, the new structure is an ionized specie, and this feature always generates perturbation in the ordering of molecular orbitals and indeed the calculations were more difficult than those performed for vanadium, therefore another possibility was considered. Hydrogen bonds have been studied by our group in other semiconductor behavior cases [64]; therefore, the next study is based on the probability that the titanium complex can form one interaction of this kind with the anthraflavic acid molecule. The hydrogen bridge is localized between the terminal hydrogen atom of the acid and the apical oxygen atom of the phthalocyanine. The shape of the molecule is shown in Figure 8. The length between the hydrogen and oxygen atoms is 1.82 Å, the Wiberg index is 0.109 and the energy contribution found taking advantage of the Grimme module application is 7.1 kcal/mol; all this data demonstrate it is a strong dispersion interaction. Making a comparison between this configuration and that which is similar to the one found for the vanadium complex by means of the Boltzmann distribution analysis, the case of the hydrogen bond is favored by 5.5 kcal/mol, and the specie remains neutral. The HOMO and LUMO of TiOPc-Anthraflavine are shown in Figure 8b, and an overview of the results is shown in Table 3. In the same way as that obtained for VOPc-Anthraflavine, the HOMO is located in the acid and the LUMO in the structure of the phthalocyanine, so the flow of electrons in the doped semiconductor would be towards this molecule, through the bridge of hydrogen.

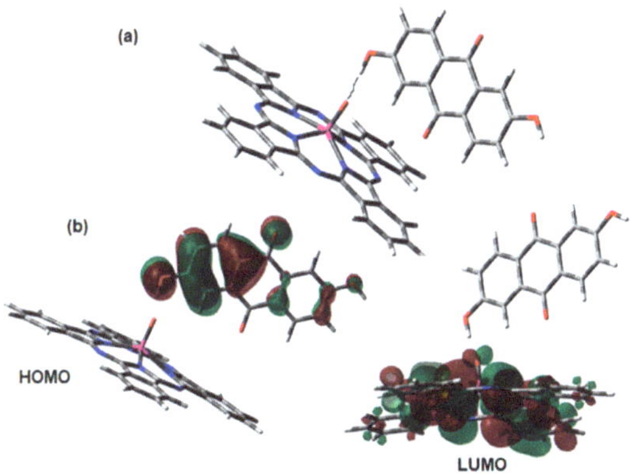

Figure 8. (**a**) Hydrogen bond between the complex of phthalocyanine of titanium and anthraflavic acid without ionization. The dotted line signals the hydrogen bond. (**b**) Frontier molecular orbitals of the TiOPc-Anthraflavine.

Table 3. HOMO, LUMO and theorical gap of the MOPc-Anthraflavine structures.

Doped Semiconductor	HOMO (eV)	LUMO (eV)	GAP (eV)
VOPc-Anthraflavine	−5.52	−4.22	1.3
TiOPc-Anthraflavine	−5.41	−3.21	2.2

Table 3 presents the gap values obtained through theoretical calculations, and when comparing them with the optical gaps experimentally obtained (see Table 2), it is observed that for the VOPc-Anthraflavine species, the theoretical value is lower. This is because DFT calculations consider isolated molecules without any type of interaction. Regarding the TiOPc-Anthraflavine species, the opposite behavior was observed: the theoretical gap is larger than that experimentally obtained. This is because the proton of anthraflavic acid is practically "loose", interacting with both the oxygen of anthraflavic acid and the titanium atom to establish a "banana" bond [65]. This bond has three atoms and a single electron pair, so oxygen and titanium are pseudo charged and there is ionic density in this vicinity. Finally, it is necessary to mention that, although there are differences between the values of the optical gaps and the theoretical gaps, both are in the same order of magnitude, and are also in the range of organic semiconductors.

3.4. Electrical Characterization

In order to evaluate the electrical behavior of the MOPc-Anthraflavine films, simple solid-state devices were manufactured, using TCS electrodes. The TCS/MOPc-Anthraflavine/Ag flexible devices were fabricated using anthraflavic acid-doped TiOPc and VOPc as the active layer. The electrical behavior obtained for each flexible device was compared with that obtained for rigid devices manufactured on the FTO-coated glass slide: FTO/MOPc-Anthraflavine/Ag. The results are presented in the current–voltage (I–V) graphs of Figure 9 under different lighting conditions to detect photoconductive properties of the samples. It is important to consider that Ag, which acts as cathode in all devices, has a work function (Φ) of 4.2 eV, and is the electrode that provides electrons to the semiconductor through its LUMO. Based on its Φ and the LUMO value reported in Table 3, the electron transit to the device is expected to be more efficient with VOPc-Anthraflavine. In graphs I–V this behavior is actually observed, and even the device on FTO (Figure 9d) is the one with the highest transported current. This may be due to the lower energy barrier between the silver cathode and the LUMO of the VOPc-Anthraflavine semiconductor, but also due to the lower energy gap between the HOMO and the LUMO of this same species.

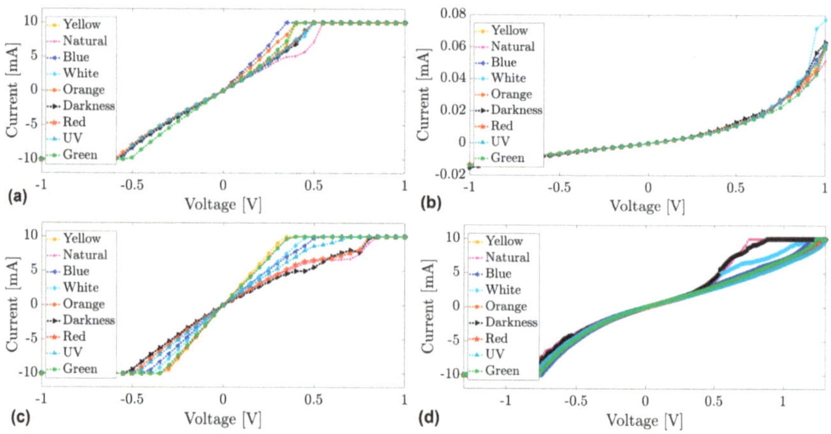

Figure 9. I–V curves from (**a**,**c**) TCS/MOPc-Anthraflavine/Ag and (**b**,**d**) FTO/MOPc-Anthraflavine/Ag devices under different lighting conditions.

From Figure 9, the I–V curves of the samples TiOPc-Anthraflavine on TCS (Figure 9a), VOPc-Anthraflavine on TCS (Figure 9c) and VOPc-Anthraflavine on FTO (Figure 9d) present a saturation current of 10 mA. The samples have the same conduction behavior, regardless of the type of material used and it can also be seen that the saturation current is reached at different voltages for each type of light. Table 4 shows the voltage at which the saturation current is reached for the different lighting conditions.

As already mentioned, samples TiOPc-Anthraflavine on TCS, VOPc-Anthraflavine on TCS and VOPc-Anthraflavine on FTO present a saturation current of 10 mA regardless of the material used; however, as can be seen in Table 4, each sample reaches saturation at different values of the applied voltage, which also changes for different lighting conditions. Specifically, the sample TiOPc-Anthraflavine on TCS saturates at lower voltage levels than samples VOPc-Anthraflavine on TCS and VOPc-Anthraflavine on n-silicon, which means it has lower resistance to current flow. In contrast, sample VOPc-Anthraflavine on n-silicon saturates at higher voltage levels compared to devices on TCS, which means it has higher resistance to current flow. In this sense, it is well known that the generated currents due to edge effects, may be important contributions to the total current in semiconductors, giving misleading results and causing variations of current [66–68]. However, it has also been reported that for semiconductors with lower resistivity, the edge effect plays a minor role on normal surfaces [68]. Since the samples have an average static resistance of less than 100 ohms, the currents generated by the edge effect will be neglected.

Table 4. Activation voltage of the saturation current under different lighting conditions.

Light	Sample		
	VOPc-Anthraflavine on FTO	TiOPc-Anthraflavine on TCS	VOPc-Anthraflavine on TCS
Natural	0.75 V	0.55 V	0.85 V
Darkness	0.88 V	0.5 V	0.85 V
White	1.17 V	0.5 V	0.5 V
UV	1.3 V	0.5 V	0.7 V
Blue	1.2 V	0.35 V	0.55 V
Green	1.23 V	0.4 V	0.4 V
Yellow	1.25 V	0.45 V	0.35 V
Orange	1.26 V	0.4 V	0.4 V
Red	1.29 V	0.5 V	0.8 V

At another context, the I–V curve of the sample TiOPc-Anthraflavine on FTO (Figure 9b) does not present a saturation behavior in the voltage range used. Since its behavior is closer to the characteristic curve of a diode, the following presents the approximation of the I–V curves of the sample TiOPc-Anthraflavine on FTO by the Shockley diode equation [69] defined by Equation (2).

$$i(v) = I_s\left(\exp\left(\frac{1}{nV_T}v\right) - 1\right) \tag{2}$$

From Equation (2), $i(v)$ is the current through the diode, I_s is the saturation current, v is the applied voltage, n is the emission coefficient of the device and V_T is the voltage due to temperature. Since the parameters I_s, n and V_T are unknown for the sample TiOPc-Anthraflavine on FTO, they were estimated using the I–V data from Figure 9b and applying the nonlinear least squares method [70] to Equation (3).

$$i(v) = I_s(\exp(av) - 1) \tag{3}$$

Since $a = \frac{1}{nV_T}$, the parameter estimation of Equation (3) consists of determining the parameters I_s and a that fit the equation to the I–V data. Table 5 shows the estimated parameters of the Shockley diode Equation (3) for the different lighting conditions in Figure 9b of the sample TiOPc-Anthraflavine on n-silicon.

Table 5. Parameters of the Shockley diode equation under different lighting conditions.

Light	TiOPc-Anthraflavine on FTO	
	a	I_s
Natural	1.6193	0.0114
Darkness	1.7601	0.0114
White	2.8563	0.0042
UV	1.8700	0.0098
Blue	1.7473	0.0111
Green	1.6628	0.0109
Yellow	1.8589	0.0096
Orange	1.8055	0.0101
Red	1.7227	0.0109
Average	1.9	0.0099

The results in Table 5 show that the estimate of the saturation current I_s in all cases approaches 0.01, while the parameter $a = \frac{1}{nV_T}$ changes for each type of lighting and on average is equal to 1.9. Besides, the threshold voltage of a diode denoted by V_T is the minimum voltage necessary to reach the conduction zone of the diode [69]. The threshold voltage can be determined graphically, by fitting a straight line in the conduction zone of the diode and the intersection of the straight line with the voltage axis which represents the threshold voltage V_T. A linear approximation at v^* of the Shockley diode Equation (3) is the tangent straight line at v^* to the curve (2) given by Equation (4):

$$y = i'(v^*)x + \left(i(v^*) - i'(v^*)v^*\right) \tag{4}$$

where $i'(v^*)$ denotes the first derivate of Equation (2) evaluated in v^*. Based on the Shockley diode Equation (2) and using the tangent line (3) at different points v^* as the straight line in the conduction zone of the diode, Figure 10 shows approximations of the threshold voltage V_T of the device.

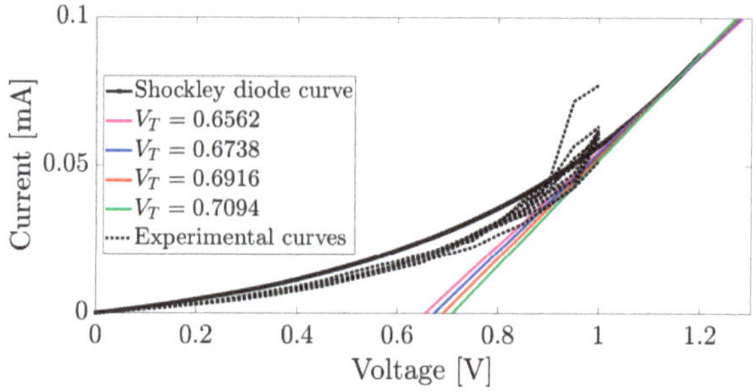

Figure 10. Approximation of the threshold voltage V_T of the device, $I_s = 0.01$, $a = 1.9$.

Based on the approximations of the Shockley diode equation and the tangent straight lines in Figure 10, the threshold voltage of sample TiOPc-Anthraflavine on FTO changes for different lighting conditions. In this sense, for the voltage range used and the different types of lighting, it can be established that the threshold voltage of the sample can vary between 0.6 and 0.8 volts. From the study on the electrical behavior of the MOPc-Anthraflavine semiconductor films, it is observed that it is the devices with the VOPc-Anthraflavine film that present the greatest transport of electrical charges. However, the substrate influences charge transport more significantly in the device with the TiOPc-Anthraflavine film, although the use of recycled Tetrapak electrodes in both films is definitely a viable option in device manufacturing.

4. Conclusions

We have demonstrated that Tetrapak waste can be used to produce a novel class of flexible electrodes. For this purpose, semiconductor films of MOPcs of vanadium and titanium, doped with anthraflavic acid, were deposited on the electrodes. The IR study confirmed the chemical structure of the semiconductor films and their non-degradation when deposited on the electrodes. The topography and structure of the films show a high RMS above 120 nm; however, they also present an acceptable crystallinity, dominated by the β-phase of phthalocyanines. Regarding its mechanical parameters, the maximum stress of the films was obtained between 7.91 and 8.58 MPa and the hardness was 11.3 HK. Regarding the optical behavior of the films, the fundamental gap energy obtained from UV-vis spectral data are found to be between 1.81 and 2.29 eV and 1.98 and 2.39 eV, for indirect and direct electronic transitions, respectively. These values are in the range of the organic semiconductors and the DFT results help to support this behavior although they reveal that the interaction between the anthraflavic acid and the metallic phthalocyanine complexes can follow two different chemical pathways in which the neutral nature of the resultant cluster always remains. The first one occurs in the case of the vanadium complex and is represented by a deprotonated form of the anthraflavic acid which makes a direct bond with the metal atom. In other context, the titanium complex arises from the formation of a strong hydrogen bond between the coordinated oxygen atom of the phthalocyanine moiety and the terminal hydrogen atom of the anthraflavic acid; in both cases, the theoretical band gap result relatively resembles the experimental one and the electric charge carriers in the device can be considered to be electrons. On the other hand, electrical experiments reveal that there are changes in the conductivity of the devices and that they do not have photoconductive properties, because the current flow does not increase under different types of light; however, when the samples TiOPc-Anthraflavine on TCS (Figure 9a), VOPc-Anthraflavine on TCS (Figure 9c) and VOPc-Anthraflavine on FTO (Figure 9d) are exposed to different types of light, the saturation current at 10 mA is reached at different voltage values. Regarding sample TiOPc-Anthraflavine on FTO, it does not have the same saturation behavior as the previous ones but shows acceptable conductive properties with a threshold voltage between 0.6 and 0.8 V and less variation in the behavior of the I–V curves when exposed to different types of light. In general, the results of this work demonstrate that low-cost, efficient and flexible electrodes can be fabricated using recycled Tetrapak, and also that these electrodes are excellent for optoelectronic devices performance. It is important to mention that edge effects were not considered in this work and an interesting future work could reveal the existence of currents due to the edge effect, and thus, the projected task would be the experimentation of the same samples using guard-rings which avoid edge effects.

Author Contributions: For research articles. Conceptualization, M.E.S.V.; methodology, M.E.S.V., L.A.C.C., R.S., L.B.-D. and E.T.D.; software, M.E.S.V., L.A.C.C. and R.S.; validation, M.E.S.V., L.A.C.C., R.S. and E.T.D.; formal analysis, M.E.S.V., E.T.D., L.B.-D. and R.S.; investigation, L.A.C.C. and E.T.D.; resources, M.E.S.V., and R.S.; data curation, M.E.S.V., L.A.C.C., R.S., L.B.-D. and E.T.D.; writing—original draft preparation, M.E.S.V., L.A.C.C., R.S., L.B.-D. and E.T.D.; writing—review and editing, M.E.S.V., L.A.C.C., R.S. and E.T.D.; visualization, M.E.S.V., L.A.C.C. and E.T.D.; supervision, M.E.S.V.; project administration, M.E.S.V.; funding acquisition, M.E.S.V. All authors have read and agreed to the published version of the manuscript.

Funding: María Elena Sánchez Vergara gratefully acknowledges financial support from COMECYT, FICDTEM-2023-65. The authors thank the LUME-UNAM staff for their technical support with SEM characterization.

Institutional Review Board Statement: Not applicable.

Informed Consent Statement: Not applicable.

Data Availability Statement: Data are contained within the article.

Acknowledgments: The authors are grateful for the technical support of Valeria López Valenzuela and thank Adriana Tejeda for his help in X-ray diffraction measurements. The authors are grateful to Citlalli Rios for the English correction.

Conflicts of Interest: The authors declare no conflicts of interest.

References

1. Hu, P.; He, X.; Jiang, H. Greater than 10 cm^2 V^{-1} s^{-1}: A breakthrough of organic semiconductors for field-effect transistors. *InfoMat* **2021**, *3*, 613–630. [CrossRef]
2. Zou, S.-J.; Shen, Y.; Xie, F.-M.; Chen, J.-D.; Li, Y.-Q.; Tang, J.-X. Recent advances in organic light-emitting diodes: Toward smart lighting and displays. *Mater. Chem. Front.* **2019**, *4*, 788–820. [CrossRef]
3. Quinn, J.T.E.; Zhu, J.; Li, X.; Wang, J.; Li, Y. Recent progress in the development of n-type organic semiconductors for organic field effect transistors. *J. Mater. Chem. C* **2017**, *5*, 8654–8681. [CrossRef]
4. Anuar, A.; Roslan, N.A.; Abdullah, S.M.; Bawazeer, T.M.; Alsoufi, M.S.; Alsenany, N.; Supangat, A. Improving the photo-current of poly(3-hexylthiophene): [6,6]-phenyl C71 butyric acid methyl-ester photodetector by incorporating the small molecules. *Thin Solid Films* **2020**, *703*, 137976. [CrossRef]
5. Heremans, P.; Gelinck, G.H.; Müller, R.; Baeg, K.-J.; Kim, D.-Y.; Noh, Y.-Y. Polymer and organic nonvolatile memory devices. *Chem. Mater.* **2010**, *23*, 341–358. [CrossRef]

6. Kumar, A.; Meunier-Prest, R.; Bouvet, M. Organic heterojunction devices based on phthalocyanines: A new approach to gas chemosensing. *Sensors* **2020**, *20*, 4700. [CrossRef]
7. Williams, G.; Sutty, S.; Klenkler, R.; Aziz, H. Renewed interest in metal phthalocyanine donors for small molecule organic solar cells. *Sol. Energy Mater. Sol. Cells* **2014**, *124*, 217–226. [CrossRef]
8. Kiran, M.R.; Ulla, H.; Satyanarayan, M.; Umesh, G. Optoelectronic properties of hybrid diodes based on vanadyl-phthalocyanine and zinc oxide. *Superlattices Microstruct.* **2017**, *112*, 654–664. [CrossRef]
9. Kiran, M.R.; Ulla, H.; Satyanarayan, M.; Umesh, G. Optoelectronic properties of hybrid diodes based on vanadyl-phthalocyanine and zinc oxide nanorods thin films. *Opt. Mater.* **2019**, *96*, 109348. [CrossRef]
10. Guo, X.; Liu, J.; Cao, L.; Liang, Q.; Lei, S. Nonvolatile memory device based on copper polyphthalocyanine thin films. *ACS Omega* **2019**, *4*, 10419–10423. [CrossRef]
11. Aziz, F.; Ahmad, Z.; Abdullah, S.M.; Sulaiman, K.; Sayyad, M. Photovoltaic effect in single-junction organic solar cell fabricated using vanadyl phthalocyanine soluble derivative. *Pigment. Resin Technol.* **2015**, *44*, 26–32. [CrossRef]
12. Ji, Z.; Shang, L.; Lu, C.; Wang, L.; Guo, J.; Wang, H.; Li, D.; Liu, M. Phototransistors and Photoswitches from an Ultraclosely pi-Stacked Organic Semiconductor. *IEEE Electron. Device Lett.* **2012**, *33*, 1619–1621. [CrossRef]
13. Kalashnyk, N.; Rochford, L.A.; Li, D.; Smogunov, A.; Dappe, Y.J.; Jones, T.S.; Guillemot, L. Unraveling giant Cu(110) surface restructuring induced by a non-planar phthalocyanine. *Nano Res.* **2017**, *11*, 2605–2611. [CrossRef]
14. Basova, T.V.; Kiselev, V.G.; Klyamer, D.D.; Hassan, A. Thin films of chlorosubstituted vanadyl phthalocyanine: Charge transport properties and optical spectroscopy study of structure. *J. Mater. Sci. Mater. Electron.* **2018**, *29*, 16791–16798. [CrossRef]
15. Ziolo, R.F.; Griffiths, C.H.; Troup, J.M. Crystal structure of vanadyl phthalocyanine, phase II. *J. Chem. Soc. Dalton Trans.* **1980**, 2300–2302. [CrossRef]
16. Bao, Z.; Lovinger, A.J.; Brown, J. New air-stable *n*-channel organic thin film transistors. *J. Am. Chem. Soc.* **1998**, *120*, 207–208. [CrossRef]
17. Li, L.; Tang, Q.; Li, H.; Hu, W. molecular orientation and interface compatibility for high performance organic thin film transistor based on vanadyl phthalocyanine. *J. Phys. Chem. B* **2008**, *112*, 10405–10410. [CrossRef] [PubMed]
18. Huang, L.; Liu, C.; Qiao, X.; Tian, H.; Geng, Y.; Yan, D. Tunable Field-Effect Mobility Utilizing Mixed Crystals of Organic Molecules. *Adv. Mater.* **2011**, *23*, 3455–3459. [CrossRef]
19. Wang, L.; Liu, G.; Zhu, F.; Pan, F.; Yan, D. Electrical instability in vanadyl-phthalocyanine thin-film transistors. *Appl. Phys. Lett.* **2008**, *93*, 173303. [CrossRef]
20. Figueroa-González, E.; Oliva, A.; Rodriguez-Gonzalez, V.; Gomez-Solis, C.; Garcia, C.; Oliva, J. Using recycled tetrapak and Ag/BaMoO4 nanoparticles to make efficient and flexible solid state supercapacitors. *J. Energy Storage* **2021**, *47*, 103544. [CrossRef]
21. Catania, F.; Oliveira, H.d.S.; Lugoda, P.; Cantarella, G.; Münzenrieder, N. Thin-film electronics on active substrates: Review of materials, technologies and applications. *J. Phys. D Appl. Phys.* **2022**, *55*, 323002. [CrossRef]
22. Irimia-Vladu, M.; Troshin, P.A.; Reisinger, M.; Shmygleva, L.; Kanbur, Y.; Schwabegger, G.; Bodea, M.; Schwödiauer, R.; Mumyatov, A.; Fergus, J.W.; et al. Edible Electronics: Biocompatible and Biodegradable Materials for Organic Field-Effect Transistors. *Adv. Funct. Mater.* **2010**, *20*, 4017. [CrossRef]
23. Ko, J.; Nguyen, L.T.H.; Surendran, A.; Tan, B.Y.; Ng, K.W.; Leong, W.L. Human hair keratin for biocompatible flexible and transient electronic devices. *ACS Appl. Mater. Interfaces* **2017**, *9*, 43004–43012. [CrossRef] [PubMed]
24. Irimia-Vladu, M.; Troshin, P.A.; Reisinger, M.; Schwabegger, G.; Ullah, M.; Schwoediauer, R.; Mumyatov, A.; Bodea, M.; Fergus, J.W.; Razumov, V.F.; et al. Environmentally sustainable organic field effect transistors. *Org. Electron.* **2010**, *11*, 1974–1990. [CrossRef]
25. Liu, X.; Shi, M.; Luo, Y.; Zhou, L.; Loh, Z.R.; Oon, Z.J.; Lian, X.; Wan, X.; Chong, F.B.L.; Tong, Y. Degradable and dissolvable thin-film materials for the applications of new-generation environmental-friendly electronic devices. *Appl. Sci.* **2020**, *10*, 1320. [CrossRef]
26. Mirjalili, A.; Dong, B.; Pena, P.; Ozkan, C.S.; Ozkan, M. Upcycling of polyethylene terephthalate plastic waste to microporous carbon structure for energy storage. *Energy Storage* **2020**, *2*, e201. [CrossRef]
27. Teng, L.; Ye, S.; Handschuh-Wang, S.; Zhou, X.; Gan, T.; Zhou, X. Liquid metal-based transient circuits for flexible and recyclable electronics. *Adv. Funct. Mater.* **2019**, *29*, 1808739. [CrossRef]
28. Lee, G.; Kang, S.; Won, S.M.; Gutruf, P.; Jeong, Y.R.; Koo, J.; Lee, S.; Rogers, J.A.; Ha, J.S. Fully biodegradable microsupercapacitor for power storage in transient electronics. *Adv. Energy Mater.* **2017**, *7*, 1700157. [CrossRef]
29. Hwang, S.; Kim, D.; Tao, H.; Kim, T.; Kim, S.; Yu, K.J.; Panilaitis, B.; Jeong, J.; Song, J.; Omenetto, F.G.; et al. Materials and fabrication processes for transient and bioresorbable high-performance electronics. *Adv. Funct. Mater.* **2013**, *23*, 4087–4093. [CrossRef]
30. Lei, T.; Guan, M.; Liu, J.; Lin, H.-C.; Pfattner, R.; Shaw, L.; McGuire, A.F.; Huang, T.-C.; Shao, L.; Cheng, K.-T.; et al. Biocompatible and totally disintegrable semiconducting polymer for ultrathin and ultralightweight transient electronics. *Proc. Natl. Acad. Sci. USA* **2017**, *114*, 5107–5112. [CrossRef]
31. Platnieks, O.; Barkane, A.; Ijudina, N.; Gaidukova, G.; Thakur, V.K.; Gaidukovs, S. Sustainable tetra pak recycled cellulose/Poly(Butylene succinate) based woody-like composites for a circular economy. *J. Clean. Prod.* **2020**, *270*, 122321. [CrossRef]
32. Adinaveen, T.; Vijaya, J.J.; Sivakumar, R.; Kennedy, L.J. Structural and electrochemical investigation of waste newspa-per-based electrodes for supercapacitor applications. *Mater. Sci.-Pol.* **2016**, *34*, 302–314. [CrossRef]

33. Al-Ghamdi, S.A.; Hamdalla, T.A.; Darwish, A.A.A.; Alzahrani, A.O.M.; El-Zaidia, E.F.M.; Alamrani, N.A.; Elblbesy, M.A.; Yahia, I.S. Preparation, Raman spectroscopy, surface morphology and optical properties of TiPcCl2 nanostructured films: Thickness effect. *Opt. Quantum Electron.* **2021**, *53*, 514. [CrossRef]
34. Song, C.; Li, Y.; Gao, C.; Zhang, H.; Chuai, Y.; Song, D. An OTFT based on titanium phthalocyanine dichloride: A new p-type organic semiconductor. *Mater. Lett.* **2020**, *270*, 127666. [CrossRef]
35. Zhang, X.-F.; Wang, Y.; Niu, L. Titanyl phthalocyanine and its soluble derivatives: Highly efficient photosensitizers for singlet oxygen production. *J. Photochem. Photobiol. A Chem.* **2010**, *209*, 232–237. [CrossRef]
36. Del Caño, T.; Parra, V.; Rodríguez-Méndez, M.; Aroca, R.; De Saja, J. Characterization of evaporated trivalent and tetravalent phthalocyanines thin films: Different degree of organization. *Appl. Surf. Sci.* **2005**, *246*, 327–333. [CrossRef]
37. Ouyang, J.; Chu, C.; Chen, F.; Xu, Q.; Yang, Y. High-conductivity poly(3,4-ethylenedioxythiophene): Poly(styrene sulfonate) film and its application in polymer optoelectronic devices. *Adv. Funct. Mater.* **2005**, *15*, 203–208. [CrossRef]
38. Cranston, R.R.; Lessard, B.H. Metal phthalocyanines: Thin-film formation, microstructure, and physical properties. *RSC Adv.* **2021**, *11*, 21716–21737. [CrossRef]
39. Hassan, A.K.; Gould, R.D. Structural studies of thermally evaporated thin films of copper phthalocyanine. *Phys. Status Solidi (A)* **1992**, *132*, 91–101. [CrossRef]
40. Szybowicz, M.; Bała, W.; Fabisiak, K.; Paprocki, K.; Drozdowski, M. The molecular structure ordering and orientation of the metallophthalocyanine CoPc, ZnPc, CuPc, and MgPc thin layers deposited on silicon substrate, as studied by micro-Raman spectroscopy. *J. Mater. Sci.* **2011**, *46*, 6589–6595. [CrossRef]
41. Klyamer, D.D.; Basova, T.V. Effect of the structural features of metal phthalocyanine films on their electrophysical properties. *J. Struct. Chem.* **2022**, *63*, 997–1018. [CrossRef]
42. Touka, N.; Benelmadjat, H.; Boudine, B.; Halimi, O.; Sebais, M. Copper phthalocyanine nanocrystals embedded into polymer host: Preparation and structural characterization. *J. Assoc. Arab. Univ. Basic Appl. Sci.* **2013**, *13*, 52–56. [CrossRef]
43. Madhuri, K.P.; John, N.S.; Angappane, S.; Santra, P.K.; Bertram, F. Influence of iodine doping on the structure, morphology, and physical properties of manganese phthalocyanine thin films. *J. Phys. Chem. C* **2018**, *122*, 28075–28084. [CrossRef]
44. Islam, Z.U.; Tahir, M.; Syed, W.A.; Aziz, F.; Wahab, F.; Said, S.M.; Sarker, M.R.; Ali, S.H.M.; Sabri, M.F.M. Fabrication and photovoltaic properties of organic solar cell based on zinc phthalocyanine. *Energies* **2020**, *13*, 962. [CrossRef]
45. Klyamer, D.; Sukhikh, A.; Nikolaeva, N.; Morozova, N.; Basova, T. Vanadyl Phthalocyanine Films and Their Hybrid Structures with Pd Nanoparticles: Structure and Sensing Properties. *Sensors* **2020**, *20*, 1893. [CrossRef]
46. Aziz, F.; Sulaiman, K.; Al-Rawi, W.K.; Ahmad, Z.; Sayyad, M.; Karimov, K.S.; Wei, L.; Tahir, M. Enhancement of electrical properties of vanadyl phthalocyanine derivative by PCBM. *Pigment. Resin Technol.* **2015**, *44*, 148–156. [CrossRef]
47. Wee, G.; Mak, W.F.; Phonthammachai, N.; Kiebele, A.; Reddy, M.V.; Chowdari, B.V.R.; Gruner, G.; Srinivasan, M.; Mhaisalkar, S.G. Particle size effect of silver nanoparticles decorated single walled carbon nanotube electrode for supercapacitors. *J. Electrochem. Soc.* **2010**, *157*, A179–A184. [CrossRef]
48. Gong, J.; Li, J.; Xu, J.; Xiang, Z.; Mo, L. Research on cellulose nanocrystals produced from cellulose sources with various polymorphs. *RSC Adv.* **2017**, *7*, 33486–33493. [CrossRef]
49. Zou, T.; Wang, X.; Ju, H.; Zhao, L.; Guo, T.; Wu, W.; Wang, H. Controllable molecular packing motif and overlap type in organic nanomaterials for advanced optical properties. *Crystals* **2018**, *8*, 22. [CrossRef]
50. Özçeşmeci, I.; Sorar, I. Synthesis and some optical results of long chain substituted phthalocyanines. *Turk. J. Chem.* **2018**, *42*, 21–28. [CrossRef]
51. Sukhikh, A.; Klyamer, D.; Bonegardt, D.; Basova, T. Octafluoro-Substituted Phthalocyanines of Zinc, Cobalt, and Vanadyl: Single Crystal Structure, Spectral Study and Oriented Thin Films. *Int. J. Mol. Sci.* **2023**, *24*, 2034. [CrossRef] [PubMed]
52. Konarev, D.V.; Nakano, Y.; Khasanov, S.S.; Kuzmin, A.V.; Ishikawa, M.; Otsuka, A.; Yamochi, H.; Saito, G.; Lyubovskaya, R.N. Magnetic and Optical Properties of Layered $(Me_4P^+)[M^{IV}O(Pc^{\bullet 3-})]^{\bullet -}(TPC)_{0.5}{\cdot}C_6H_4Cl_2$ Salts (M = Ti and V) Composed of π-Stacking Dimers of Titanyl and Vanadyl Phthalocyanine Radical Anions. *Cryst. Growth Des.* **2017**, *17*, 753–762. [CrossRef]
53. Ridhi, R.; Neeru; Gautam, S.; Saini, G.; Tripathi, S.; Rawat, J.; Jha, P. Study of the effect of orbital on interaction behaviour of SWCNT- metal phthalocyanines composites with ammonia gas. *Sens. Actuators B Chem.* **2021**, *337*, 129767. [CrossRef]
54. Bonegardt, D.V.; Klyamer, D.D.; Atilla, D.; Gürek, A.G.; Basova, T.V. Thin films of poly(oxyethylene)-substituted phthalocyaninato zinc(II) and oxotitanium(IV) complexes: Synthesis, structure and sensor response to ammonia. *J. Mater. Sci. Mater. Electron.* **2021**, *32*, 5955–5964. [CrossRef]
55. Kadish, K.M.; Smith, K.M.; Guilard, R. *The Porphyrin Handbook*; Elsevier BV: Amsterdam, The Netherlands, 2003; pp. 201–232.
56. Atilla, D.; Gürek, A.G.; Basova, T.V.; Kiselev, V.G.; Hassan, A.; Sheludyakova, L.A.; Ahsen, V. The synthesis and characterization of novel mesomorphic octa- and tetra-alkylthio-substituted lead phthalocyanines and their films. *Dye. Pigment.* **2011**, *88*, 280–289. [CrossRef]
57. El-Saady, A.A.; Roushdy, N.; Farag, A.A.M.; El-Nahass, M.M.; Basset, D.M.A. Exploring the molecular spectroscopic and electronic characterization of nanocrystalline Metal-free phthalocyanine: A DFT investigation. *Opt. Quantum Electron.* **2023**, *55*, 662. [CrossRef]
58. Vergara, M.E.S.; Heredia, L.F.V.; Hamui, L. Influence of the Coordinated Ligand on the Optical and Electrical Properties in Titanium Phthalocyanine-Based Active Films for Photovoltaics. *Materials* **2023**, *16*, 551. [CrossRef]

59. Bandas, C.; Popescu, M.I.; Orha, C.; Nicolaescu, M.; Pop, A.; Lazau, C. Development of Hybrid Electrodes Based on a Ti/TiO_2 Mesoporous/Reduced Graphene Oxide Structure for Enhanced Electrochemical Applications. *Coatings* **2023**, *13*, 1359. [CrossRef]
60. Frisch, M.J.; Trucks, G.W.; Schlegel, H.B.; Scuseria, G.E.; Robb, M.A.; Cheeseman, J.R.; Scalmani, G.; Barone, V.; Petersson, G.A.; Nakatsuji, H.; et al. *Gaussian 16, Revision A.03*; Gaussian Inc.: Wallingford, CT, USA, 2017. Available online: https://gaussian.com/ (accessed on 7 September 2023).
61. Becke, A.D. Density-functional exchange-energy approximation with correct asymptotic behavior. *Phys. Rev. A* **1988**, *38*, 3098–3100. [CrossRef]
62. Perdew, J.P.; Wang, Y. Accurate and simple analytic representation of the electron-gas correlation energy. *Phys. Rev. B* **1992**, *45*, 13244–13249. [CrossRef]
63. Basiuk, E.V.; Basiuk, V.A.; Toscano, R.A.; Gómez-Lara, J. A 1:1:2 complex between 1,4,8,11-tetraazacyclotetradecane (cyclam), anthraflavic acid, and water. *J. Chem. Crystallogr.* **2000**, *30*, 339–343. [CrossRef]
64. Vergara, M.E.S.; Hamui, L.; Hernandez, M.F.B.; Rios, C.; Salcedo, R. Combined experimental and theoretical study of conjugated ferrocene semiconductors and the effect of doping on their opto-electrical and structural properties. *J. Mol. Struct.* **2022**, *1262*, 132998. [CrossRef]
65. Palke, W.E. Double bonds are bent equivalent hybrid (banana) bonds. *J. Am. Chem. Soc.* **1986**, *108*, 6543–6544. [CrossRef]
66. Ruzin, A. Edge effect in ohmic contacts on high-resistivity semiconductors. *Nucl. Instruments Methods Phys. Res. Sect. A Accel. Spectrom. Detect. Assoc. Equip.* **2016**, *806*, 356–359. [CrossRef]
67. Tove, P.A. Methods of avoiding edge effects on semiconductor diodes. *J. Phys. D Appl. Phys.* **1982**, *15*, 517–536. [CrossRef]
68. Tove, P.; Hyder, S.; Susila, G. Diode characteristics and edge effects of metal-semiconductor diodes. *Solid-State Electron.* **1973**, *16*, 513–521. [CrossRef]
69. Dennis, J.E., Jr.; Schnabel, R.B. *Numerical Methods for Unconstrained Optimization and Nonlinear Equations*; Society for Industrial and Applied Mathematics: Philadelphia, PA, USA, 1996.
70. Boylestad, R.L.; Nashelsky, L. *Electronic Devices and Circuit Theory*; Pearson Education: London, UK, 2018.

materials

MDPI

Article

Research of the Pre-Processing Strategy Influence on the Tribological Properties of PEI Processed by Fused Filament Fabrication Technology

Gerhard Mitaľ, Ivan Gajdoš * and Emil Spišák

Department of Technology, Materials and Computer Supported Production, Faculty of Mechanical Engineering, Technical University of Košice, Mäsiarska 74, 04001 Košice, Slovakia; gerhard.mital@tuke.sk (G.M.); emil.spisak@tuke.sk (E.S.)
* Correspondence: ivan.gajdos@tuke.sk; Tel.: +421-55-602-3518

Abstract: The aim of this study was to investigate the effect of Fused Filament Fabrication (FFF) layer generation strategies on abrasive wear resistance and compare the material loss of PEI (polyetherimide) plastic specimens based on different specimen building strategies. The study also compares a newly proposed path generation strategy (parallel paths in layers with 0.25 mm displacement of alternate layers) with samples from a previous study where samples were printed without displacement of alternate layers, i.e., layers stacked perpendicularly to each other. The primary focus was on the weight loss due to abrasive wear before and after the test. The tests were conducted on a tribometer constructed according to ASTM G65/16 standards using dry sand. Two printing directions were examined: X (longitudinal) and Z (portrait) orientations. For X construction, three orientations of deposition path generation were utilized, resulting in three samples for each orientation (nine samples in total for X construction). The same approach was applied to Z construction, resulting in another nine samples. In total, 18 samples were produced and tested. The deposited infill path width was 0.5 mm, and the layer thickness used in printing was 0.254 mm. Garnet abrasive $Fe_3Al_2(SiO_4)_3$ was employed in this test. Analysis of the experimental data revealed a relationship between the construction method (X and Z orientations) and the variation in different orientations (1X–3X and 1Z–3Z). The research results can be categorized as overall and partial. The overall results indicate poorer wear resistance for 1X–3X and 1Z–3Z specimens, while the partial results illustrate the findings within each individual specimen.

Keywords: polyetherimide; fused filament fabrication; 3D printing; path generation strategy; abrasive wear; tribological properties

check for updates

Citation: Mitaľ, G.; Gajdoš, I.; Spišák, E. Research of the Pre-Processing Strategy Influence on the Tribological Properties of PEI Processed by Fused Filament Fabrication Technology. *Materials* **2023**, *16*, 4527. https://doi.org/10.3390/ma16134527

Academic Editor: Sergei Tarasov

Received: 19 May 2023
Revised: 15 June 2023
Accepted: 20 June 2023
Published: 22 June 2023

1. Introduction

FFF as one of the additive manufacturing technologies has undergone a great boom in the last decades and has found applications ranging from hobby applications to highly demanding industrial applications. Compared to conventional technologies, FFF (Fused Filament Fabrication) offers the possibility of producing parts by adding material instead of removing it, but it also has its limitations, especially when it comes to small batch production or small part sizes [1,2]. While many studies have been devoted to the actual printing process and mechanical properties of printed parts, due to the limitations resulting from the nature of FFF technology [3–5], the tribological properties of printed models have only been studied marginally and on a limited range of materials [6]. The most widely and frequently utilized materials for FFF were also the subject of numerous studies, investigating their tribological properties. Among such materials we can include acrylonitrile butadiene styrene (ABS) [7–12], polyactic acid (PLA) [13–15], polycarbonate/acrylonitrile butadiene styrene (PC-ABS) [16], polyamide (PA) [17], polyether ether ketone (PEEK) [18], and polyethersulfone (PES) [19].

A study by Jadwiga Pisula et al. [7] investigated the contact wear of spur gears made by the FFF method from ABS, polyetherimide (PEI), and PEEK, finding that the greatest wear was observed for gears made from PEI material. The hardness of the material decreased due to the operating temperature of the loaded gear. Studies by Sencer Sureyya Karabeyoglu et al. [9], Investigated the dependence of texture and filling density on wear and wear rate, friction coefficient, wear mechanism, and microscopic wear characteristics of parts printed from ABS. The results revealed that a grid pattern of high infill density and a honeycomb pattern of low infill density showed the lowest wear rate and lowest coefficient of friction compared to the rectilinear pattern. Infill pattern and density affected the wear rate behavior of specimens directly.

Wenzheng Wu et al. [19], Investigated fiber-reinforced thermoplastic polymer composites based on PES and basalt fibers. By improving the interfacial bonding properties between the fiber and resin matrix with a silane coupling agent, the mass wear and the wear scar width decreased by 80% and 45%, respectively, compared with those for pure PES. Some studies indicated the nature of plastic part behavior during abrasive wear such as S. Perepelkina et al. [14], Where analysis of experimental data showed that changing the 3D printing (filling) settings during part construction made of PLA had a significant effect not only on the strength and stiffness of the parts but also on the surface quality, which influences the tribological properties of the tribological couples. The results of these investigations make it possible to select the optimal 3D printing settings depending on the desired tribological properties of the parts, such as friction and wear coefficients, the ability of the thermoplastic material to soften by heating, and to form under pressure.

Singh R. et al. [17], Stated that the use of functional prototypes based on fused deposition modeling (FDM) for wear resistance applications is currently limited mainly due to selective availability of filament material. Several thermoplastic-based fillers have been reinforced by various research groups to improve mechanical properties. Current research with three particle sizes added suggests that the pin reinforced with double particle size shows better wear properties, followed by triple particle size and ABS reinforcement with the first particle type.

In the experiment of Lin L. et al. [18], A PEEK coating material reinforced with carbon fibers (CF) was prepared on a pure PEEK substrate. The results show that the friction coefficient and wear rate are strongly dependent on the direction of wear action with respect to the orientation of the print paths. The values are much lower when the wear direction was perpendicular to the carbon fibers. Comparative analyses of the worn surfaces show that an effective distribution of shear stress on the carbon fibers can contribute to a large reduction in friction and wear.

Tribological properties of PEI were the object of research in studies by Gu, Y. et al. [20] and Bijwe et al. In these cases, PEI was processed and utilized by other technologies than 3D printing. Observed wear performance of PEI was significantly affected by operating parameters such as load, speed, temperature, and sliding duration and depends largely on the type of wear. Neat PEI polymer performed best in abrasive and erosive wear modes [21].

The application of PEI as a high-performance polymer material in FFF additive manufacturing is gaining ground and research into the properties and performance of the parts made of PEI by FFF is becoming crucial. Much of the research is focused on mechanical properties [5,22,23]. On the other hand, tribological properties of 3D printed parts made of PEI are investigated only marginally.

A study by Mitaľ et al. [24] investigated the influence of processing parameters on wear performance of 3D printed parts. Abrasion resistance of PEI material showed the suitability of the chosen build orientation and deposition strategy for the production of parts by additive manufacturing, depending on the required tribological properties such as friction coefficient and wear behavior at 3D printing technological parameters such as layer height of 0.254 mm and filament width of 0.5 mm. The results of the study show that the wear continuity and frictional force depend on the trajectory traversed in the model orientation of the fabrication. The magnitude of wear (material loss) ranged from 0.451% to 0.809%. It was shown that the weight loss of the specimen under load was on average greater for the chosen fiber orientation strategy in the Z direction than in the X direction [20].

The purpose of this research study is to enhance the knowledge about abrasion resistance of PEI 3D parts, especially the influence of the path generation strategy. To fill this gap, in the present work, the effect of the sample building orientation and in-layer deposition strategy has been studied. This work is envisaged to help the designer in considering both the manufacturing settings through FFF along with the real abrasion resistance of parts made of PEI by FFF.

2. Materials and Methods

2.1. Measurements and Equipment

The testing apparatus (Figure 1) contains a tribological pair consisting of a rotating disc and a specimen mounted in a holder.

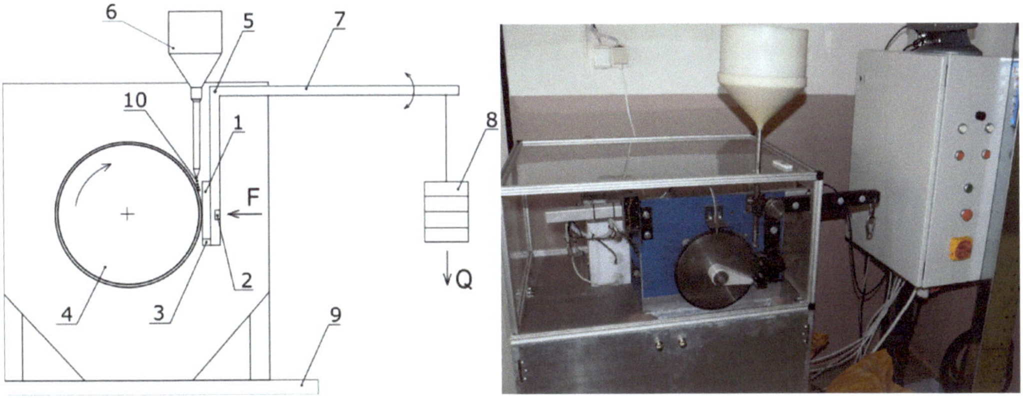

Figure 1. Principle of friction and wear measurement (G65/16)—**left**; tribometer—**right**.

In a tribometer with a rotating rubber disc mounted on a base (9), abrasive wear conditions are created by particles (10) which are discharged from a reservoir (6) between the sample (1) and the rotating disc (4).

During testing, the wear rate of the surface layers is monitored as a function of the sliding path and load. At the same time, the effect of different abrasive particles on the wear of the specimens can be investigated. It is possible to change the load magnitude and the rotation speed of the wheel (so-called sliding speed). The load magnitude is generated by the load unit (5, 7, 8). The device allows samples with different physical and mechanical properties to be tested. During testing, the time history of the specimen temperature, normal force, friction force, and sliding distance can be continuously recorded. The device can be set to automatically terminate the test after a certain number of rotations. The test process can be controlled and monitored via computer.

Prior to the testing process, it is possible to adjust the magnitude of the normal force by fixing and positioning the weight on the lever, whereby the normal force is measured using the force sensor (2) which is located in the arm (7). The arm holding the test specimen is pivotally mounted by means of a pin (5). To measure the frictional force between the specimen (1) and the rotating disc (4), a frictional force sensor (3) is located in the load (8). The control of the sliding speed (rotation) is realized based on the desired value of the time course of the movement or the number of cycles. The advantage of this technical solution is that it has the ability to record the friction force, the normal force, and the temperature of the sample and to convert this data into a tabular form. Figure 2 shows a schematic of the flow and extent of data acquisition and processing.

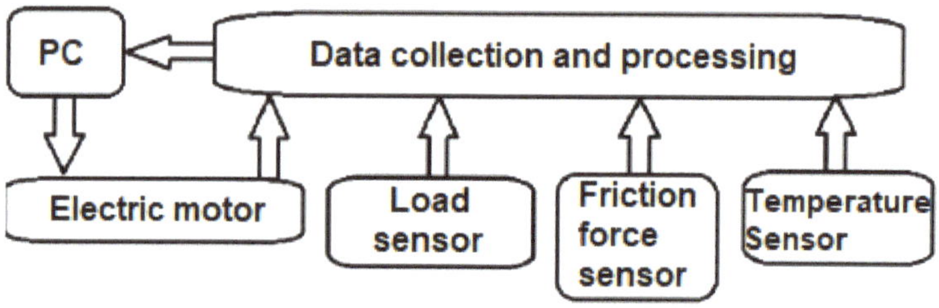

Figure 2. Scheme of data processing during the test.

2.2. Experimental Parameters

Experimental measurements were carried out according to ASTM G65/16 (Standard Test Method for Measuring Abrasion Using the Dry Sand/Rubber Wheel Apparatus) at an ambient temperature of 21 °C and relative humidity of 64%.

To determine the wear rate as a function of specimen build direction and orientation of the deposition paths, 9 tests were performed for specimens with X orientation and 9 tests for specimens with Z orientation, using garnet $Fe_3Al_2(SiO_4)_3$ abrasive. The same rubber disc was used for all tests to eliminate any differences in the chemical composition or hardness of the rubber. As a basis for determining wear, the weight loss of the samples was determined by weighing the samples on a precision balance with an accuracy of 0.0001 g before and after the test. The abrasive test results are reported as weight loss in grams for a particular test procedure according to the following formula [25]:

$$mass\ loss = \frac{weight\ loss}{density\left(\frac{g}{cm^3}\right)} \times 1000 \tag{1}$$

The advantage of this technical solution is that it allows the recording of friction and normal force and temperature of the sample during the test. The machine parameter settings during the test are shown in Table 1.

Table 1. ASTM G65-16 test parameters.

Disc diameter	(mm)	229
Rotations per minute	(RPM)	278
Sample size	(mm)	70 × 20 × 6
Load	(N)	30
Path	(m)	200

The three-point abrasive process involves loose particles that are free to move (abrasive). Garnet abrasive was used in this experiment. Garnet sand is a natural material [24]. The composition of the abrasive used is shown in Table 2. Figure 3 shows a snapshot of the abrasive used.

Table 2. Chemical composition of garnet abrasive [24].

Garnet $Fe_3Al_2(SiO_4)_3$	SiO_2	FeO	Fe_2O_3	Al_2O_3	CaO	MgO	MnO
	(%) 41.34	(%) 9.72	(%) 12.55	(%) 20.36	(%) 2.97	(%) 12.35	(%) 0.85

Figure 3. Garnet abrasive.

2.3. Pre-Processing and Production of Experimental Samples

The most common material used in the production of models using FDM technology is ABS thermoplastic, PC polycarbonate and derivatives, polyphenyl sulfone PPSU/PPSF, and ULTEM 9085™ material [26].

The samples were made from PEI polymer (ULTEM 9085™) on a Fortus 400mc™ 3D printer at 380 °C nozzle temperature according to the scheme shown in Figure 4. All crucial parameters (nozzle temperature, chamber temperature, and layer height) were applied according to default process setting values valid for ULTEM9085™ in Stratassys Insight 10.1™ slicing software. Printing conditions and parameters are summarized in Table 3.

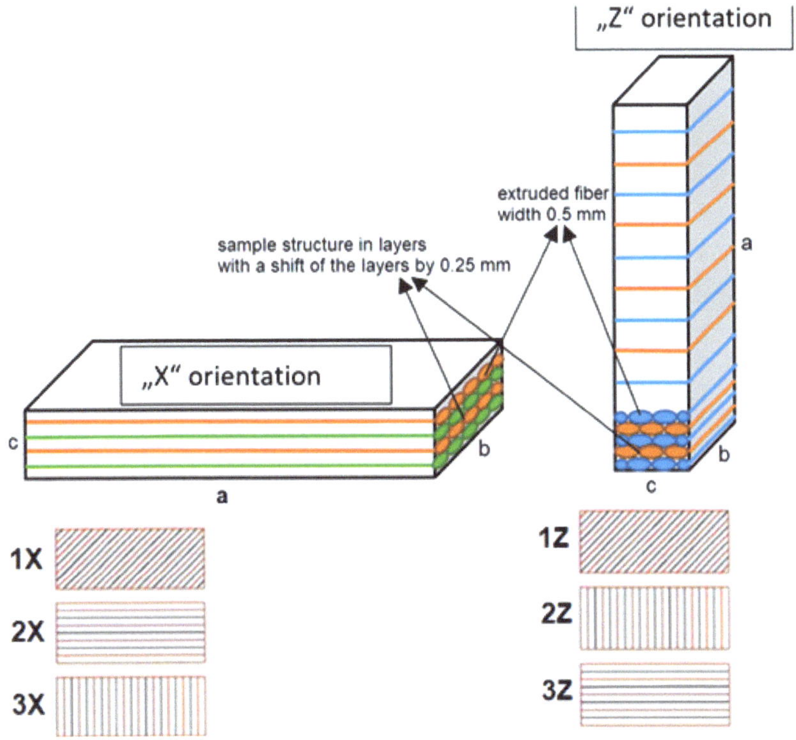

Figure 4. Layer deposition strategy in X and Z orientation (**top**) and orientation of deposited fibers in alternate layers (**bottom**).

Table 3. Slicing and printing parameters.

Parameter	Value
Printing material	ULTEM 9085™
Slicing height	0.254 mm
Width of infill deposition path	0.5 mm
Width of outline deposition path	0.25 mm
No. of outlines in even layer	1
No. of outlines in odd layer	0
Infill deposition path angle	0°, 45°, 90°
Printing head temperature	380 °C
Building chamber temperature	145 °C

PEI is a high-temperature-resistant amorphous plastic with an amber-transparent coloration. The material meets the highest requirements for heat resistance (permanent operating temperatures up to 170 °C are possible depending on the application and type), and at the same time, thanks to its chemical composition, slows down combustion without the addition of further flame retardants, with very low smoke generation. It has a very high strength, which can be increased by adding glass or carbon fibers. The wide range of material properties is complemented by above-average dimensional stability and a low tendency to creep. All these properties make the material an ideal candidate for the automotive and aerospace industries. Thermoplastic is also popularly used in the telecommunications and medical technology sectors.

The specimens were in the shape of a 70 × 20 × 6 mm block (a × b × c), and the contact surface in contact with the rotating disc was the wall of the specimen with a dimension of 70 × 20 mm (Figure 4 top). The strategies used in the sample construction are shown in Figure 4, where the alternating layers (XY plane) when printing the sample were at the same angle but with a shift of the printing nozzle by 0.25 mm which provided a better filling of the sample volume and a denser structure [24]. The same construction procedure was used for both orientations.

Figure 4 (lower half) depicts the method used for printing samples in alternating layers. In this experiment, the alternating layers were printed at the same angle during sample formation, but the axis of the deposition path was shifted by 0.25 mm in a direction perpendicular to the deposition direction. In Figure 4 (top half), the odd layers are denoted in blue and green to enhance clarity in the schematic representation. Samples 1X, 2X, and 3X were deposited with the structure's orientation labeled as X, while samples 1Z, 2Z, and 3Z were deposited with the structure's orientation indicated as Z.

In the abrasive wear resistance test, three types of track fit orientation were applied to the specimen structure in the X and Z directions to investigate the weight loss of the material. Table 4 shows the specimen designations and the angles at which the specimen was formed.

Table 4. Angle of deposition path alignment to the testing surface.

Sample Type	Strategy	Sample Type	Strategy
1X	45°	1Z	45°
2X	0°	2Z	90°
3X	90°	3Z	0°

3. Results and Discussion

The experimental results were evaluated as the value of weight loss and friction force waveform as a function of the traveled path for different orientations of specimen construction by FFF process. Two orientations of specimen construction (X and Z orientation) were made for this experiment. For each orientation, three identical methods of path deposition and layer formation were used, up to the desired specimen size. Three samples were made from each method, thus a total of 18 samples were tested (Figure 5).

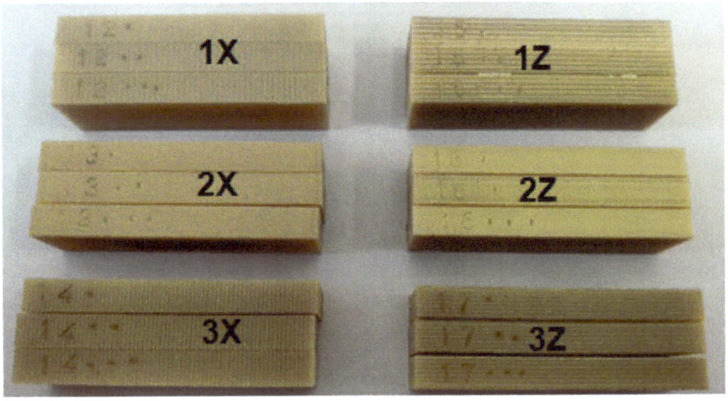

Figure 5. Samples used in the experiment.

3.1. Comparison of 1X–3X Samples with the Same Path Orientation without Displacement in the Intermediate Layers of Samples 2–4

The experimental results are presented as the weight loss of the sample as a function of the deposition orientation. Figure 6 shows the average weight loss of the samples used with X orientation.

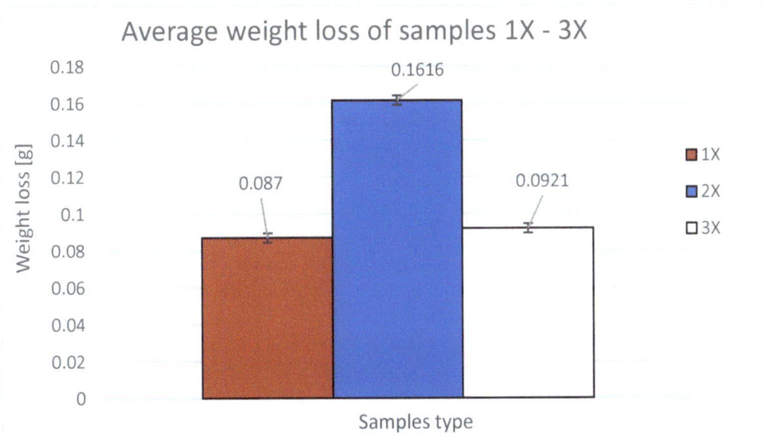

Figure 6. Weight losses and standard deviation of 1X, 2X and 3X samples.

Figure 7 shows the orientation of the path deposition in odd and even layers in samples from study [24]. These orientations are compared with newly produced samples with a denser structure. The denser structure was achieved by shifting the internalized layers by 0.25 mm, which ensured a higher weight of the printed samples in the X orientation and thus a denser interlayered sample structure.

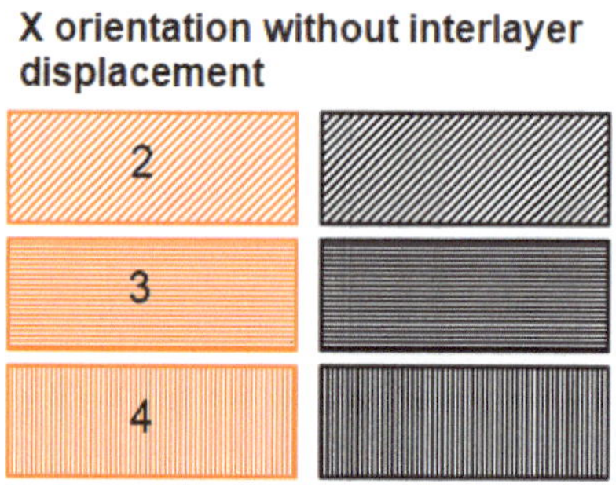

Figure 7. Path formation strategy in even (black) and odd (orange) structural layers of investigated samples with X orientation of deposited fibers [24].

Figure 8 displays a comparison of mass differences between specimens in the X orientation and the material loss due to abrasive wear. The blue color represents the loss of samples 1X (new samples) and 2 (from the previous study). Within the 45° orientation, the specimen with a perpendicular build system in the alternate layers (printing paths directly above each other) proved to be more durable, exhibiting a 2.64% decrease in wear compared to specimen 1X, with a loss difference of 0.0023 g. The yellow color illustrates the losses of specimens 2X (new specimens) and 3 (from the previous study). Within the 0° orientation, the specimen with a perpendicular printing method in alternate layers (printing paths directly above each other) demonstrated higher durability, with a 17.08% decrease in wear compared to the 2X specimen, resulting in a loss difference of 0.0276 g. The green color showcases the losses of the 3X specimens (new specimens) and the 4 specimens (from the previous study). Within the 90° orientation, the specimen with perpendicular printing in alternate layers (printing paths directly above each other) was found to be more resistant, exhibiting a 30.4% decrease in wear compared to the 3X specimen, resulting in a loss difference of 0.0286 g. Based on this comparison, it was observed that there was a decrease in abrasive wear resistance for the specimens with a 0.25 mm offset in the alternate layers when printed in the X orientation. Thus, specimens with the alternating layers printed perpendicularly on top of each other (without interlayer offset) demonstrated higher resistance in all print path orientations.

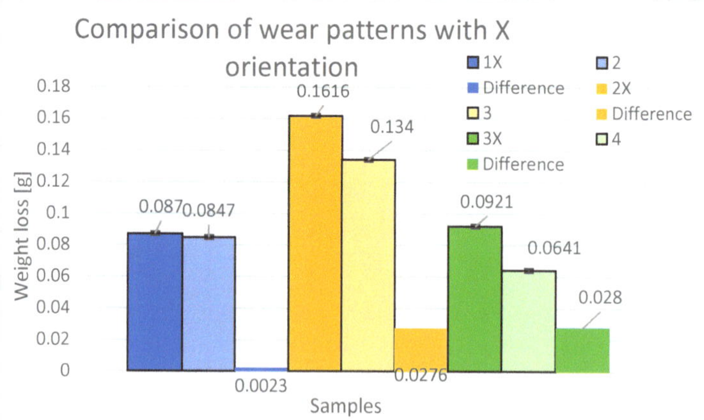

Figure 8. Comparison and difference and standard deviation of weight loss of 1X, 2X, and 3X samples with samples with the same orientation but without displacement in alternate layers of samples 2, 3, and 4 [24].

The friction force was also evaluated as a dependent variable in the evaluation of the experimental data as a function of the path of the rubberized disc. The frictional force was read directly from the tribometer output data, and its evolution depends on several factors such as the mechanical properties of the material, contact pressure and friction, chemical interactions of the materials in the three contact tribosystem (disc, abrasive particles, specimen), and the contact time length. The resulting friction force curves are shown in Figures 9–11 and are processed into separate graphical dependencies for better clarity.

The graphical dependencies of the samples with 1X–3X orientation show that they are less resistant to wear compared to samples 2–4. Samples 1X and 2 in the test showed similar material loss characteristics. This means that if the layers are applied at an angle of 45° from the direction of the abrasion flow that this does not have much effect on the wear, whether the deposition paths are overlapping or shifted with a 0.25 mm rasterization offset.

The highest frictional force was at the beginning of the abrasion test, which was due to the material putting up the most resistance to the frictional force at the beginning. It can be seen from the dependencies that the coefficient of friction from the initial value at the beginning of the experiment increases and stabilizes after a path of 40–50 m, where it then oscillates around a certain value until the end of the experiment. The lowest coefficient of friction at the beginning of the experiment can be observed for the Type 1X samples, which is also the lowest at the end along with the coefficient of friction for the Type 3 and 3X samples.

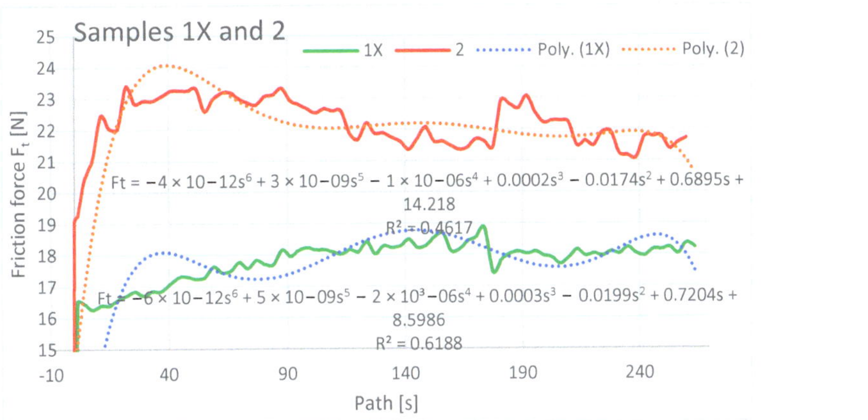

Figure 9. Plot of coefficient of friction versus path for samples 1X and 2.

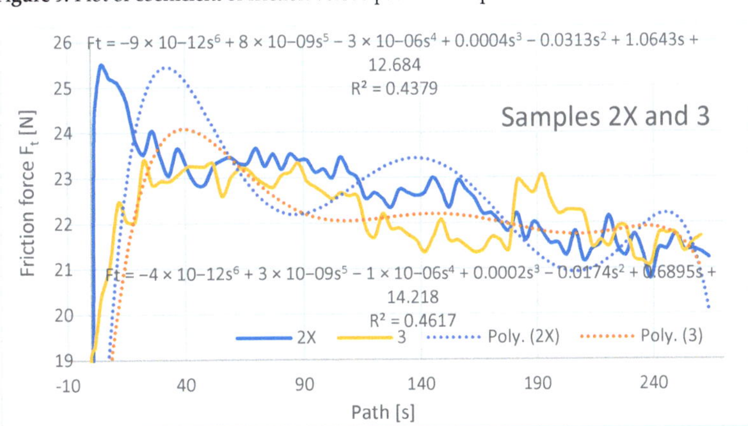

Figure 10. Plot of coefficient of friction versus path for samples 2X and 3.

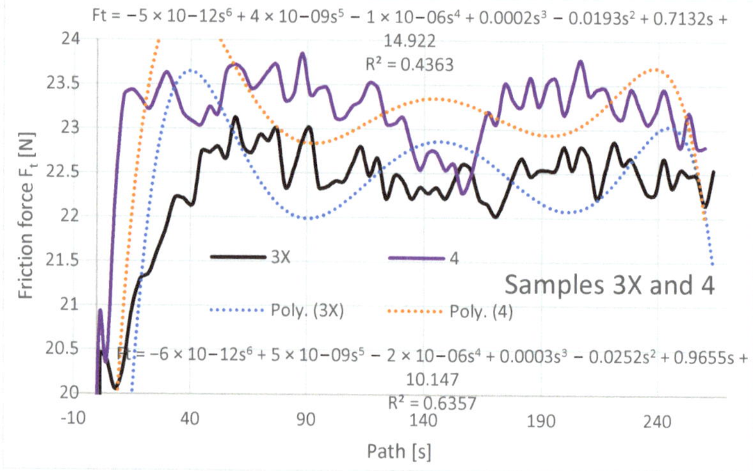

Figure 11. Plot of coefficient of friction versus path for samples 3X and 4.

Within the graphical characterization of the friction forces, regression mathematical models of the given dependencies were also worked out, which indicate that there is negligible mathematical dependence between the compared variables, i.e., that it is mathematically meaningless to investigate the compared variables and to create mathematical models.

3.2. Comparison of 1Z–3Z Samples with the Same Fiber Orientation without Displacement in the Intermediate Layers of Samples 7, 8, and 10

The experimental results are presented as the weight loss of the sample as a function of the deposition orientation. Figure 12 shows the average weight loss of the samples used with Z orientation.

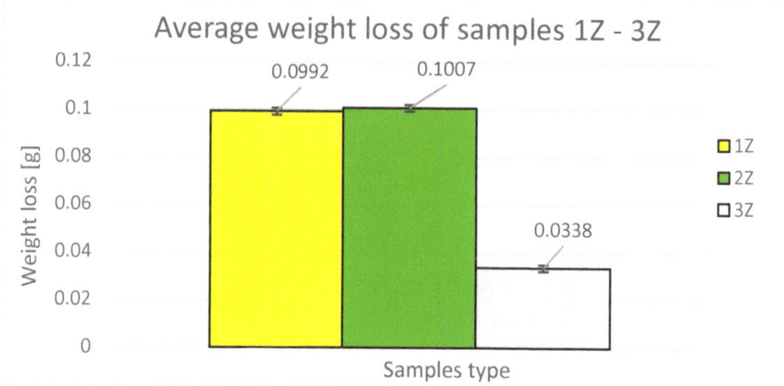

Figure 12. Weight losses and standard deviation of 1Z, 2Z, and 3Z samples.

Figure 13 shows the orientation of the path deposition in odd and even layers in samples from study [24]. These orientations are also compared with new samples with a denser structure, where the denser structure was achieved by shifting the internalized layers by 0.25 mm, which provided a lower weight of the printed samples in the Z orientation and thus a more densely bonded sample structure.

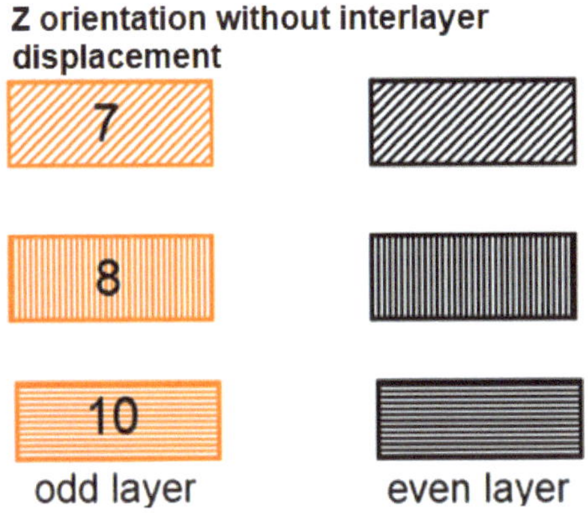

Figure 13. Path formation strategy in even (black) and odd (orange) structural layers of investigated samples with Z orientation of deposited fibers [24].

The graphical representation in Figure 14 provides a comparison of weight differences between specimens in the Z orientation and the material loss due to abrasive wear. The blue color indicates the loss of samples 1Z (new samples) and 7 (from the previous study). Within the 45° orientation, the specimen with perpendicular printing in alternating layers (printing paths directly above each other) demonstrated lower durability, with a 15.22% increase in wear compared to specimen 1Z and a loss difference of 0.0151 g.

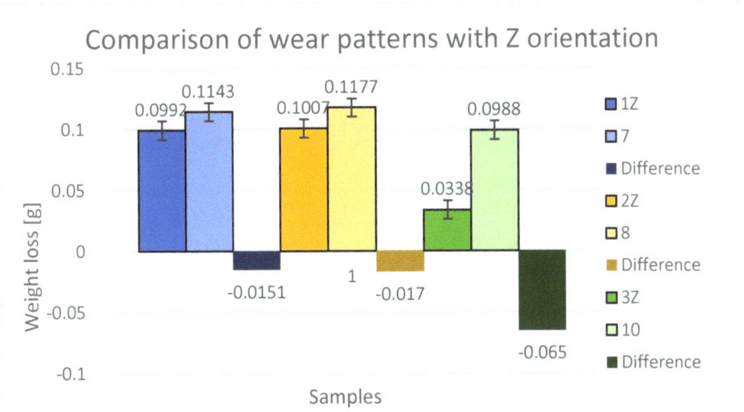

Figure 14. Comparison and difference of weight losses and standard deviation of samples 1Z, 2Z, and 3Z with samples with the same orientation but without displacement in the intermediate layers of samples 7, 8, and 10.

The yellow color represents the losses of specimens 2Z (new specimens) and 8 (from the previous study). Within the 90° orientation, the specimen with a perpendicular construction system in alternate layers (fibers directly and with each other) also exhibited lower durability, with a 15.88% increase in wear compared to specimen 2Z, resulting in a loss difference of 0.017 g.

The green color shows the losses of specimens 3Z (new specimens) and 10 (from the previous study). Within the 0° orientation, the specimen with a perpendicular printing method in alternating layers (printing paths directly above each other) demonstrated even lower durability, with a 192.31% increase in wear compared to specimen 3Z, and a loss difference of 0.065 g.

Based on this aforementioned comparison, it was found that specimens built in the Z orientation, with a 0.25 mm offset of the alternating layers of fibers, exhibited higher abrasive wear resistance compared to specimens with a perpendicularly superimposed fiber deposition construction system. Among the trio of 1Z, 2Z, and 3Z, specimen 3Z demonstrated the lowest weight loss. For specimens 1Z and 2Z, weight loss was approximately the same. This indicates that the specimens produced in the Z orientation experienced a lower weight loss and thus a lower wear.

In the evaluation of the experimental data, the friction force was also evaluated as a dependent variable in the Z orientation depending on the traveled path of the rubberized disc. The friction force was read directly from the tribometer output data. The resulting friction force curves are shown separately in Figures 15–17.

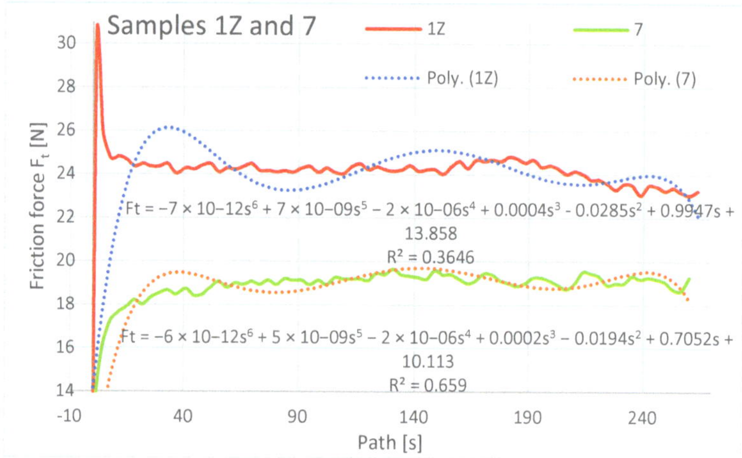

Figure 15. Plot of coefficient of friction versus path, samples 1Z and 7.

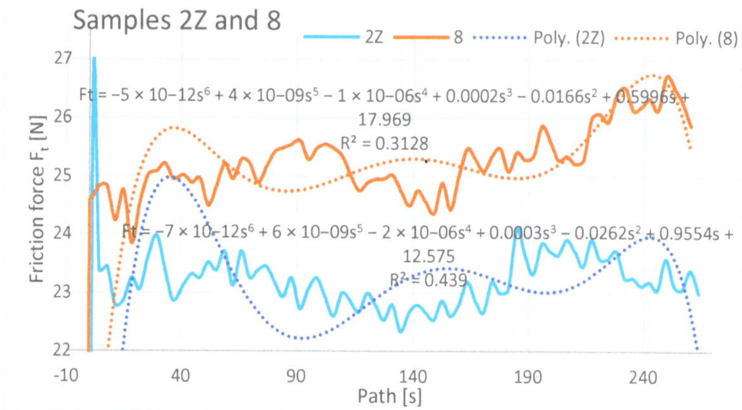

Figure 16. Plot of coefficient of friction versus path, samples 2Z and 8.

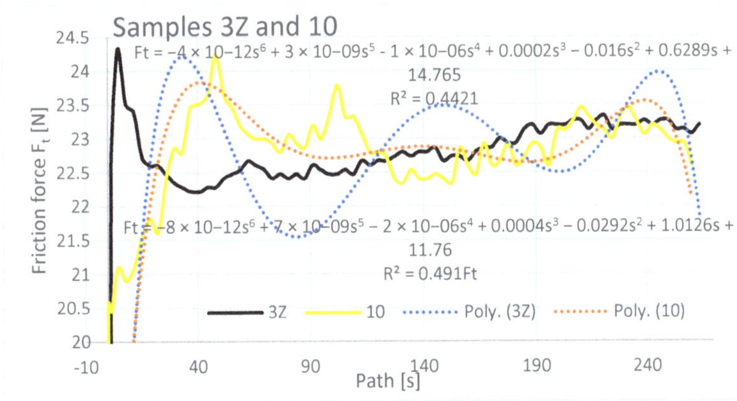

Figure 17. Graph of coefficient of friction versus path, samples 3Z and 10.

The graphical analysis of specimens with orientations 1Z to 3Z reveals their higher resistance to wear compared to specimens 7, 8, and 10. As weight loss is directly proportional to frictional force, specimens 3Z and 10 exhibited similar weight loss characteristics in the frictional force test. This suggests that depositing the layers at 0° does not significantly impact friction and wear. Similarly, a comparable friction and wear pattern occurs when the layers are deposited on top of each other with a 0.25 mm rasterization offset. Regarding samples 1Z, 7, 2Z, and 8, different friction forces were observed. The highest friction was recorded for sample 8 (ranging from 24 N to 26.5 N) and 1Z (ranging from 24.8 N to 23.3 N). Specimens with orientations 2Z and 7 exhibited lower frictional forces, while the lowest friction was observed for specimen 7 (ranging from 18 N to 19.4 N) from the previous study. In the current study, the specimen with 3Z orientation demonstrated the lowest friction.

It is important to note the history of frictional forces in the compared specimens of the X and Z series in the current study. The frictional force did not change significantly when comparing specimens with the same orientations. Initially, the highest friction force was observed at the beginning of the abrasion test, indicating that the material offered the greatest resistance to the applied force. The friction coefficient increased from the initial value at the start of the experiment and stabilized after a distance of 10–20 m. It then oscillated around a certain value until the end of the experiment, with a slight increase towards the end of the traveled path observed for samples 2Z and 8. At the beginning of the experiment, the lowest coefficient of friction was observed for the Type 1X samples, which remained low at the end along with the coefficient of friction for the Type 3 and 3X samples. Graphically characterizing the friction forces, regression mathematical models were also developed for the given dependencies. These models indicate a negligible mathematical relationship between the compared variables. From a mathematical perspective, it is not meaningful to investigate the variables compared and create mathematical models, as the polynomial functions of the sixth degree used in the mathematical description resulted in low regression and complex mathematical relationships.

3.3. *Wear Images of Specimen Surfaces at the Abrasive Inlet (Left) to the Contact Zone and the Abrasive Outlet (Right) from the Containment Zone*

In the Figures 18 and 19 shown below, we have processed the test results from a visual perspective. In the first column, the surfaces before the abrasion test are photographed. The second column contains photos of the surfaces after the input abrasion test (from left to right). The last third column contains the surface of the material at the outlet of the abrasion (from right to left). In these photos, we can see how the abrasive penetrated the different layers of the prints during the test. This can be seen most clearly in specimens with alternating filament deposition. Although the samples were cleaned with the ultrasonic cleaner after testing, we observed residues of abrasive on the surface, especially in orientation Z but also in X orientation.

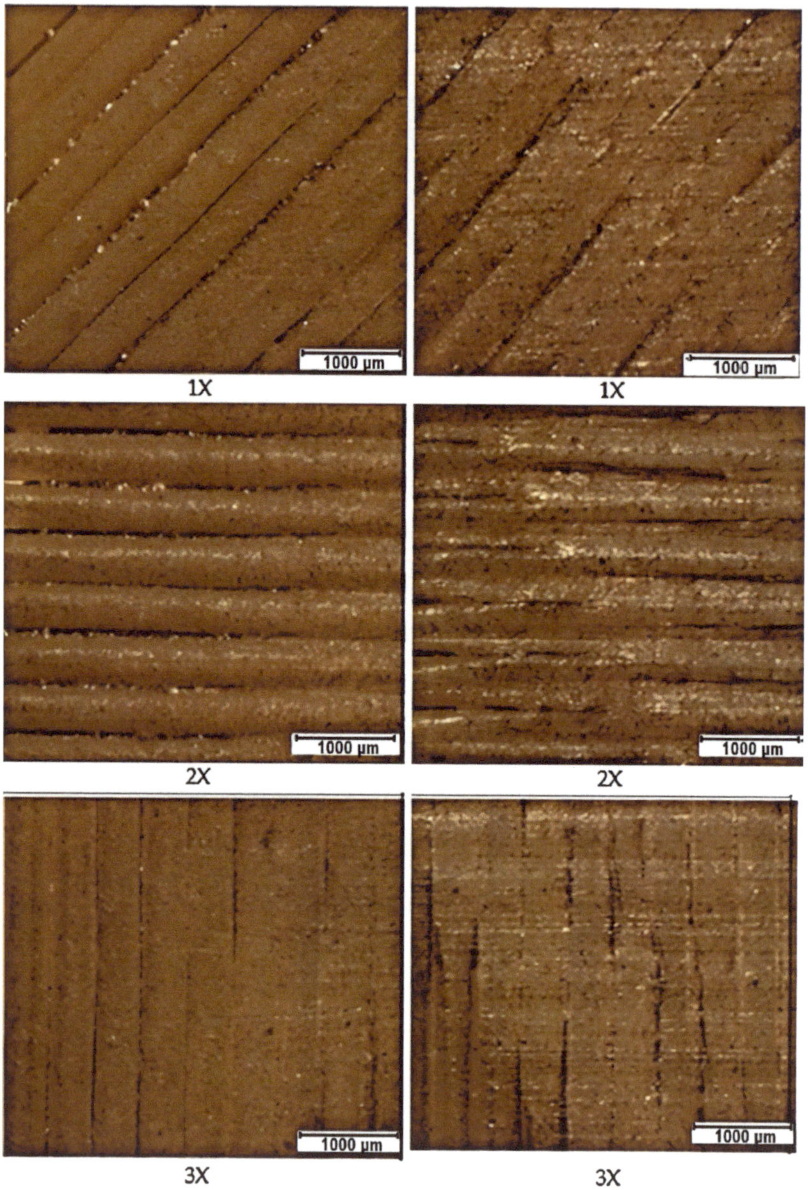

Figure 18. Microscopic images of sample surfaces with X orientation.

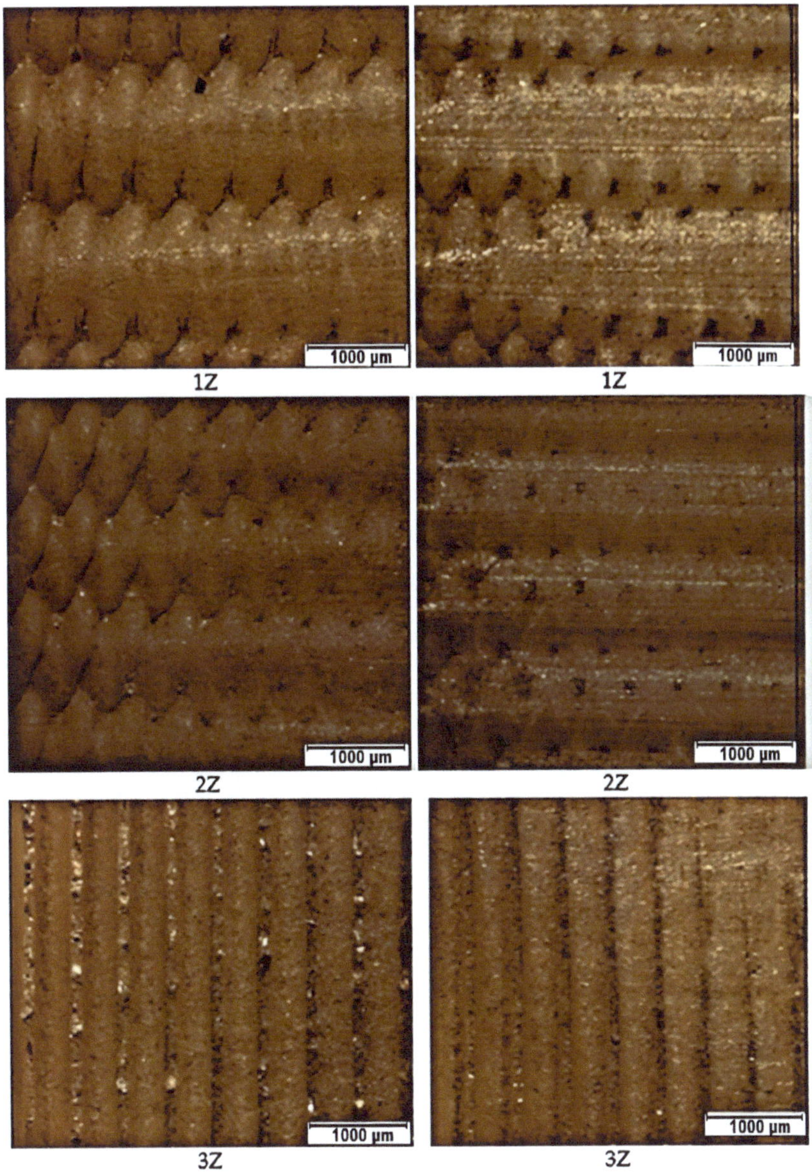

Figure 19. Microscopic images of sample surfaces with Z orientation.

4. Discussion

To assess the wear and mass loss of the material, the printing methods of X and Z specimens are compared and evaluated against each other (Table 5 and Figure 20). The specimens were built in layers with a 0.25 mm offset. Additionally, the results of the current study are compared with those of a previous study, where the specimens were built with alternating layers perpendicularly stacked on top of each other. Table 5 presents the average initial weight values of the specimens before the test, the average weight loss in grams, the cumulative loss for each orientation, and the weight loss percentage.

Table 5. Difference in weight loss within the X and Z sample building direction.

Building Direction	Sample Type	Initial Weight (g)	Weight Loss (g)	Weight Loss (%)
X	1	13.9903	0.0729	0.520
	2	14.0028	0.0847	0.605
	3	13.8603	0.134	0.967
Z	7	14.5401	0.1143	0.786
	8	14.5558	0.1177	0.809
	10	14.5100	0.0988	0.681

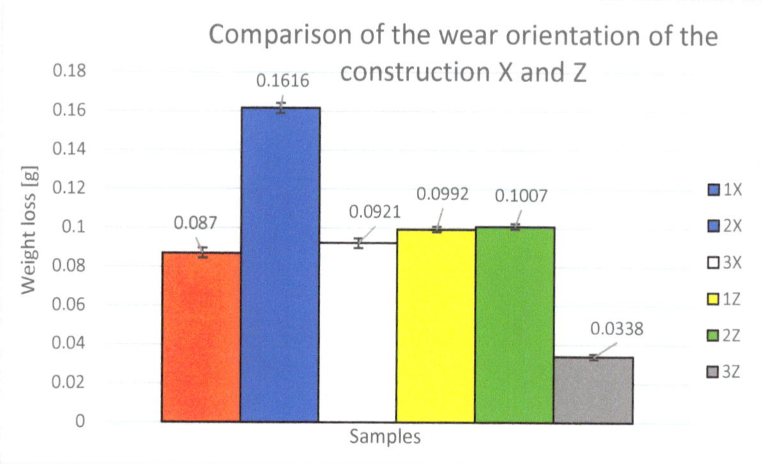

Figure 20. Graphical representation of the comparison of X and Z orientations and standard deviations.

The primary factors investigated were the weight loss in relation to the traveled path and the coefficient of friction. According to Table 5 and Figure 19, it is evident that the highest material loss was observed for the second specimen from orientation X (labeled as 2X), which was constructed at a 90° angle to the tested specimen wall. This indicates that this orientation is the least resistant to abrasive wear, with a material loss percentage of 1.199% of the total specimen weight. The second lowest material loss was recorded for specimens 2Z and 1Z, with orientations of 45° (2Z) and 90° (1Z) relative to the tested specimen wall. These specimens exhibited weight loss percentages of 0.753% and 0.734% of their respective total weights. The third lowest material loss was observed for samples 3X and 1X, with orientations of 45° (3X) and 0° (1X) relative to the tested specimen wall. The weight loss percentages for these samples were 0.668% and 0.624% of their total weights, respectively. The lowest material loss was recorded for the third specimen from the Z orientation (labeled as 3Z), which was constructed at a 0° angle to the tested specimen wall. This indicates that this orientation is the most resistant to abrasive wear, with a material loss percentage of 0.257% of the total specimen weight.

Comparison of weight loss between samples 1X and 1Z revealed a lower weight loss for sample 1X. The experimental results allowed the evaluation of weight loss and friction coefficient as a function of distance traveled.

The comparison reveals that the most abrasion-resistant specimen appears to be the specimen designated 3Z, which has a material loss value of 0.0338 g. When considering the percentage of weight loss relative to the average weight loss in the Z orientation, it corresponds to 14.46% of the average Z orientation loss and 9.921% of the X orientation loss.

The second sample with relatively low weight loss is sample 1X, with a material loss value of 0.087 g. In terms of the percentage of weight loss relative to the average weight loss in the Z orientation, it represents 37.23% of the average Z orientation loss and 25.54% of the X orientation loss.

Next, we have the third sample with relatively low weight loss, which is sample 3X with a material loss value of 0.0921 g. In terms of the percentage of weight loss relative to the average weight loss in the Z orientation, it corresponds to 39.41% of the average Z orientation loss and 27.03% of the X orientation loss.

The following is the fourth sample with increasing weight loss, which is sample 1Z with a material loss value of 0.0992 g. When considering the percentage of weight loss relative to the average weight loss in the Z orientation, it represents 42.45% of the average loss in the Z orientation and 29.12% of the X orientation loss.

The sample labeled 2Z has a relatively low weight loss compared to sample 1Z, with a difference of 19% and a material loss value of 0.1007 g. In terms of the percentage of weight loss relative to the average weight loss in the Z orientation, it corresponds to 43.09% of the average Z orientation loss and 29.56% of the X orientation loss.

The last specimen with the highest percentage of weight loss due to abrasive wear is the specimen labeled 2X, with a material loss value of 0.1616 g. When considering the percentage of weight loss relative to the average weight loss in the Z orientation, it represents 69.15% of the average loss in the Z orientation and 47.43% of the X orientation loss.

The evaluation of the specimen construction system, with alternate layer displacement of 0.25 mm and without alternate layer displacement, reveals the influence of layer orientation on material wear characteristics and the rate of wear by abrasive particles, allowing for predictions regarding their performance in terms of surface and subsurface layer abrasion. Based on this evaluation, specimens with the Z orientation experience less wear and weight loss compared to the X orientation. The Z orientation exhibits a 45.785% lower weight loss compared to the X orientation.

The highest weight loss was observed in sample 3 of the X-oriented samples, where the print paths in both layers were parallel (0°) to the tested sample's wall. Samples 1 and 2, with deposition paths forming a 45° angle, exhibited weight losses 36.79–45.6% lower than sample 3. Sample 4, with the same deposition strategy for all layers (90°), showed a weight loss reduction of 52.16% compared to sample 3. Sample 5, with alternating layers of perpendicular paths and paths parallel to the test specimen's wall (90° and 0°), had the lowest weight loss compared to sample 3, up to 52.91% lower.

Based on the results of the weight loss, it can be concluded that using the same strategy in both layers (sample 2, 45°) or a strategy parallel to the direction of wear in both layers (sample 3) led to higher weight loss. On the other hand, employing a strategy in the straight direction for both layers (sample 4, 0°) or a combined strategy resulted in decreased weight loss in the layers (sample 5, 0° and 90°).

Compared with the construction system in the previous study, where the Z orientation had a higher average weight loss of 0.5315 g, indicating a 26.91% increase in wear compared to the X orientation with an average wear of 0.4188 g, significant differences were observed in the behavior of layers deposited using different methods and construction systems towards abrasive wear. Perpendicular layers exhibited higher wear resistance in the X orientation, with a 26.91% increase in wear compared to the Z orientation. In contrast, layers deposited with a 0.25 mm offset in the alternating layers showed the opposite behavior, with X-oriented layers showing a lower abrasive wear resistance.

Based on this evaluation, recommendations are made for 3D printing parts that will experience contact with other surfaces. Specimens with offset alternate layers and X orientation have lower abrasion resistance compared to specimens with a perpendicular and self-aligned construction system in the interlayers. Conversely, specimens built in the Z orientation with a 0.25 mm interlayer offset appear to be more resistant to abrasion.

5. Conclusions

The first objective of this work was to investigate and demonstrate the impact of the FFF path generation strategy with a 0.25 mm offset in the alternate layers on the abrasion resistance of the material. It also aimed to compare the wear or material loss among different specimens and between the specimen construction methods in the X and Z orientations. The second objective was to compare the results obtained from the abrasion testing of specimens exposed to a 0.25 mm displacement of alternate layers with the specimens from a previous experiment where the layers were not displaced during construction, i.e., they were stacked perpendicularly on top of each other.

While numerous studies focus on testing the mechanical properties of plastic components, there is a lack of research examining parts printed using FFF technology. This study aims to fill this gap by investigating how the chosen construction orientation and path generation strategy affect the abrasion resistance of 3D printed PEI parts. The objective is to analyze the influence of the construction orientation and coating strategy in the FFF additive manufacturing technique on the evolution of the friction coefficient and wear rate of ULTEM9085™ thermoplastic samples. This will be assessed by measuring the weight loss of the material according to the ASTM G65-16 standard. The PEI material offers the potential to use printed products in demanding applications, making it important to conduct experimental measurements of its abrasive wear resistance.

A total of 18 PEI plastic samples were made by the FFF (FDM) method. The samples were fabricated with two orientations of sample construction and different methods of print path generation. From each path generation method, three samples were made from which average mass loss values were obtained and subsequently worked with in our study.

Based on the data collected and analyzed, a recommendation can be made for parts fabricated using FFF technology with a 0.25 mm offset in alternate layers, particularly when they will be operating in environments where contact with other surfaces or surface abrasion may occur. In such cases, it is advised to construct the parts using the Z orientation construction method. Specifically, the recommended orientation is similar to that used for the 3Z specimen, where the specimen is built in a portrait orientation, and the print paths are oriented at a 0° angle to the specimen wall under test or the location prone to abrasion.

A comparison between specimens with a 0.25 mm offset in alternate layers and specimens constructed with perpendicular alternate layers revealed differences in the durability of the construction systems used. When comparing the X orientation constructions, it was observed that the construction with a 0.25 mm displacement in alternate layers was 22.92% less durable than the system with perpendicular layering. However, in the Z orientation with a 0.25 mm offset in alternate layers, the reverse construction system proved to be up to 127.43% more wear-resistant than the perpendicular layering.

In terms of the total mass loss of material after wear, when considering both orientations and expressed as a percentage, the construction with a 0.25 mm offset in layers exhibited 65.44% greater resistance compared to the construction system with perpendicular layers. These findings can serve as a foundation for future research on the durability of elastomeric and abrasion-resistant products manufactured using FFF technology for component construction. It is anticipated that the abrasion resistance of such materials will present significant challenges in their application [27,28].

Author Contributions: Conceptualization, G.M. and I.G.; methodology, I.G.; software, E.S.; validation, E.S.; formal analysis, G.M.; investigation, G.M.; resources, I.G.; data curation, G.M.; writing—original draft preparation, I.G. and G.M.; writing—review and editing, G.M. and E.S.; visualization, I.G.; supervision, E.S. and G.M. All authors have read and agreed to the published version of the manuscript.

Funding: This work was supported by the Slovak Research and Development Agency under the contract no. APVV-21-0418. The authors would like to thank the VEGA grant agency for supporting this research work and co-financing the project VEGA 1/0384/20.

Institutional Review Board Statement: Not applicable.

Informed Consent Statement: Not applicable.

Data Availability Statement: Not applicable.

Conflicts of Interest: The authors declare no conflict of interest.

References

1. Levy, G.N.; Schindel, R.; Kruth, J.P.; Leuven, K.U. Rapid manufacturing and rapid tooling with layer manufacturing (LM) technologies–state of the art and future perspectives. *CIRP Ann. Manuf. Technol.* **2003**, *52*, 589–609. [CrossRef]
2. Dey, A.; Roan Eagle, I.N.; Yodo, N. A Review on Filament Materials for Fused Filament Fabrication. *J. Manuf. Mater. Process.* **2021**, *5*, 69. [CrossRef]
3. Singh, J.; Kumar Goyal, K.; Kumar, R. Effect of filling percentage and raster style on tensile behavior of FDM produced PLA parts at different build orientation. *Mater. Today Proc.* **2022**, *63*, 433–439. [CrossRef]
4. Mahmoud Moradi, M.; Aminzadeh, A.; Rahmatabadi, D.; Rasouli, S.A. Statistical and Experimental Analysis of Process Parameters of 3D Nylon Printed Parts by Fused Deposition Modeling: Response Surface Modeling and Optimization. *J. Mater. Eng. Perform.* **2021**, *30*, 5441–5454. [CrossRef]
5. Gajdoš, I.; Slota, J.; Spišák, E.; Jachowicz, T.; Tor-Swiatek, A. Structure and tensile properties evaluation of samples produced by Fused Deposition Modeling. *Open Eng.* **2016**, *6*, 86–89. [CrossRef]
6. Norani, M.; Chua Abdullah, M.; Abdollah, M.; Amiruddin, H.; Redza Ramli, F.; Tamaldin, N. Mechanical and tribological properties of FFF 3D-printed polymers: A brief review. *J. Tribol.* **2021**, *29*, 11–30.
7. Pisula, J.; Budzik, G.; Turek, P.; Cieplak, M. An Analysis of Polymer Gear Wear in a Spur Gear Train Made Using FDM and FFF Methods Based on Tooth Surface Topography Assessment. *Polymers* **2021**, *13*, 1649. [CrossRef]
8. Bankupalli, N.; Srinivasa Rao, D.; Vamsi Krishna, T.S. Role of butadiene content on tribological properties of polymeric components fabricated by FDM. *Mater. Today Proc.* **2020**, *38*, 3019–3026. [CrossRef]
9. Karabeyoglu, S.S.; Eksi, O.; Yaman, P.; Kucukyildirim, B.O. Effects of infill pattern and density on wear performance of FDM-printed acrylonitrile-butadiene-styrene parts. *J. Polym. Eng.* **2021**, *41*, 854–862. [CrossRef]
10. Norani, M.N.M.; Abdollah, M.F.B.; Abdullah, M.I.H.C.H.; Amiruddin, H.; Ramli, F.R.; Tamaldin, N. Correlation of tribo-mechanical properties of internal geometry structures of fused filament fabrication 3D-printed acrylonitrile butadiene styrene. *Ind. Lubr. Tribol.* **2020**, *72*, 1259–1265. [CrossRef]
11. Kaur, G.; Singari, R.M.; Kumar, H. A review of fused filament fabrication (FFF): Process parameters and their impact on the tribological behavior of polymers (ABS). *Mater. Today Proc.* **2021**, *51*, 854–860. [CrossRef]
12. Gurrala, P.K.; Regalla, S.P. Friction and Wear Behavior of ABS Polymer Parts made by Fused Deposition Modeling (FDM). In Proceedings of the International Conference on Advances in Tribology [icat14], Abstract id NO.–"300", Calicut, Kerala, India, 21−24 February 2014.
13. Yaman, P.; Ekşi, O.; Karabeyoğlu, S.S.; Feratoğlu, K. Effect of build orientation on tribological and flexural properties of FDM-printed composite PLA parts. *J. Reinf. Plast. Composites.* **2023**. [CrossRef]
14. Perepelkina, S.; Kovalenko, P.; Pechenko, R.; Makhmudova, P.K. Investigation of Friction Coefficient of Various Polymers Used in Rapid Prototyping Technologies with Different Settings of 3D Printing. *Tribol. Ind.* **2017**, *39*, 519–526. [CrossRef]
15. Phogat, A.; Chhabra, D.; Sindhu, V.; Ahlawat, A. Analysis of wear assessment of FDM printed specimens with PLA, multi-material and ABS via hybrid algorithms. *Mater. Today Proc.* **2022**, *62*, 37–43. [CrossRef]
16. Mohameda, O.A.; Masooda, S.H.; Bhowmikb, J.L.; Somersc, A.E. Investigation on the tribological behavior and wear mechanism of parts processed by fused deposition additive manufacturing process. *J. Manuf. Process.* **2017**, *29*, 149–159. [CrossRef]
17. Singh, R.; Kumar, S.; Bedi, P.; Hashmi, M.S.J. On wear of 3D printed Al2O3 reinforced Nylon6 matrix based functional prototypes. *Mater. Today Proc.* **2020**, *33*, 1477–1482. [CrossRef]
18. Lin, L.; Ecke, N.; Huang, M.; Pei, X.-O.; Schlarb, A.K. Impact of nanosilica on the friction and wear of a PEEK/CF composite coating manufactured by fused deposition modeling (FDM). *Compos. Part B Eng.* **2019**, *177*, 107428. [CrossRef]
19. Wu, W.; Li, Z.; Lin, G.; Ma, J.; Gao, Z.; Qu, H.; Zhang, F. Additive manufacturing of continuous BF-reinforced PES composite material and mechanical and wear properties evaluation. *J. Mater. Sci.* **2022**, *57*, 12903–12915. [CrossRef]
20. Gu, Y.; Fei, J.; Zheng, X.; Li, M.; Huang, J.; Qu, M.; Zhang, L. Graft PEI ultra-antiwear nanolayer onto carbon spheres as lubricant additives for tribological enhancement. *Tribol. Int.* **2020**, *153*, 106652. [CrossRef]
21. Bijwe, J.J.; Indumathi, J.; Rajesh, J.J.; Fahim, M. Friction and wear behavior of polyetherimide composites in various wear modes. *Wear* **2001**, *249*, 715–726. [CrossRef]
22. Ding, S.; Zou, B.; Wang, P.; Ding, H. Effects of nozzle temperature and building orientation on mechanical properties and microstructure of PEEK and PEI printed by 3D-FDM. *Polym. Test.* **2019**, *78*, 105948. [CrossRef]
23. El Magri, A.; Vanaei, S.; Vaudreuil, S. An overview on the influence of process parameters through the characteristic of 3D-printed PEEK and PEI parts. *High Perform. Polym.* **2021**, *33*, 862–880. [CrossRef]
24. Mitaľ, G.; Gajdoš, I.; Jezný, T.; Spišák, E.; Majerníková, J. Analysis of the selected technological parameters' influence on tribological properties of products manufactured by FFF tech-nology. *Appl. Sci.* **2021**, *12*, 3853. [CrossRef]
25. *ASTM G65-16*; Standard Test Method for Measuring Abrasion Using the Dry Sand/Rubber Wheel Apparatus 1. ASTM International: West Conshohocken, PA, USA, 2016.

26. Coranič, T.; Gašpár, Š.; Paško, J. Structural modification impact analysis of selected components of shielding equipment using rapid prototyping technology. *MM Sci. J.* **2018**, *2018*, 2183–2187. [CrossRef]
27. Khalaf, E. A comparative study for the main properties of silica and carbon black Filled bagasse-styrene butadiene rubber composites. *Polym. Polym. Compos.* **2023**, *31*, 09673911231171035. [CrossRef]
28. Suder, J.; Mlotek, J.; Panec, A.; Fojtík, F. Design of Printing Parameter Settings Methodology for FFF Printing of Waterproof Samples from a Flexible Material. *Acta Mech. Slovaca* **2023**, *27*, 58–64. [CrossRef]

materials MDPI

Article

Evaluation of Cross-Linked Polyamide 6 Micro-Indentation Properties: TAIC Concentration and Electron Radiation Intensity

Martin Ovsik *, Michal Stanek and Martin Bednarik

Faculty of Technology, Tomas Bata University in Zlin, Vavreckova 5669, 760 01 Zlín, Czech Republic
* Correspondence: ovsik@utb.cz; Tel.: +420-576-035-100

Abstract: Nowadays, technical practice puts emphasis on improving selected material properties of polymers which could lead to new applications. Material properties can be modified in numerous ways, among which is radiation treatment. This study looks into the influence of beta radiation on several properties of polyamide 6, e.g., indentation hardness, modulus and creep. Main changeable parameters were the concentration of triallyl isocyanurate (TAIC), which promotes cross-linking, and intensity of radiation. The concentration was in the range from 2 to 6 wt.%, while the radiation dose was 0, 66, 99 and 132 kGy. The treated materials were measured for indentation hardness, modulus and creep. Degree of cross-linking was verified by thermo-mechanical analysis (TMA), while degradation processes was investigated by Fourier-transform infrared spectroscopy (FTIR). The results indicate that electron radiation positively affects the tested material properties. The best results were seen in polyamide with 6 wt.% of TAIC, which demonstrated a 38% improvement in mechanical properties after exposure to 132 kGy. This improvement in properties affects the final parts and their application (e.g., in the automotive industry—engine parts; in electrical engineering—insulation of wires and cables; and in industry—pipes for underfloor heating, etc.).

Keywords: polyamide 6; cross-linking; electron radiation; micro-indentation test; indentation hardness; structural properties

check for updates

Citation: Ovsik, M.; Stanek, M.; Bednarik, M. Evaluation of Cross-Linked Polyamide 6 Micro-Indentation Properties: TAIC Concentration and Electron Radiation Intensity. *Materials* **2023**, *16*, 2391. https://doi.org/10.3390/ma16062391

Academic Editor: Wolfgang Linert

Received: 20 February 2023
Revised: 14 March 2023
Accepted: 15 March 2023
Published: 16 March 2023

1. Introduction

This study is focused on PA 6 modified with differing concentration of cross-linking agent (CLA) irradiated with three varying doses. Industry puts emphasis on material improvement, one of the methods of which is irradiation. The literature sources laid out in this research focus on the characterisation of radiation influence on mechanical properties of polyamide 6. On the other hand, these articles lack sufficiently complex information regarding this topic. Some correlation between ionising beta radiation, polymer morphology and changes in mechanical properties are hard to find, even though this information is necessary for industrial application of irradiated polyamides with CLA.

Overall, resistance of polyamides to ionising radiation is intermediate. It is lower in type –CONH–(CH_2)n– polyamides, which have a higher number of methyl groups within peptide bonds. As a result of ionising radiation, both cross-linking and degradation of chains commences. If the cross-linking prevails over degradation, then the irradiated material can be improved [1].

Construction of 3D networks is not too strong. The presence of oxygen leads to scission of polyamide chains and the creation of peroxides. Kaindle and Graul show [2], that free radicals are produced during irradiation, especially in Structures (1) and (2).

$$-\mathrm{CONH}-\mathrm{C^{*}H}-\mathrm{CH_2}- \quad (1)$$

$$-\mathrm{C^{*}O}-\mathrm{NH}-\mathrm{CH_2}- \quad (2)$$

Radicals are also produced in C=N and C=C double bonds. Rexer stated [3] that the dose that leads to the creation of gel in the presence of oxygen is 350 kGy. Irradiation in vacuum leads to structure changes that result in improved mechanical and thermo-mechanical properties. The addition of poly functional monomers, such as TAIC, can cause the polyamides to cross-link after exposure to relatively low doses of radiation, even in the presence of oxygen. Finding the way in which these doses influence selected mechanical properties is the topic of this study. Hydrogen on carbon neighbouring with nitrogen in amide group is taken out and water, carbon monoxide and dioxide and methane are created. Three allyl groups of TAIC can react due to created macro-radicals in polyamide, creating a 3D network [3].

Aside from cross-linking, the irradiation of polyamides in the presence of oxygen (typical interceptor of radicals) also leads to fragmentation, disproportion and oxidation, resulting in scission of chains in the microscopic world and embrittlement in the macroscopic world. An initial study of the oxidation behaviour of polyamides was conducted by Sharkley, Mochel, Levantovskaya, Lock and Sagar. The mechanism was based on primary engagement of methyl groups neighbouring with –NH-groups. The radicals created on these groups subsequently react with oxygen, which leads to the creation of more radicals, which can either isomerise or undergo other reactions, including both the cross-linking and scission of polymer chains. All of these reactions eventually lead to the creation of final carbonyl or carboxyl groups [4,5]. The scission of polymer chains leads to a reduction in the number of free amino groups [6].

Radiation cross-linking of polyamides has been researched in numerous studies [6–10], but these studies do not contain sufficient amounts of complex information, and it is hard to obtain the required interactions among electron radiation and morphological changes from them.

The authors of this study examined the influence of electron radiation on the morphological, thermal and mechanical properties of polyamide 6. The material was exposed to radiation of up to 150 kGy. Properties were improved in the specimen irradiated by 75 kGy, but more than that led to degradation [11]. Shin et al. investigated PA6 irradiated by beta radiation with maximum intensity 200 kGy. The results showed that mechanical properties improved, and especially elastic modulus rose by 47% in comparison with unaltered PA6 [12]. These findings are consistent with the results of this study.

Shin et al. [13] studied the effect of beta radiation on the morphological, rheological and mechanical properties of polymer blend composed of PA 6 and PP. It was found that electron radiation improved properties, although they were not measured in the surface layer.

Bradler et al. [14] tested commonly accessible virgin and filled PAs with 30 and 35 wt.% glass fibres and 5 wt.% TAIC. These samples were irradiated with 50, 100, 150 and 200 kGy. Contrary to the aforementioned study, in this case, the variation in TAIC concentration was smaller, while the radiation doses varied more widely.

Studies [15,16] have shown that irradiation can lead to improved mechanical properties, and that the addition of cross-linking agent, e.g., TAIC, enhances this effect. These studies confirm the findings presented in this work.

Other authors [17–20] have investigated the effect of beta radiation on the mechanical and thermal properties of the observed polymer types. It was found that finding the optimal dose of radiation always leads to improved material properties. Numerous polymers are investigated in this study. This study focuses on varying concentrations of CLA in PA 6, in order to obtain a more detailed description of this problem.

The measured results construct a basis upon which a more detailed study of this problem can be built. Finding deeper correlations of various dependencies requires a higher degree of repeatability during measurements, which is more demanding of time and finances.

Overall, this study deals with the influence of varying concentrations of cross-linking agents (2, 4 and 6%) on the micro-mechanical properties of irradiated polyamide 6. Individual samples were irradiated by electron radiation with differing intensities (66, 99 and 132 kGy). The main goal of this work is to find the optimal concentration of CLA and the optimal radiation dose at which to improve micro-mechanical properties. Suitable processing conditions can lead to more economical processes, thus generating better profit. The degree of irradiation was evaluated by thermo-mechanical analysis (TMA), while the degree of polyamide degradation was investigated by Fourier-transform infrared spectroscopy (FTIR).

2. Materials and Methods

The tested polyamide 6 was injection moulded with three different concentrations of CLA (0, 2, 4 and 6 wt.%). These samples were then irradiated by an electron beam with differing intensities (66, 99 and 132 kGy). Following manufacturing, the samples were measured with respect to their mechanical properties, i.e., indentation hardness, modulus and creep. Degree of irradiation was evaluated by thermo-mechanical analysis (TMA), while the occurrence of the degradation processes taking place was proved by means of Fourier-transform infrared spectroscopy (FTIR). Each sample combination was measured ten times, and these results were subsequently statistically evaluated by arithmetic mean and standard deviation. Table 1 presents the designation of individual samples.

Table 1. Designation of test samples.

Sample	CLA Concentration [%]	Radiation Dose [kGy]
A	0	-
B	2	66, 99, 132
C	4	66, 99, 132
D	6	66, 99, 132

2.1. Material

Polyamide 6 with commercial designation PA 6 Frianyl B63 VN made by Frisset (Utzenfeld, Germany) was chosen as the experimental material. This material has a broad range of application in industrial practice, and is often used for complex parts. Polyamide 6 is a semi-crystalline polymer with high strength. This material was modified through the addition of TAIC in the following concentrations: 2, 4 and 6 wt.%.

2.2. Sample Preparation

Test samples were manufactured by injection moulding in an Arburg Allrounder 470 H injection machine, which was made by Arburg (Losburg, Germany). Test samples were prepared in accordance with the ČSN EN ISO 179-1 standard in the shape of a rectangular block with dimensions 4 × 80 × 10 mm. The injection moulding parameters were set according to the manufacturer's recommendations, as provided on the material sheet. This information can also be found in Table 2.

Table 2. Technological parameters of the injected samples.

Injection Parameter	Unit	Value
Pressure	MPa	80
Velocity	mm/s	50
Length of a dose	mm	26
Temperature under the hopper	°C	70
Cooling time	s	20
Holding pressure	MPa	68
Holding pressure duration	s	10
Heat zone settings		
Zone n. 1	°C	220
Zone n. 2	°C	230
Zone n. 3	°C	245
Zone n. 4	°C	265

Polyamide 6 is characterised by its tendency to absorb water, which means it is necessary to dry the material before injection moulding. The drying process was performed in a Thermolift 100-2 dryer made by Arburg (Losburg, Germany). The parameters of the drying process were as follows: drying temperature 120 °C and time of drying 4 h. Material was dried and injected in the form of granules, which were provided by the manufacturer. Transport of the material from the dryer to the injection machine was carried out using a pneumatic transporter in a manner such that contact with additional humidity was prevented.

2.3. Sample Irradiation

Prepared samples were irradiated in cooperation with BGS Beta Gamma Service GmbH & Co., KG company (Saal an der Donau, Germany). Radiation was applied in a high-voltage accelerator of the Rhodotron type with a maximal energy output of 10 MeV. Radiation doses were set within the range commonly found in practice (66, 99 and 132 kGy), and irradiation was performed at ambient temperature in a normal atmosphere. Polyamide 6 was enhanced by TAIC in 2, 4 and 6 wt.%. TAIC enables chemical bonding of free radicals to applied CLA, leading to a certain degree of cross-linking.

2.4. Micro-Indentation Properties

Micro-indentation properties were measured using an MCT^2 instrumented hardness meter provided by CSM (Graz, Austria). Parameters for the measurement were set as follows: applied load 1 N, loading and de-loading speed 2 N/min, maximum load duration 90 s. A Vickers four-sided pyramid with tip angle 136° was chosen as the indentation tip. This measurement was realised in accordance with the ČSN EN ISO 14577-1 standard, and indentation hardness, modulus and creep were evaluated based on this standard. All obtained parameters were evaluated using the method of Oliver and Pharr (Figure 1). The derivation of the measured mechanical properties can be seen below.

Indentation hardness H_{IT} measures the resistance to permanent deformation or damage, where F_{max} is the maximum force and A_p is the projected contact area (theoretical or calibrated) [20–22].

$$H_{IT} = \frac{F_{max}}{A_p} \quad (3)$$

$$A_p = 23.96 \cdot h_c^2 \quad (4)$$

As described in the ISO 14577 standard, the reduced modulus, E_r, is used to account for the fact that the elastic displacements occur in both the indenter and the sample. The instrumented elastic modulus in the test material, E_{IT}, can be calculated from E_r using the following formula [20–22]:

$$E_{IT} = E^* \cdot (1 - {\nu_s}^2) \quad (5)$$

$$E^* = \frac{1}{\frac{1}{E_r} - \frac{1-\nu_i^2}{E_i}} \quad (6)$$

$$E_r = \frac{\sqrt{\pi}}{2 \cdot C\sqrt{A_p}} \quad (7)$$

where ν_s is the Poisson's ratio of the sample (polymer 0.3 to 0.4), and E_i and ν_i are the elastic modulus (diamond 1141 GPa) and Poisson's ratio (0.07), respectively, of the indenter.

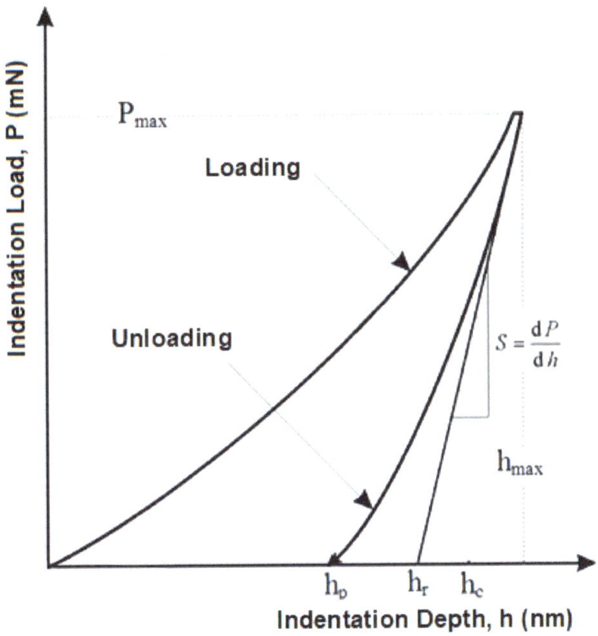

(**a**)

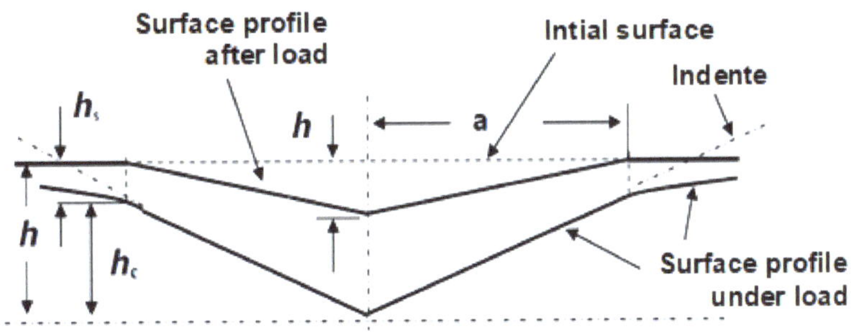

(**b**)

Figure 1. Principle of indentation: (**a**) load vs. indentation depth; (**b**) schematic representation of the indentation processes showing the decreasing of the indentation depth during loading (according to Oliver and Pharr).

The instrumented hardness test allows the measurement of changes in imprint depth under constant load. The results can be used to calculate relative depth of imprint, i.e., creep. Indentation creep can be calculated using the following equation [20,22]:

$$C_{IT} = \frac{h_2 - h_1}{h_1} 100 \tag{8}$$

where h_1 is the depth of imprint in time t_1 after the testing load is reached, and h_2 is the depth of imprint in time t_2 that is reached at maximal testing load P_{max} (Figure 2) [22].

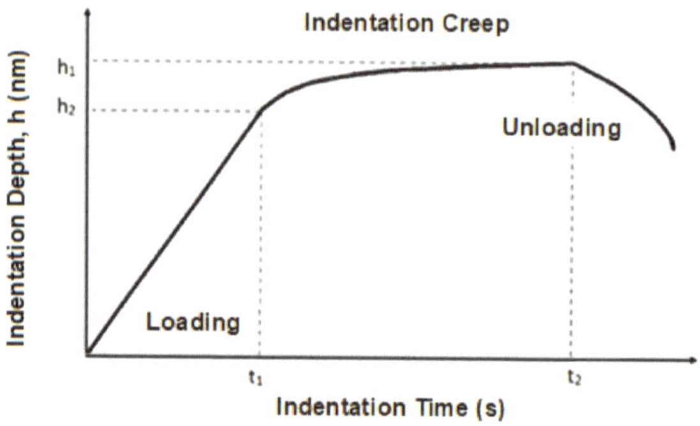

Figure 2. Indentation creep.

2.5. Thermo-Mechanical Analysis (TMA)

Measurement of the degree of polyamide 6 cross-linking was performed by thermo-mechanical analysis using a DMA made by Mettler Toledo (Langacher, Greifensee, Switzerland). The measurement was performed on three test samples with dimensions 10 × 4 × 10 mm for all TAIC concentrations and radiation doses. The process parameters are shown in Table 3 and Figure 3. Deformation calculation was carried out according to Equation (9).

$$\frac{\text{deformation of the sample } (\mu\text{m})}{\text{thickness of the sample } (\mu\text{m})} \times 100 = \text{deformation } (\%) \quad (9)$$

Table 3. Parameters of TMA measurements.

Technological Parameter	Unit	Value
Starting temperature	°C	180
Dwelling time at temperature	min	10
Load	N	1
Speed of heating	K/min	1
Final temperature	°C	250

2.6. Fourier-Transform Infrared Spectroscopy (FTIR)

Measurement of FTIR spectra was carried out on an AVATAR 320 made by Nicolet (Watertown, MA, USA). The method used was ATR with ZnSe crystal (zinc selenide). The background was scanned before and after each sample, and subsequently deducted in order to obtain more accurate measurements. This was done on three test samples with each TAIC concentration and radiation dose.

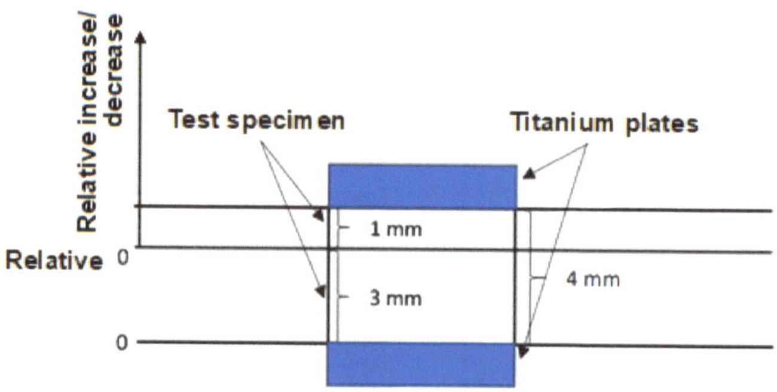

Figure 3. TMA measurement principle.

3. Results

The goal of this study was to show the effect of CLA concentration and radiation intensity on the indentation hardness, modulus, and creep of polyamide 6. The main goal was to find the optimal concentration of CLA and radiation intensity to be able to achieve a significant improvement in the mechanical properties of the surface layer of a commonly used construction polymer (polyamide 6). Micro-hardness was measured as a part of the measurements of micro-mechanical properties. This test is sufficiently sensitive to catch even small changes in observed properties.

3.1. Micro-Indentation Properties

The most important area of this research is the measurement of micro-mechanical properties, which was performed using contemporary technology, i.e., via the instrumented hardness test. This test provides a complex description of the mechanical behaviour of polymer materials in a non-destructive way. The material characteristics evaluated were indentation hardness, modulus and creep. These values can be obtained from the characteristics given in Figure 4.

Figure 4a–c present the indentation characteristics of indentation force as a function of depth of indentation for individual concentrations of CLA and radiation doses. This dependence provides initial information about polymer behaviour and represents a basis for the calculation of indentation hardness and modulus. It can be stated that more leftwards dependencies signify improved mechanical properties, because the indentation reached a lower depth.

Figure 4a–c show the indentation characteristics of indentation depth as a function of indentation time. When the maximum indentation force (1 N) is reached, it stays the same for a predetermined duration, during which the indentation depth increases. The indentation depth is determined based on these changes in indentation depth. The angle of the curve representing this process is used to determine the creep behaviour of the individual materials.

Hardness is one of the most important parameters for describing the mechanical behaviour of the tested surface layer of polyamide 6. With increasing hardness, the product can be used in more demanding applications, as it will present higher resistance against scratching and tearing.

The indentation hardness results indicate that both the concentration of CLA TAIC and the radiation intensity have a significant effect on indentation hardness (Figure 5). The indentation hardness of sample A was 154 MPa. The indentation hardness in sample B increased with increasing radiation intensity, which was capped at 132 kGy. The most significant increase was found in the sample irradiated with 66 kGy (190 MPa). In subsequent samples, further irradiation led to smaller, although still significant, increases. Indentation hardness was 196 MPa in the sample irradiated by 99 kGy and 200 MPa in the sample irradiated by 132 kGy. There was a 30% difference between sample A and sample B irradiated by 132 kGy. The greatest increase in indentation hardness for sample C was found in the sample irradiated by 66 kGy (209 MPa). Samples irradiated at greater intensity displayed quite similar indentation hardness values. This value was 36% higher for the sample irradiated by 66 kGy than it was for sample A. Sample D was measured as possessing an indentation hardness of 213 MPa following irradiation by 66 kGy. In comparison with the base material, this was a 38% increase. Higher intensities led to decrease in indentation hardness values, which was due to the degradation processes occurring with high radiation doses.

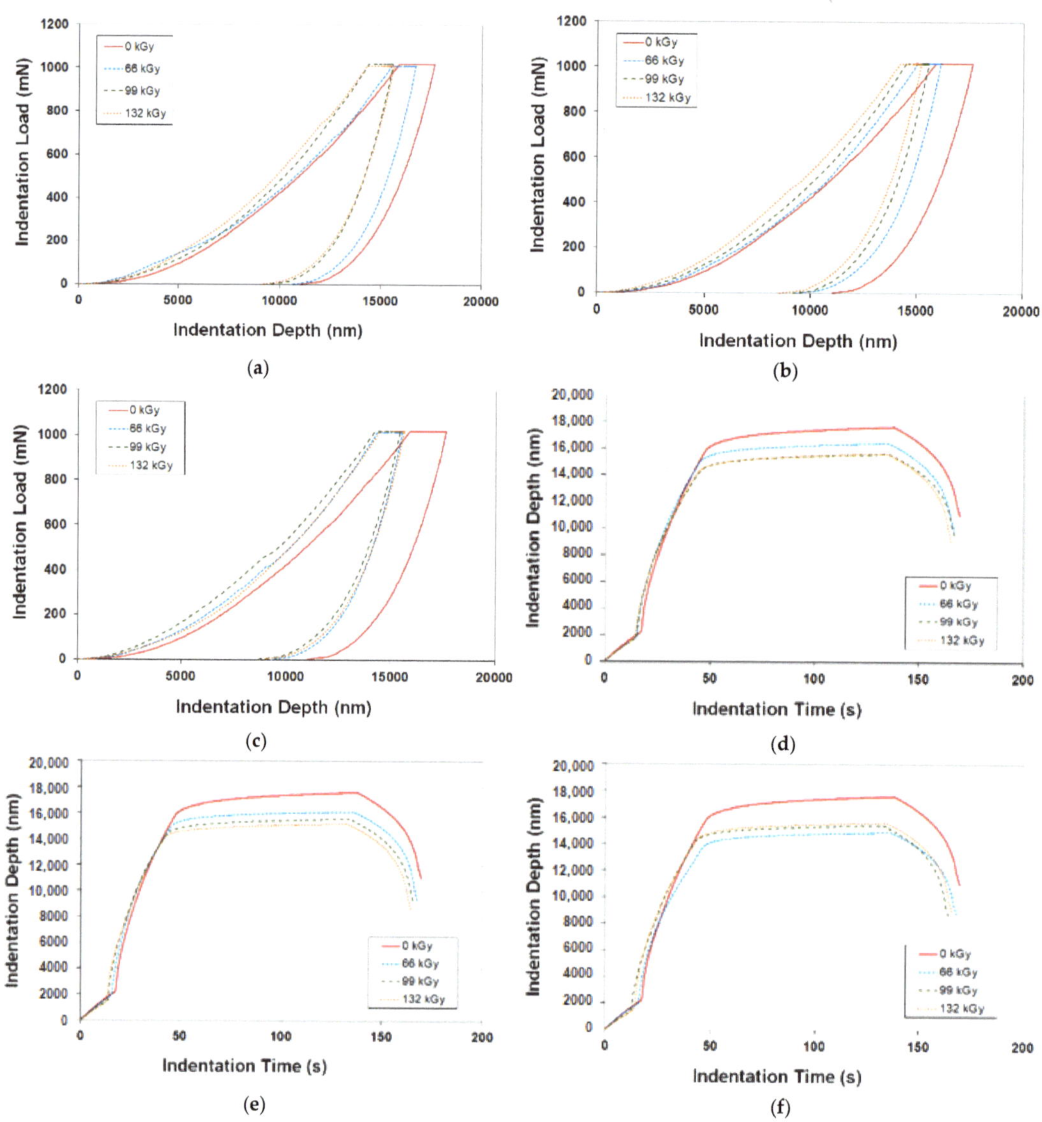

Figure 4. Indentation characteristics: (**a**) indentation force vs. depth of indentation—concentration of TAIC 2%; (**b**) indentation force vs. depth of indentation—concentration of TAIC 4%; (**c**) indentation force vs. depth of indentation—concentration of TAIC 6%; (**d**) depth of indentation vs. time of indentation—concentration of TAIC 2%; (**e**) depth of indentation vs. time of indentation—concentration of TAIC 4%; (**f**) depth of indentation vs. time of indentation—concentration of TAIC 6%.

The indentation hardness results indicate that the optimal radiation for polyamide 6 improvement was 66 kGy, while the optimal concentration of TAIC was 4 or 6 wt.%. Higher radiation intensity made indentation hardness more likely to decrease to increase.

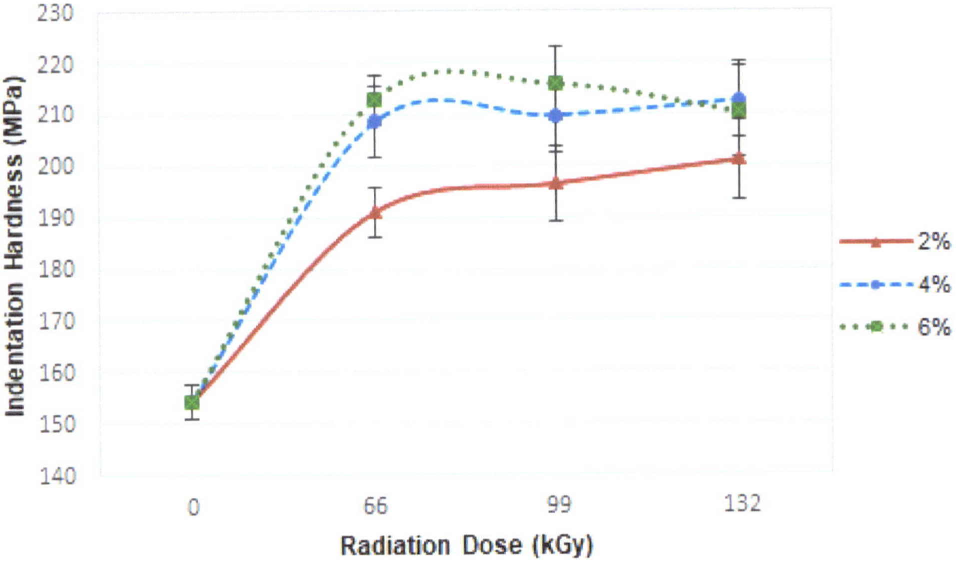

Figure 5. Dependence of indentation hardness on radiation intensity and TAIC concentration.

A further critical parameter for characterising the mechanical behaviour of materials is elastic modulus. Great material resistance can be achieved by improving the elastic modulus of the tested polyamide 6.

The results of elastic modulus indicate that both CLA concentration and radiation intensity have a significant effect on indentation modulus (Figure 6). The indentation modulus of sample B irradiated by 66 kGy was 5.5 GPa. Similar tendencies were found in samples irradiated by 99 and 132 kGy. The difference between the virgin (0 kGy) polyamide 6 (4.6 GPa) and the irradiated one (66 kGy) was 20%. The highest indentation modulus value obtained by samples C (5.8 GPa) was found for the sample irradiated by 132 kGy. The difference between the base material and sample C irradiated by 132 kGy was 26%. The best improvement in indentation modulus (5.8 GPa) was found in sample D irradiated by 99 kGy. In comparison with the virgin material, this represented a 26% increase. Indentation modulus decreased at 132 kGy, mostly likely as a result of degradation processes occurring due to the high intensity of the radiation.

Indentation creep is an important parameter for characterising material under long-term load (use). On the basis of indentation creep, it is possible to estimate how the product will behave in practice over a longer period of time.

Indentation creep tendency was similar for all radiation intensities (Figure 7). The base material had a creep of 10.5%. The most interesting development in indentation creep was measured in samples irradiated by 66 kGy. On the other hand, the indentation creep values became worse with increasing intensity. However, in the end, samples D demonstrated the best results, especially when irradiated with 66 kGy, following which the value of indentation creep was 7.7%. The difference between virgin material and that irradiated by 66 kGy was 36%.

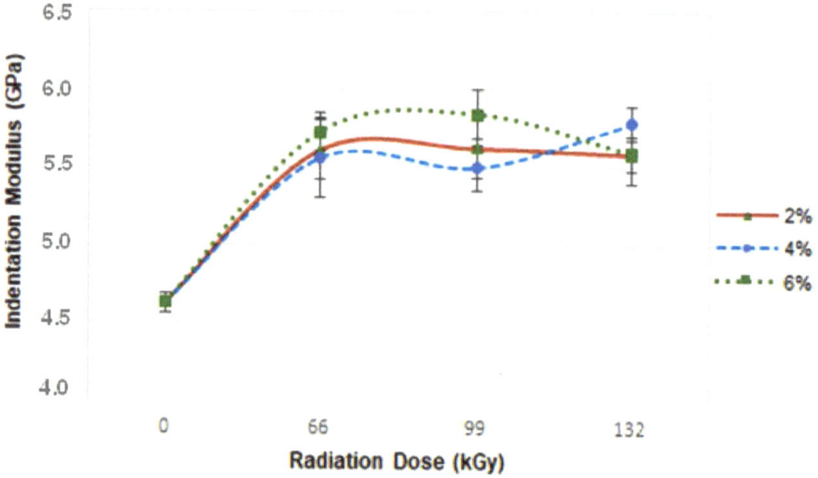

Figure 6. Dependence of indentation modulus on radiation intensity and TAIC concentration.

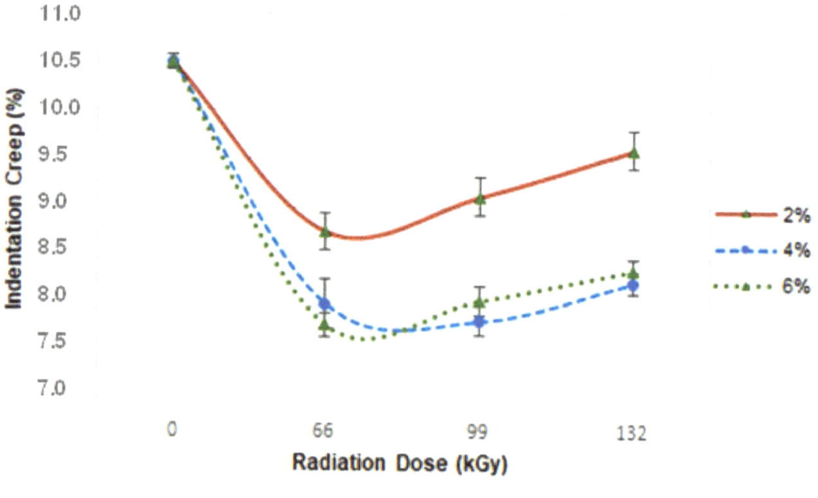

Figure 7. Dependence of indentation creep on radiation intensity and TAIC concentration.

This significant change in the behaviour of polyamide 6 was caused by the cross-linking of the polymer. The cross-linking of polymers is a chemical process, during which transverse bonds are created in the polymer structure. Infinite 3D structure, i.e., spatial network or gel, is produced due to these transverse bonds. Cross-linking induced by radiation is a result of the numerous free radical polymers moving around, especially in amorphous regions. During cross-linking, two opposite events occur simultaneously, i.e., cross-linking and degradation (scission of the main chain). The cross-linking of a polymer and its degree of cross-linking depend on many factors, including crystallinity, glass transition temperature, chemical structure of the polymer, etc.

3.2. Thermo-Mechanical Analysis (TMA)

TMA was performed to confirm the degree of irradiation of the test samples. In the case of TMA, two parameters were measured, i.e., thermal expansion and the flow of the non-cross-linked part. These two parameters influence one another, and are dependent on the currently dominating one. The curve can either go up due to thermal expansion or down due to thermal degradation. Similar results of deformations were found in all test samples, as can be seen in Figures 8–10. The curves in the graphs were generated from the median values of three measurements. In order to provide better a picture of the measured trend, deformation was determined at different temperatures (230, 240 and 248 °C).

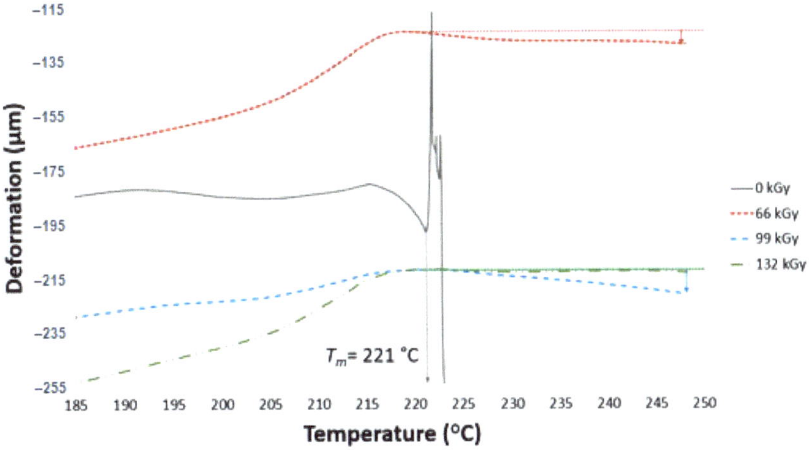

Figure 8. TMA of sample B.

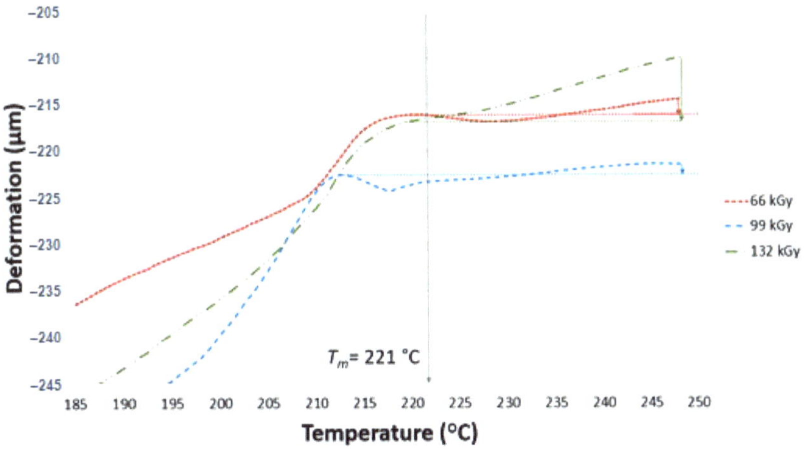

Figure 9. TMA of sample C.

Figure 8 displays the deformation of samples A and B. As can be seen, the virgin sample (0 kGy) exhibited a melting temperature of 221 °C. On the other hand, the irradiated polyamide 6 showed different mechanical behaviour from the virgin material. The interrupted line represents a tangent that indicates where the deformations were taken from at all three temperatures (230, 240 and 248 °C). The deformation of the test sample started to rapidly increase when the temperature increased beyond 217 °C, and this phenomenon remained in effect until a temperature of 248 °C. The test sample exposed to 66 kGy showed improved mechanical behaviour with increasing temperature. Deformation was −0.09% at 230 °C, −0.1% at 240 °C and −0.12% at 248 °C.

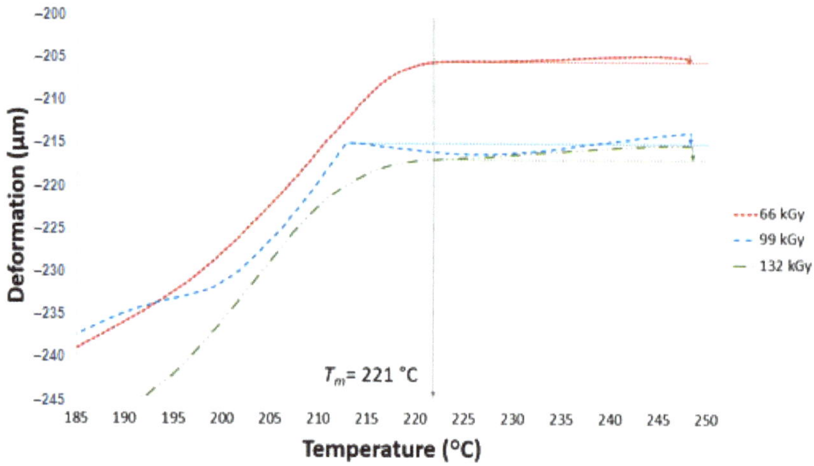

Figure 10. TMA of sample D.

The test sample irradiated by 99 kGy exhibited differences similar to those in the previous sample. The deformation was −0.06% at 230 °C, −0.15% at 240 °C and −0.23% at 248 °C. This slightly higher deformation in comparison with the sample irradiated by 66 kGy could be a result of insufficient cross-linking in the entire volume.

The highest radiation intensity (132 kGy) resulted in the lowest deformation of all of the test samples B. The measured deformation was −0.02% at 230 °C, −0.01 at 240 °C and −0.03% at 248 °C. On the basis of the mechanical behaviour angle, this radiation dose resulted in improved cross-linking compared to earlier test samples.

Figure 9 presents the course of TMA analysis for sample C. Deformation grew by −0.2% with the increase in temperature from 222 °C to 230 °C and after this temperature, thermal expansion began to dominate, as manifest by the curve growth. Deformation was 0.01 at 240 °C, while the thermal expansion was 0.04% at 248 °C.

Polyamide 6 irradiated by 99 kGy exhibited a slight decrease at 218 °C, which was probably caused by the incorrect clamping of the test sample between the titan discs. Higher temperatures enabled the domination of thermal expansion to be observed, which was 0% at 230 °C, 0.02% at 240 °C and 0.03% at 248 °C.

PA 6 exposed to 132 kGy exhibited the dominance of thermal expansion from a temperature of 220 °C. Thermal expansion was 0.05% at 230 °C, 0.12% at 240 °C and 0.17% at 248 °C. There was no deformation in this case. Value of thermal expansion as high as those measured makes it likely that the polyamide 6 was adequately cross-linked, and there was need to neither add more TAIC nor increase radiation intensity.

Figure 10 presents the thermo-mechanical analysis of sample D. This sample, irradiated by 66 kGy, exhibited the best balance of thermal expansion and flow of the non-cross-linked part. Thermal expansion was 0% at 230 °C, 0.02% at 240 °C and 0.01% at 248 °C. No significant deformation was observed, here.

The test sample irradiated by 99 kGy showed deformation from a temperature of 213 °C. This deformation prevailed until 227 °C, where thermal expansion started, as well. At 230 °C, deformation was still dominating, at −0.03%. At 240 °C, deformation was zero and thermal degradation was 0.01%. Thermal expansion at 248 °C was 0.03%.

PA 6 irradiated by the highest radiation dose showed thermal expansion of 0.01% at 230 °C, 0.04% at 240 °C and 0.04% at 248 °C.

The thermo-mechanical analysis results for sample C exposed to 132 kGy led to significant cross-linking. Samples D cross-linked after exposure to all radiation doses, and therefore they all demonstrated significant improvement in their mechanical property values.

3.3. Fourier-Transform Infrared Spectroscopy (FTIR)

Figures 11–13 display the FTIR analysis of the test samples. FTIR measurements were performed in order to evaluate the degree of degradation of polyamide 6 when exposed to varying intensities of radiation. The curves in the spectra were created on the basis of the median of three measurements for each test sample. The presence of chemical groups was determined via infrared spectroscopy. A comparison of IR spectra demonstrated slight material degradation, represented by the 1736 cm^{-1} belt, which is typical for carbonyl groups. Degradation is commonly also indicated by hydroxyl groups, as shown by 3500 cm^{-1} belts, but there was no trace of these groups in this area. The vibrations at 3290, 3074, 1635, 1537, 964 and 667 cm^{-1} demonstrate the presence of amide groups. The vibrations at 2920, 2852, 1463, 1352 and 1259 cm^{-1} demonstrate the presence of –CH2 groups, also indicating the presence of aliphatic chains. The intensity of these belts decreases with increasing radiation dose, which indicates decreased mobility of the –CH2 groups. The belt located in the 667 cm^{-1} area is a vibration of amide in crystalline phase α and β. Crystalline phase α is characterised by vibration with a wave number of 1201 cm^{-1}, which can be attributed to the deformation vibration with a fan-like shape of –CH2. The branch of this belt increased in the irradiated test samples. The belt with the wave number 1697 cm^{-1}, which is connected with CLA TAIC, was lacking in both test samples. This indicates that the addition was most likely consumed in reactions connected with PA 6 irradiation.

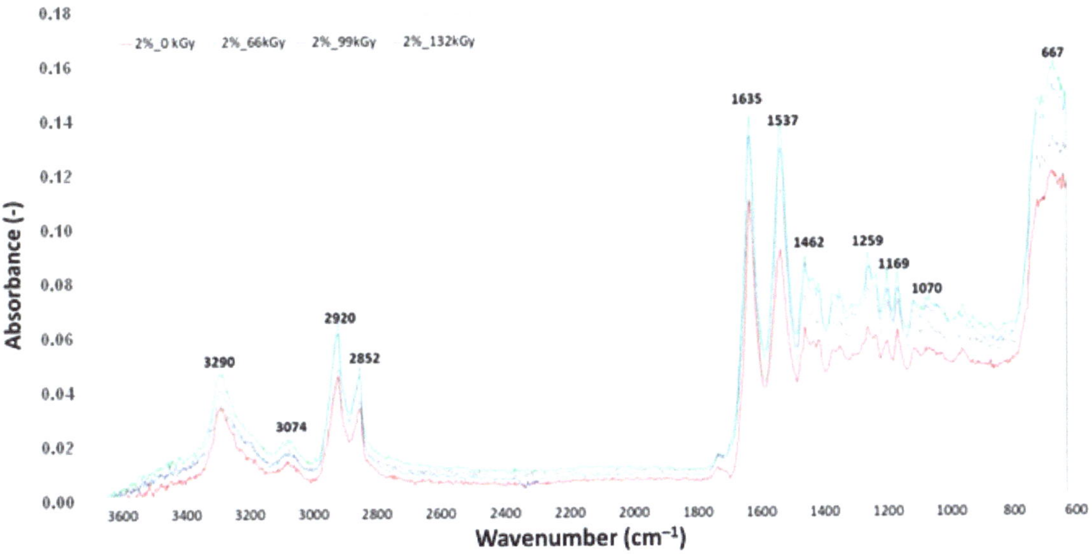

Figure 11. Fourier-transform infrared spectroscopy of sample B.

Table 4 shows the values of absorbance at wave number 1736 cm^{-1}, which is common for carbonyl groups, and represents the degradation of PA 6. The absorbance marked in green is the lowest with a value of 0.00613, which corresponds with the results obtained for micro-mechanical properties, where sample C irradiated by 132 kGy showed the highest values of indentation hardness and elastic modulus. At the same time, the TMA results indicate that this material has the highest degree of cross-linking. The absorbance value marked in red (0.01653) belongs to test sample B irradiated by 99 kGy. This test sample showed the highest degradation.

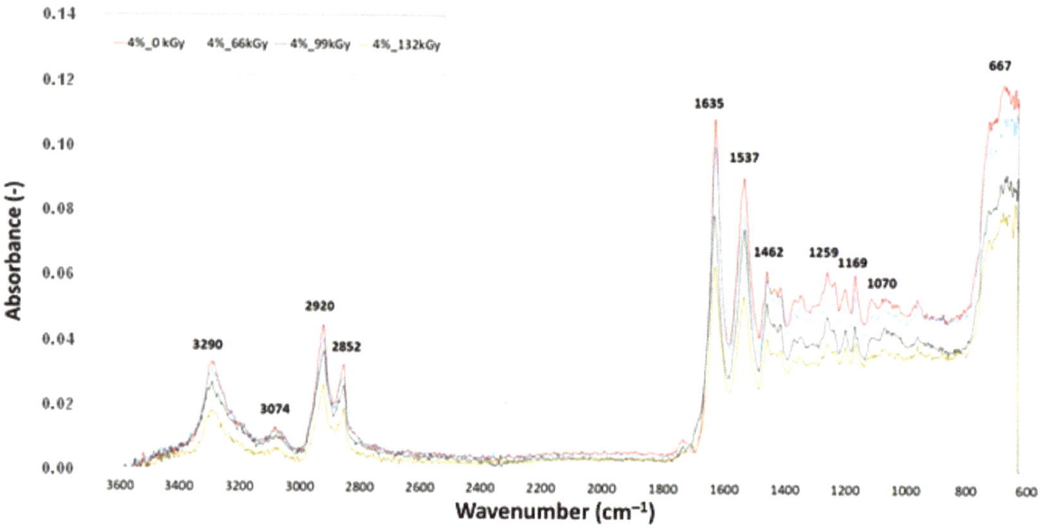

Figure 12. Fourier-transform infrared spectroscopy of sample C.

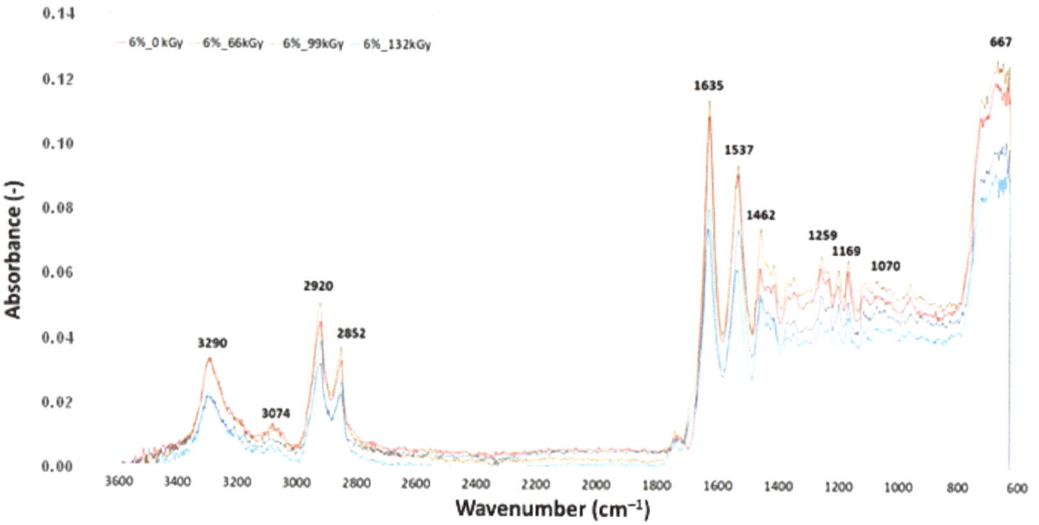

Figure 13. Fourier-transform infrared spectroscopy of sample D.

Table 4. Values of absorbance at wavenumber 1736 cm^{-1}.

Specimen	Absorbance
A, 0 kGy	0.00979
B, 66 kGy	0.01230
B, 99 kGy	0.01653
B, 132 kGy	0.01448
C, 66 kGy	0.01219
C, 99 kGy	0.01119
C, 132 kGy	0.00613
D, 66 kGy	0.00988
D, 99 kGy	0.01225
D, 132 kGy	0.00848

4. Conclusions

In this study, the effect of CLA concentration on the indentation hardness, modulus and creep of irradiated PA 6 (66, 99 and 132 kGy) was studied. of the occurrence of cross-linking was evidenced by the thermal deformation and expansion values obtained by TMA analysis. Proof of degradation was further obtained by FTIR spectroscopy.

The micro-mechanical property results indicate that the best combinations of radiation and CLA were those in sample C irradiated by 132 kGy and in sample D irradiated by 66 kGy. The combinations resulted in the highest values of indentation hardness and indentation modulus. On the other hand, the lowest values of indentation creep were found in samples with the same TAIC concentration irradiated by 66 kGy. On the basis of the TMA measurements, it was found that the highest cross-linking of PA 6 was reached in sample C irradiated by 132 kGy. From an economic perspective, it is more advantageous to use a lower radiation intensity (e.g., sample D irradiated by 66 kGy), thus achieving high mechanical properties, but also economic relief due to fewer passes under the electron accelerator being required.

FTIR spectroscopy showed a slight degradation of the test samples, indicated by presence of carbonyl groups in vibration belt 1736 cm^{-1}. The absence of the belt with the wave number 1697 cm^{-1}, indicating the presence of TAIC, shows that the added CLA was consumed during the irradiation process.

Since CLA is crucial for the irradiation process, its absence means that the polymer is degrading upon irradiation. It is important to adopt correct content of this agent. It is also important to choose the correct dose of radiation.

Author Contributions: Conceptualisation, M.O.; methodology, M.O. and M.B.; formal analysis, M.S. and M.B.; data curation, M.O. and M.S.; writing—original draft preparation, M.O.; visualisation, M.O.; project administration, M.O.; funding acquisition, M.S. All authors have read and agreed to the published version of the manuscript.

Funding: This article was written with the support of the project TBU in Zlin Internal Grant Agency: No. IGA/FT/2023/005.

Institutional Review Board Statement: Not applicable.

Informed Consent Statement: Not applicable.

Data Availability Statement: The data presented in this study are available on request from the corresponding author.

Conflicts of Interest: The authors declare no conflict of interest.

References

1. Brocka, Z. *Werkstoff- und Einsatzpotential Strahlenvernetzter Thermoplaste, Lehrsthulfür Kunststofftechnik (LKT)*; Universität Erlangen-Nürnberg: Nürnberg, Germany, 2008.
2. Kaindl, K.; Graul, E.H. *Strahlenchemie*; Dr. Alfred Hüttig Verlag: Heildelberg, Germany, 1967.
3. Rexer, E.; Wuckel, L. *Chemische Veränderungen von Stoffen durch Energiereiche Strahlung*; VEB Deutscher Verlag fürGrundstoffindustrie: Leipzig, Germany, 1965.
4. Grigg, M.N. Thermo-Oxidative Degradation of Polyamide 6. Ph.D. Thesis, Philosophy, Queensland University of Technology, Brisbane, QLD, Australia, 2006.
5. Karstens, T.; Rossbach, V. Thermo-oxidative degradation of polyamide 6 and 6, 6. Kinetics of the formation and inhibition of UV/VIS-active chromophores. *Die Makromol.* **1989**, *190*, 3033–3053. [CrossRef]
6. Dadbin, S.; Frounchi, M.; Goudarzi, D. Electron beam induced crosslinking of nylon 6 with and without the presence of TAC. *Polym. Degrad. Stab.* **2005**, *89*, 436–441. [CrossRef]
7. Pramanik, N.K.; Haldar, R.S.; Bhardwaj, Y.K.; Sabharwal, S.; Niyogi, U.K.; Khandal, R.K. Radiation processing of Nylon 6 by e-beam for improved properties. *Radiat. Phys. Chem.* **2009**, *78*, 199–205. [CrossRef]
8. Porubska, M.; Janigova, I.; Jomova, K.; Hodak, I.C. The effect of electron beam irradiation on properties of virgin and glass fiber-reinforced polyamide 6. *Radiat. Phys. Chem.* **2014**, *102*, 159–166. [CrossRef]
9. Adem, E.; Burillo, G.; del Castillo, L.F.; Vasquez, M.; Avalos-Borja, M.; Marcos-Fernandez, A. Polyamide-6: The effects on mechanical and physiochemical properties by electron beam irradiation at different temperatures. *Radiat. Phys. Chem.* **2014**, *97*, 165–171. [CrossRef]

10. Burillo, G.; Adem, E.; Munoz, E.; Vasquez, M. Electron beam irradiated polyamide-6 at different temperatures. *Radiat. Phys. Chem.* **2013**, *84*, 140–144. [CrossRef]
11. El-Nemr, K.F.; Hassan, M.M.; Ali, M.A. Effect of electron beam irradiation on mechanical and thermal properties of waste polyamide copolymer blended with nitrile-butadiene rubber. *Polym. Adv. Technol.* **2010**, *21*, 735–741. [CrossRef]
12. Shin, B.Y.; Kim, J.H. Rheological and mechanical properties of polyamide 6 modified by electron-beam initiated mediation process. *Radiat. Phys. Chem.* **2015**, *112*, 88–96. [CrossRef]
13. Shin, B.Y.; Ha, M.H.; Han, D.H. Morphological, Rheological, and Mechanical Properties of Polyamide 6/Polypropylene Blends Compatibilized by Electron-Beam Irradiation in the Presence of a Reactive Agent. *Materials* **2016**, *9*, 342. [CrossRef] [PubMed]
14. Bradler, P.R.; Fischer, J.; Wallner, G.M.; Lang, R.W. Characterization of irradiation crosslinked polyamides for solar thermal applications—Fatigue properties. *Compos. Sci. Technol.* **2019**, *175*, 55–59. [CrossRef]
15. Porubska, M.; Szollos, O.; Janigova, I.; Jomova, K.; Chodak, I. Crosslinking of polyamide-6 initiated by proton beam irradiation. *Radiat. Phys. Chem.* **2017**, *133*, 52–57. [CrossRef]
16. Nino, C.G.; Vidal, J.; Del Cerro, M.; Royo-Pascual, L.; Murillo-Ciordia, G.; Castell, P. Effect of Gamma Radiation on the Processability of New and Recycled PA-6 Polymers. *Polymers* **2023**, *15*, 613. [CrossRef] [PubMed]
17. Cui, C.W.; Zhang, Y. Effect of Electron Beam Irradiation on the Mechanical and Thermal Properties of Ternary Polyamide Copolymer. *Macromol. Res.* **2018**, *26*, 359–364. [CrossRef]
18. Yang, Y.; Xi, C.; Lu, N.; Gao, F. *Injection Molding Process Control, Monitoring and Optimization*; Carl Hanser Verlag GmbH &, Co., KG: München, Germany, 2016; ISBN 978-1-56990-592-0.
19. Banerjee, S.S.; Janke, A.; Gohs, U.; Fery, A.; Heinrich, G. Some nanomechanical properties and degree of branching of electron beam modified polyamide 6. *Eur. Polym. J.* **2017**, *88*, 221–230. [CrossRef]
20. Manas, D.; Mizera, A.; Navratil, J.; Manas, M.; Ovsik, M.; Sehnalek, S.; Stoklasek, P. The electrical, mechanical and surface properties of thermoplastic polyester elastomer modified by electron beta radiation. *Polymers* **2018**, *10*, 1057. [CrossRef] [PubMed]
21. Manas, D.; Mizera, A.; Manas, M.; Ovsik, M.; Hylova, L.; Sehnalek, S.; Stoklasek, P. Mechanical properties changes of irradiated thermoplastic elastomer. *Polymers* **2018**, *10*, 87. [CrossRef] [PubMed]
22. Ovsik, M.; Manas, M.; Stanek, M.; Dockal, A.; Vanek, J.; Mizera, A.; Adamek, M.; Stoklasek, P. Polyamide surface layer nano-indentation and thermal properties modified by irradiation. *Materials* **2020**, *13*, 2915. [CrossRef] [PubMed]

materials

MDPI

Article

Experimental and Numerical Analysis of the Residual Stresses in Seamed Pipe in Dependence on Welding and Metal Forming

Pavel Solfronk *, Jiří Sobotka, Šárka Bukovská and Josef Bradáč

Department of Engineering Technology, Faculty of Mechanical Engineering, Technical University of Liberec, Studentská 1402/2, 46117 Liberec, Czech Republic
* Correspondence: pavel.solfronk@tul.cz; Tel.: +420-485-353-374

Abstract: Concerning the increasingly widespread utilization of the finite element method (FEM), the concept of the so-called virtual factory is also gaining ground, and not only in the engineering industry. This approach does not use numerical simulations of individual production technologies separately but treats the entire production process as a chain of interrelated technologies. Thus, the output data from one technology is taken as input data for the following technology. The resulting thermal and mechanical effects are then not only dealt with within one technology but always comprehensively within the production process. In the consideration of the loading and subsequent service lives of manufactured components, values of residual stresses are one of the very important characteristics. For these reasons, this paper deals with the effect of residual stresses' magnitude and distribution during the formation and the final springback of the seamed pipe end section with and without respect to the influence of the preceding welding. The resulting residual stress values from numerical simulations are subsequently compared with the actual values of residual stresses experimentally measured using X-ray diffraction.

Keywords: virtual factory; numerical simulation; X-ray diffraction; welding; metal forming; residual stresses; Sysweld; PAM-Stamp 2G; weld zone; heat affected zone

Citation: Solfronk, P.; Sobotka, J.; Bukovská, Š.; Bradáč, J. Experimental and Numerical Analysis of the Residual Stresses in Seamed Pipe in Dependence on Welding and Metal Forming. *Materials* **2023**, *16*, 2256. https://doi.org/10.3390/ma16062256

Academic Editor: Thomas Niendorf

Received: 23 January 2023
Revised: 2 March 2023
Accepted: 8 March 2023
Published: 10 March 2023

1. Introduction

The general effort to account for the effects of various technological operations within a complex whole in numerical simulations is currently a major challenge in the pre-production process. In this case, there is a considerable influence of residual stresses after welding (especially in the heat-affected zone—HAZ) on the accuracy of metal forming numerical simulations. That is why two specialized FEM software programs (Sysweld—v.2021 for welding and PAM-Stamp 2G—v2022.0 for metal forming) were linked in this paper.

Nowadays, the utilization of the FEM represents quite a common approach to solving a substantially wide range of different engineering problems. All of the theoretical aspects of the FEM in light of the engineering industry, such as math equations, basic concepts and reviews of the mathematics and mechanics of materials, are given in the work of Kim et al. [1]. Highly appropriate applications of the FEM in structural mechanics and thermal sciences can be found in [2,3]. In terms of the focus of this paper, there are important books that describe general simulations in metal forming and heat transfer [4,5].

Regarding the experimental part of this paper, it was necessary to take into account two major steps: the determination (with the help of X-ray diffraction) or prediction (using the FEM) of the residual stresses induced by the welding of the steel pipe, and numerical simulations of the subsequent metal forming, which has to take into account the previous welding process. There are quite a lot of different studies that deal with these problems. However, the vast majority of them take into account these problems separately, i.e., they describe either the welding process or the metal forming process.

Many authors have already focused on the phenomenon of residual stresses in welding technologies because residual stresses are very important issues in welded parts. A general prediction of the welding-induced residual stresses using the FEM has been investigated in [6–10]. The prediction of the residual stresses through the FEM in a laser beam, an electron beam and arc welding when considering welding technologies can be found in [11–17]. Some studies have already incorporated the prediction of the residual stresses via the FEM and its validation through experimentation. Li et al. [18] investigated the welding-induced residual stresses in P92/SU304 dissimilar metal butt-welded joints using Sysweld software and simultaneously, by the hole-drilling strain gauge method, measured the distribution of the residual stresses on the top and bottom surfaces of the two butt-welded joints. It was proven that predicted welding residual stresses revealed a relatively good agreement with the measured ones. The application of the hole drilling method to compare measured and predicted welding residual stresses was also used in the work of Puymbroeck [19], where it was taken into account its effect on the fatigue life (also studied in [20,21]). In addition to the fatigue properties, great attention is also paid to the heat treatment processes and their influence on the magnitude of the welding-induced residual stresses [22–24]. Some authors studied also the influence of the residual stresses on the distortion of welded material [25–27]. Highly interesting studies were carried out by García et al. [28] and Pavan et al. [29]; these authors applied isotropic and kinematic hardening models to estimate their influence on the welding residual stress distribution. Predicted results were compared with the experimentally measured data obtained from X-ray diffraction, and it was found that residual stress prediction accuracy depends on the hardening models as well. In light of the accurate estimation and minimalization of residual stresses in welding, there is also an effort to take advantage not only of FEM but also of machine learning (MA) and genetic algorithms (GA)—see [30]. As mentioned before, in addition to the hole-drilling method, the X-ray diffraction method is very often used to experimentally measure the magnitude of welding-induced residual stresses (e.g., [31–37]). In addition, there is also an effort to improve the efficiency of X-ray welding image defect recognition, so the deep learning network has also sometimes been used [38].

The prediction of the residual stresses is also very important for metal forming processes. However, the majority of authors, e.g., [39–45], studied this phenomenon as only welding problems or metal forming problems, thus separately and not as two consecutive steps. Some authors [46–48] dealt with the determination of residual stress in formed members and predicted results were compared with the experiments (sectioning, hole drilling method and X-ray diffraction). On the other hand, Lee et al. [49] identified manufacturing variables (and the determination of residual stresses was one of them) that are closely related to the surface defects in the exterior panels of large commercial vehicles. They also investigated the most suitable conditions to minimize such defects. However, in all cited studies, residual stresses were taken into account as a major output or as one of the major effects in light of metal forming.

Nevertheless, there are very few papers, where authors have already incorporated residual stress into their own computation of the metal-forming process. Li et al. [50] studied the effect of the machine-induced residual stresses on the springback of creep age-formed AA2050 plates. They proposed and implemented a simplified residual into a finite element model; a commercial FE software program, PAM-Stamp 2G, was used in this case. They concluded that the required springback prediction of the FE model was significantly improved when the effect of the residual stresses was taken into account. The influence of the welding-induced residual stresses on the subsequent fatigue damage analysis is discussed in [51]. In this case, Wang et al. investigated the fatigue of butt-welded joints using a damage mechanics method. First of all, they determined residual stresses through thermomechanical FEA. Subsequently, the residual stresses were superimposed on the fatigue loading, and such welding-induced plastic damage was taken as initial damage via the elasto-plastic fatigue damage model. Finally, they evaluated, predicted and measured the fatigue properties and relaxation of the residual stresses, which were in good agreement.

2. Materials and Methods

The affordability of high-end HW computing systems has enabled the rapid development of technological processes and numerical simulations at all levels of production. The application of specialized software for solving individual engineering tasks is nowadays relatively common, and FEM results show a progressive match with the real technological processes. The mentioned method shows high flexibility in different designs of the given problem in a relatively short period of time and thus significantly reduces the cost. For these reasons, there can be observed constant pressure and efforts to create the concept of a so-called Virtual Factory, where individual technological tasks will not be solved separately, but as a chain of successive technological processes with mutual iterations and the possibility of optimal correction of the production parameters leading to the production of the desired product. This concept assumes the mutual transferability of input data and obtained results in all stages of such a solution, which has turned out to be one of the main problems in the implementation and development of this concept in practice.

The paper presents the possibility of using a combination of two specialized software programs for the technological processes of welding (Sysweld) and sheet metal forming (PAM-Stamp 2G) developed by ESI Group. For the presentation of the results from experimental measurements and numerical simulations, the forming process of a straight seam welded pipe made from material S355JR with an outer diameter of 133 mm and a wall thickness of 3 mm was chosen. The material used for the production of such pipes is representative of common non-alloy structural steel with low carbon content (max. 0.24 wt%), which is suitable for all types of welding. According to the producer, Huta Pokój Profile Sp. z o.o. (Ruda Śląska, Poland), the joint of the pipe is formed by laser welding without subsequent heat treatment. The shape of the tested pipe with the formed end is shown in Figure 1. In this paper, attention is focused on the analysis of the residual stresses in the formed seam pipe with respect to the influence of both used technologies (welding and metal forming). Values of the residual stresses obtained from numerical simulations are compared with the real values, which are determined through X-ray diffraction. The whole solution process can be summarized as follows:

- Analysis of the tested material mechanical properties to obtain input data for the subsequent numerical simulations;
- Analysis of the residual stresses through X-ray diffraction;
- FEM analysis of the residual stresses for the metal forming in the environment of software PAM-Stamp 2G, marked as **standard FEM**;
- FEM analysis of the residual stresses both for welding and metal forming in the environment of software Sysweld and PAM-Stamp 2G, marked as **modified FEM**.

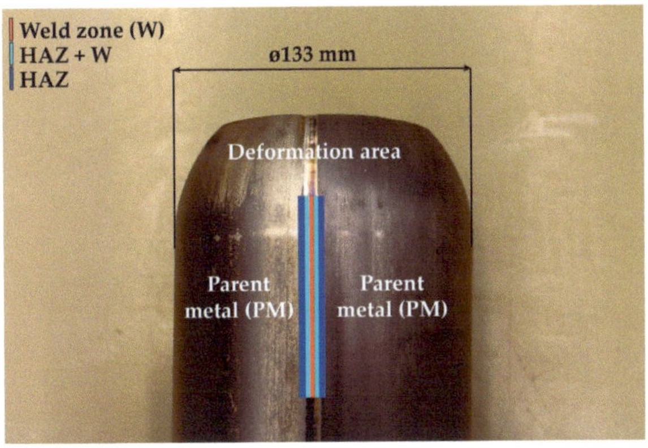

Figure 1. Overview of the analyzed zones (PM, W, HAZ + W and HAZ).

2.1. Static Tensile Test

A static tensile test was used to determine the basic material characteristics of the tested material. With a view of using the experimentally obtained data in future for the definition of material modeling in numerical simulations, these tests have already been performed for samples taken from the parent metal (marked as PM), weld zone (W), zone immediately adjacent to the weld (W + HAZ) and heat-affected zone (HAZ). The locations from where the samples were taken are shown in Figure 1. To eliminate any material anomalies, material characteristic values were determined in each instance from a set of 5 measured samples. When testing the mechanical properties of welds, it is quite difficult to divide the weld zone into precisely defined zones. For this reason, a metallographic scratch pattern of the given weld was made (see Figure 2), which shows how the individual samples were taken for the subsequent tensile testing. For the weld zone (W) and zone immediately adjacent to the weld (W + HAZ), samples were milled and ground to a final square cross-section area of 2 mm × 2 mm (see FIG. X). For the heat-affected zone (HAZ) and parent metal (PM) zones, the samples were only milled to cross-sectional areas of 3 mm × 3 mm (HAZ) and 10 mm × 3 mm (PM), respectively. These samples were then no longer modified in terms of thickness. The gauge length of the test samples was 40 mm. The whole methodology of the test execution and evaluation was in accordance with the standard EN ISO 6892-1. The static tensile test was carried out on a modernized device, TIRA Test 2300, using the software Labtest v.4 to evaluate all basic mechanical properties of tested material (proof yield strength $R_{p0.2}$, ultimate strength R_m, total ductility A_{40mm}, uniform ductility A_g and Young's modulus E). The results of the individual material quantities are summarized in Table 1. Engineering stress–strain curves from the static tensile test of tested material S355JR for the individual testing zones are graphically shown in Figure 3. This figure shows averaged curves from 5 measurements.

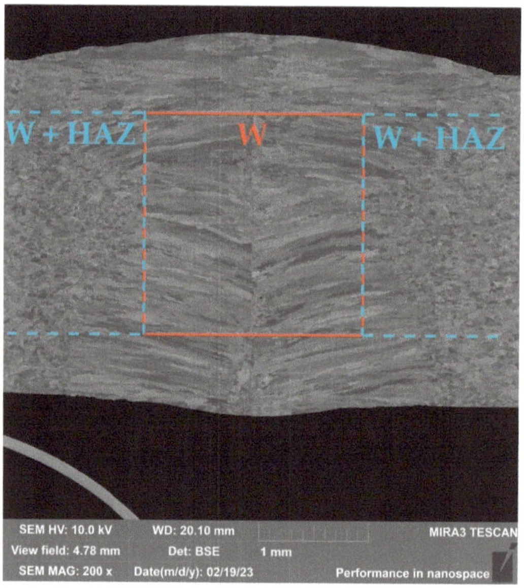

Figure 2. SEM image of the tested laser weld.

Table 1. Basic material properties of the tested material S355JR.

Area	$R_{p0.2}$ (MPa)	R_m (MPa)	A_g (%)	A_{40mm} (%)	E (MPa)
PM	392.5	491.7	12.3	20.8	204,587
HAZ	433.4	513.6	9.8	17.9	202,771
W + HAZ	474.2	522.1	4.6	8.1	203,744
W	564.7	600.2	2.2	3.2	201,196

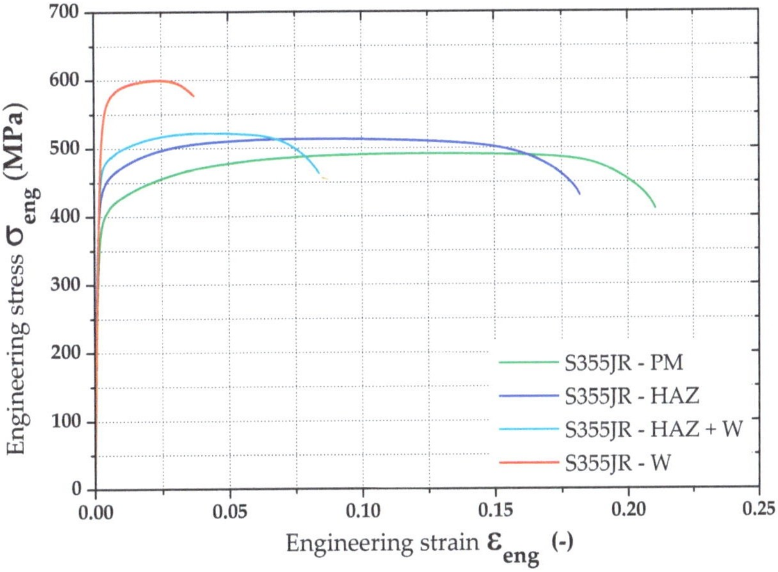

Figure 3. Engineering stress-strain curves (static tensile test) of tested sandwich material.

For a numerical simulation, stress–strain curves (hardening curves) have to be entered as a dependence of true stress (effective stress) vs. true strain (effective strain). Measured dependences of the true stress vs. true strain for the individual rolling directions, 0°, 45° and 90°, were subsequently approximated (fitted) using the Krupkowski law (power–law function) according to Equation (1).

$$\sigma = \mathrm{K}\,(\varepsilon_{\mathrm{pl}} + \varepsilon_0)^{\mathrm{n}} \tag{1}$$

where

σ—true stress (effective stress) (MPa)

K—strength coefficient (MPa)

ε_{pl}—true strain (effective strain) (-)

ε_0—offset true strain (pre-strain) (-)

n—strain hardening exponent (-)

Approximation of the stress–strain curves (hardening curves) was performed within the interval from 1% up to A_g in the software OriginPro 9.0. The advantage of defining the hardening curve by using the Krupkowski hardening law is the simple possibility of integrating this function, and thus it is easy to express the equivalent plastic work parametrically in dependence on the true strain and also the approximation constants arising from such a function. An example of the approximation result is shown in Figure 4. The determined results of approximation constants are given in Table 2.

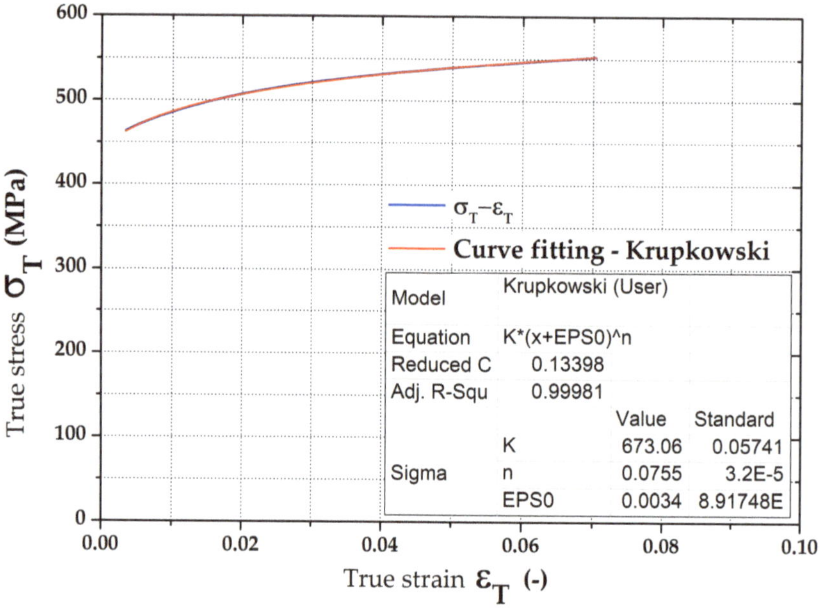

Figure 4. True stress-strain curve (STT) and application of the Krupkowski law.

Table 2. Stress–strain curve approximation constants from the static tensile test.

Zone	K (MPa)	n (-)	ε_0 (-)
PM	707.4	0.11701	0.00740
HAZ	673.1	0.07551	0.00346
W + HAZ	658.2	0.05886	0.00282
W	697.7	0.03264	0.00121

2.2. *Principle of the X-ray Method*

X-ray diffraction is based on the scattering of X-rays on crystals of a material. Based on the X-ray scattering, changes are measured in the lattice interplanar spacing (generally d-spacing), which are caused by the applied stress. These changes result in the change in Bragg angle θ (reflection of X-rays). Determined deformations are then recalculated by using equations from the theory of elasticity. Radiation scattering on the adjacent planes leads to an interference maximum in the direction of angle θ, if Bragg's law (2) is satisfied [52,53].

$$n \cdot \lambda = 2d_{hkl} \cdot \sin(\theta) \quad (2)$$

where

n—order of diffraction (-)

λ—wavelength of incident radiation (m)

d_{hkl}—interplanar spacing of the adjacent lattice planes (m)

θ—Bragg angle (-)

The residual stress values in the tested pipes were analyzed via X-ray diffraction (XRD). Device PROTO iXRD COMBO (Proto Manufacturing Inc., Windsor, ON, Canada) was used for the analysis. Tested steel was of ferritic–bainitic structure, and therefore a chromium X-ray tube was used for residual stress analysis (voltage 25 kV, current 4 mA, wavelength Kα = 2.293606 A). Following diffraction elastics, constants of $s_1 = -1.25$ TPa^{-1} and $\frac{1}{2}s_2 = 5.75$ TPa^{-1} were used to convert deformation to stresses. The diffraction angle value of planes {211} α-Fe material in an undeformed state was 156° 2θ. These values were taken from the database of the software XDR Win2000. The diffraction angles were determined by approximating the Gaussian function using the absolute peak method. The algorithm for calculating the residual stresses was $\sin^2\psi$. Table 3 shows the basic parameters of the diffraction experiment.

Table 3. X-ray diffraction—basic parameters.

Parameter	Value
Exposure time	1 s
Collimator	Ø 1 mm
X-ray oscillation	±5°
Number of angle β inclinations	11
Number of exposure repetitions for each measured diffraction maximum	15

Due to the 1 mm diameter of the collimator, feed of the work desk and oscillation of the X-ray beam, it was only possible to measure residual stresses at points with a minimum distance of 1 mm. Another problem was the curvature of pipe. To achieve a high degree of accuracy in the measured results, the method chosen had the oscillation (displacement) of the work desk occur only in the longitudinal axis of the pipe, thus eliminating its curvature. Four characteristic positions on the pipe, located at 90° intervals, were chosen for residual stress measurements. The starting point, designated as position 0°, is the center of the weld zone.

The starting point, which was marked as position 0°, is at the center of the weld zone. Around each characteristic point, 5 more points were then measured on both sides at distances of 1, 2, 4, 6 and 10 mm. The method of measuring the residual stresses via X-ray diffraction can be seen in Figures 5 and 6. Four samples were used for the XRD measurements, and the measured residual stress values are shown in Table 4 as the average value from 4 measurements. Since the tested samples were from real industrial production, it was first necessary to remove the surface soil from the pipes without affecting the magnitude of the residual stresses. The XRD measurements were always carried out in a location free of such surface defects. For all 4 measurements, identical experimental conditions (clamping method, position of the scanning point and measurement parameters) were ensured.

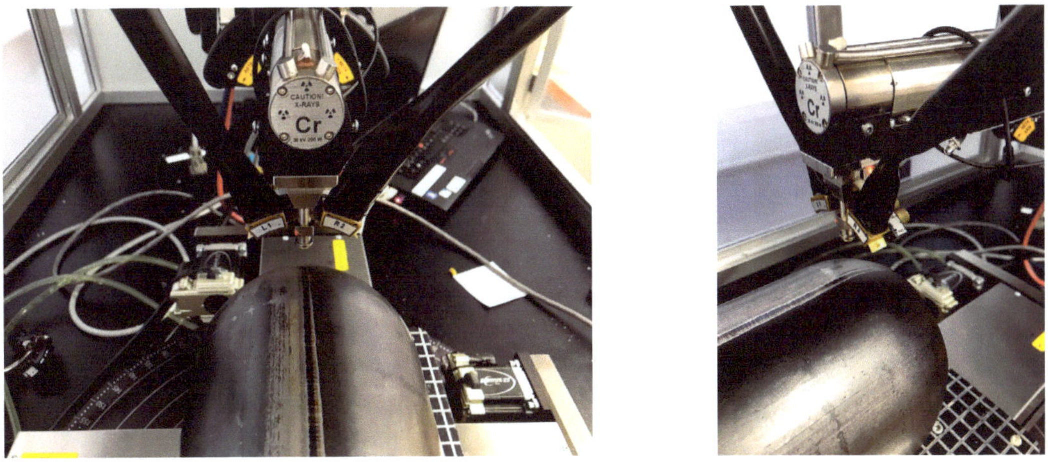

Figure 5. Measurement of the residual stresses via X-ray diffraction.

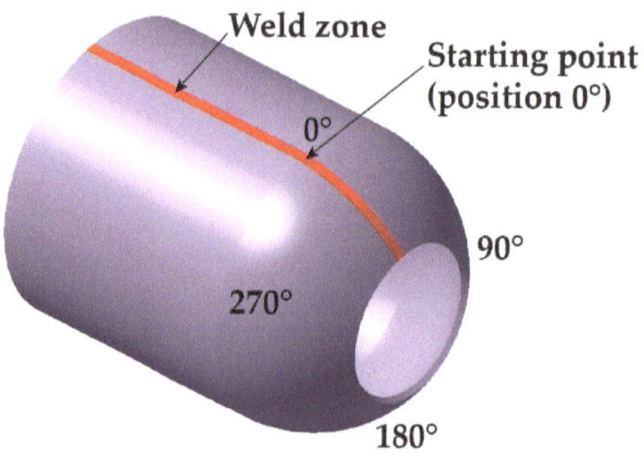

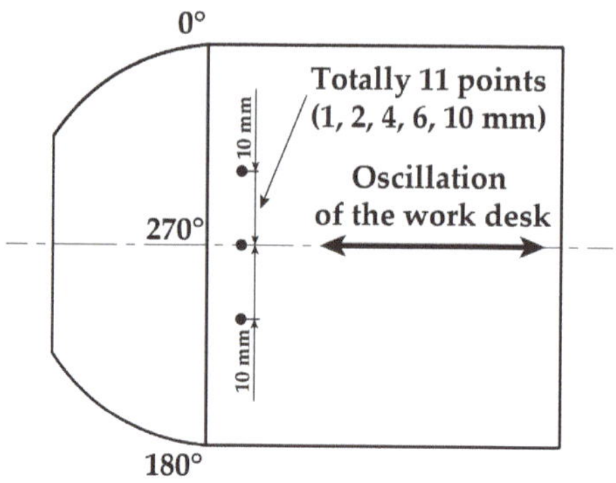

Figure 6. Used measured positions at measurement of residual stress values using XRD method.

Table 4. Measured values of the residual stresses σ_R (MPa) in the relevant points.

Distance from Center of Weld	Measured Position (°)			
	0° (Weld Zone)	90°	180°	270°
−10 mm	392	378	356	388
−6 mm	412	371	347	371
−4 mm	458	383	341	382
−2 mm	487	369	352	376
−1 mm	506	382	361	374
0	492	376	355	367
1 mm	502	370	344	369
2 mm	496	384	349	382
4 mm	442	375	339	372
6 mm	403	381	342	379
10 mm	386	376	338	370

3. Numerical Simulations

As mentioned in the previous chapter, two different approaches were used to monitor residual stresses. The so-called standard FEM computed residual stresses directly in the PAM-Stamp 2G. On the other hand, the so-called modified FEM took into account results from the previous numerical simulation of welding (software Sysweld).

3.1. Numerical Simulation of Welding

The origins of the ESI Group's Sysweld simulation program date back to the 1980s, when the first software for numerical solutions of welding processes using FEM was developed. This method is based on the approximate solution of partial differential and integral equations. Since the first version of the software was based on a stepwise solution, it always started with the thermal loading and the associated metallurgy and phase transformations, continuing with the mechanical analysis of stresses and strains. Over the years, the software has been continuously improved and extended with new modules. In its current form, it is a comprehensive system covering all interactions between thermal, metallurgical, chemical and mechanical processes.

The program links thermal, electro-kinetic and metallurgical equations. The computations of the mechanical properties are then dependent on the metallurgical history, thermal loading and plasticity. In general, the computations are performed in a given sequence, and the output of one step is the input for the following step. The advantage of this program is that all relevant physical phenomena associated with temperature effects on the material being welded are included in the computations. The material properties are dependent on the temperature, phase and chemical composition.

The program can simulate standard arc welding technologies as well as high-performance welding technologies. It allows the design of 3D volume parts as well as shell components or their combinations. To achieve high-quality results, it is very important to correctly set up the main input parameters, especially the simulation model, heat-source model and input material data. The numerical computation in the software Sysweld is based on two major analyses: thermo-metallurgical and mechanical. Thermo-metallurgical analysis has to be computed first because its basis is the mechanical analysis.

In order to perform a simulation computation for welding a pipe of defined dimensions, it is first necessary to create a simulation model that corresponds to the real situation. When creating the finite element mesh for the welding process, a commonly used model meshing strategy was chosen, wherein the weld zone (weld line), a high-density mesh, was created with sufficiently small elements to allow accurate computation. As the distance from the weld increases, the elements become progressively larger. Considering the assumed laser weld width corresponds to approx. 3 mm of sheet thickness, a high-density mesh of $0.3 \times 0.3 \times 1$ mm (circumferential direction $\times$ thickness direction $\times$ length of pipe) with a total width of 4.8 mm was created in the weld zone. The change in the mesh element's dimensions with increasing distance from the weld can be seen in Figure 7. The final simulation model contains 126,000 volume elements. The isotropic hardening model was used to compute the residual stresses.

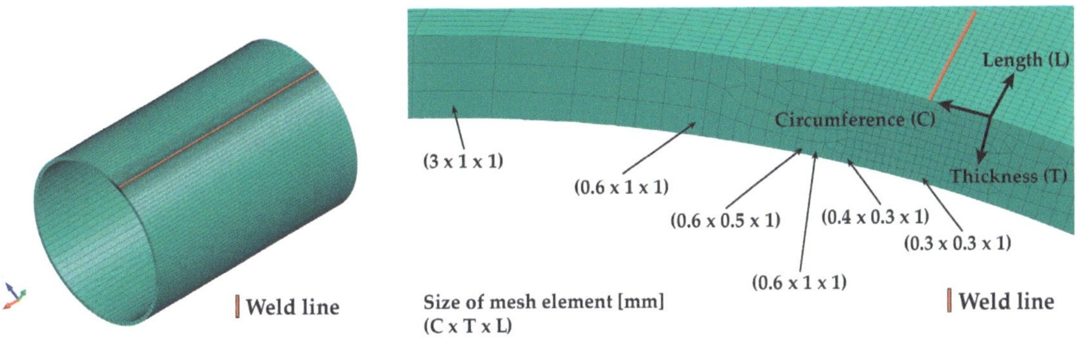

Figure 7. Welding simulation model of pipe and detail of mesh in the weld zone.

The welding process was performed on the device IPG YLS-2000-S2T (IPG Photonics, Oxford, MA, USA) under the following basic process parameters: power 1800 W, pulse duration 50/5 ms, spot size 1 mm and feed rate 0.01 $m \cdot s^{-1}$. The same conditions were adjusted for the numerical simulation of welding.

Subsequently, the model needed to define the method and area of clamping during welding, which should correspond to the real conditions of the welding process. In fact, the seamed pipe is manufactured by continuous roll bending from a strip of sheet metal. During production, the pipe is fixed during welding using rotary bending tools that allow the sheet to move only in the direction of the pipe length. Regarding this, the boundary conditions during welding were chosen such that a clamping line (in Figure 8 shown in blue) was placed at a distance of 11.3 mm from the weld line (in Figure 8 shown in red). Along the clamping line, displacement in the y and z axes was fixed. Displacement in the x-axis and rotation about all three axes at the location of the clamping line were enabled. The elements in the area between the two blue lines (clamping lines) have no limiting boundary conditions.

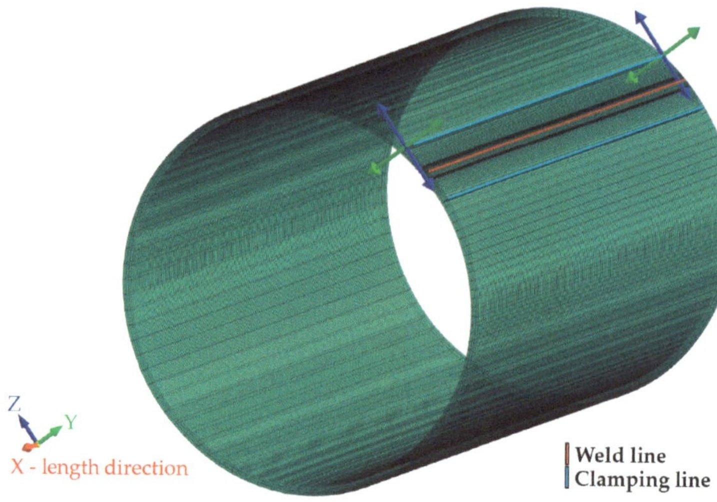

Figure 8. Heat transfer region and definition of simulation model clamping.

After setting up the simulation model, material data were assigned from the extensive library of the simulation program Sysweld, which contains most of the materials used. All material parameters are given here in terms of temperature dependence. For materials undergoing phase transformations, CCT (continuous cooling transformation) and TTT (time–temperature transformation) diagrams are available, which is also the case for the material S355JR, which was used in the submitted paper.

Another important step in the preparation of the simulation computation is setting the heat source parameters, especially the level of heat input corresponding to the real welding process. Based on the temperature loading, it is then possible to compute stresses and strains. The setting of the computation time should be sufficiently long so that the results can be obtained after the welded part has cooled down. This is especially true in the case of residual stress analysis after welding.

According to the procedure mentioned above, the simulation computation of the pipe with a butt weld was set up and subsequently performed. Thermo-metallurgical and then mechanical analysis were computed in sequence. The major aim was to determine the resulting stress state after the welding and cooling of the part. Figure 9 shows the output in the form of the residual stresses according to HMH theory after cooling down. Such theory (in numerical simulation software, often referred to as von Mises) was subsequently used for the computation of all residual stresses. The highest stress values are achieved in the weld zone and HAZ, ranging from 300–470 MPa.

For the numerical simulation of the welding process, a 200-mm-long pipe was chosen. From the result presented in Figure 9, it can be seen that the magnitude of residual stresses at the beginning and end of the pipe is different from that in the middle of the pipe. This is due to the different heat dissipation conditions during welding. Since the pipe production process is, in practice, continuous, the residual stresses after welding were analyzed in the center of the welded pipe, where the welding conditions are already stabilized. The plane section, where these stresses were measured, can be seen in Figure 9.

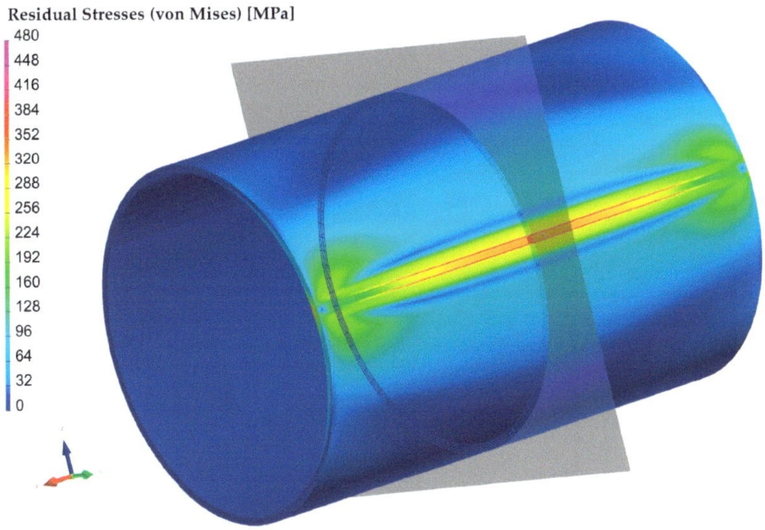

Figure 9. Distribution of the residual stresses (von Mises) after welding and cooling of tested pipe.

A graphical illustration of the residual stress distribution around the weld, obtained by the numerical simulation of welding, is shown in Figure 10.

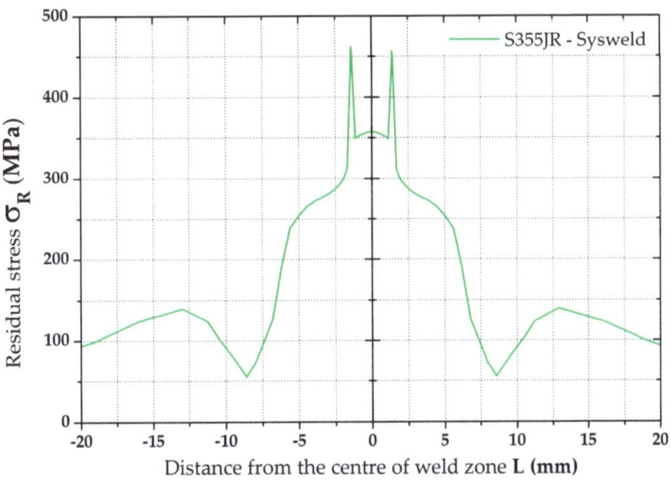

Figure 10. Distribution of residual stresses (von Mises) along section plane after welding and cooling.

For further use of the residual stresses in the subsequent computations, it was necessary to export these values to a binary file in *.erf format. When exporting, it is possible to select the state to be exported and the required values of the numerical simulation results. In this case, the geometric information of the finite element network and the own values of the residual stresses (von Mises ones) in the state after cooling (State 3600, which is equal to 3600 s) were exported. An example of such an export can be seen in Figure 11.

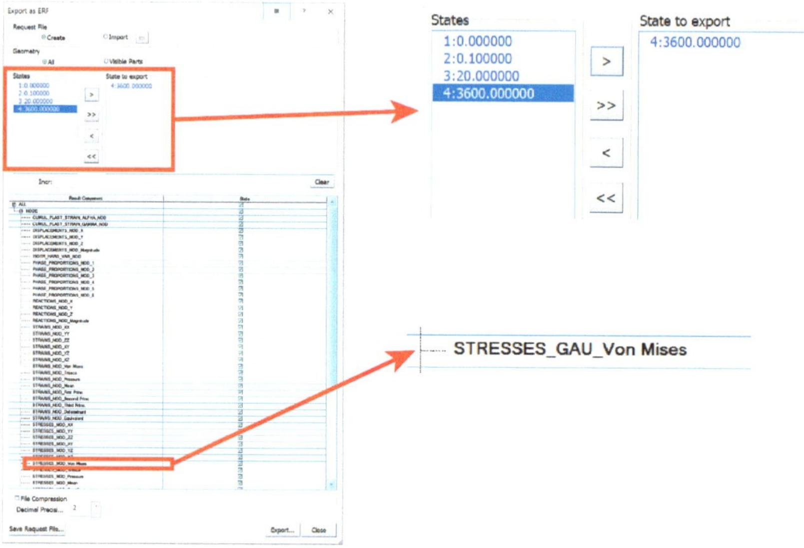

Figure 11. Export of the output data with detail about the required state and residual stress values.

3.2. Numerical Simulation of Metal Forming

The software program PAM-Stamp 2G was used for the metal forming numerical simulation of the pipe end part (see deformation area in Figure 1) with subsequent analysis of residual stresses. For the numerical computation, it was necessary to know the geometry of the tool's active surfaces, the mechanical properties of the formed material, and the technological conditions of the actual stamping process. The transition area of the tested pipe, which was used for the residual stress analysis, is actually used for the connection of pipes with a smaller diameter in the production of street lighting masts. The dimensions and shape of the forming die were provided by the manufacturer of these masts (Kooperativa Stožáry v.d. (co-opt)) in the form of a CAD file. The real pressing is carried out in such a way that one end of the pipe is fixed (propped) on the machine frame and the other end of the pipe is pressed by a tool (die) placed on a hydraulic piston. The feed rate of the tool during real pressing is 20 mm·s^{-1}. The magnitude of the working stroke is about 35 mm. The criterion to terminate the pressing process was obtaining a pipe inner diameter of 91 mm.

In the first step of the numerical simulation, the material cards of the formed pipe were defined. The pipe was divided into four regions (PM, HAZ, W + HAZ and W) in which the mechanical properties were investigated as described in the previous chapters. Based on the measured and calculated values (shown in Tables 1 and 2), material cards were defined in the PAM-Stamp 2G environment. To define a material card, Young's modulus, Poisson's ratio, the density (specific mass) and the stress–strain curve had to be entered. The values of Poisson's ratio and density were taken from the commonly available physics tables. The remaining values needed to fully define the material card were experimentally measured. In addition, the yield criteria had to be selected in the material card. The PAM-Stamp 2G software allows the selection of several major yield criteria. In addition to the commonly used yield criteria, according to Hill (isotropic and anisotropic approach), it is also possible to select so-called advanced yield criteria, e.g., Vegter, Barlat or Yoshida yield criteria. Due to the nature of the formed part (more precisely, the pipe) and the requirement of results that were in accordance with the practice, the isotropic yield criterion according to Hill (rather, Hill 48) was chosen. An example of the material card for the parent metal is shown in Figure 12. Based on the measured data presented in Tables 1 and 2, the remaining material cards for all tested zones (W, W + HAZ and HAZ) were created analogously.

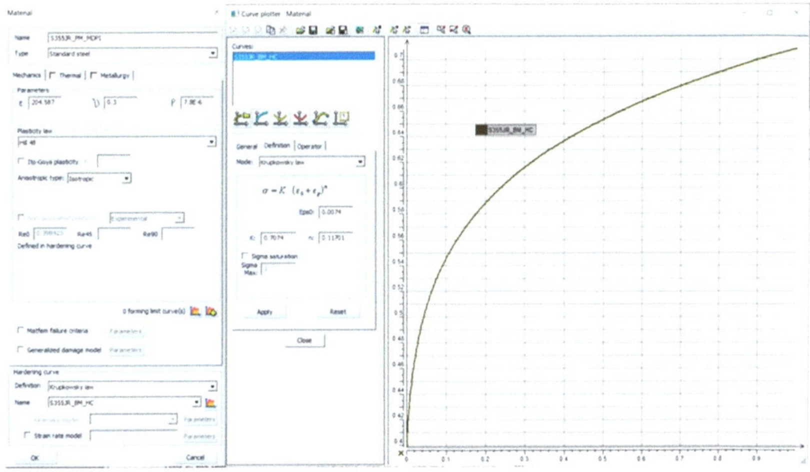

Figure 12. Setting of the stress–strain curve and yield criterion Hill 48.

3.2.1. **Standard FEM**—Analysis of the Residual Stresses Only by PAM-Stamp 2G

The next step of the metal forming process' numerical simulation was the geometric definition of tools by importing so-called active tool surfaces in the format *.igs with the subsequent automatic discretization of these surfaces. Surface elements with a maximum edge size of 2 mm were chosen for the discretization of the tool surfaces.

A very important step in determining the accuracy and computational time of the metal forming numerical simulation is creating the finite element model of the tested formed pipe. For this purpose, the so-called "Blank editor", implemented into PAM-Stamp 2G, was used. Regarding the thickness of the pipe (3 mm), volume elements with a constant plane size of 1 × 1 mm were chosen in the Blank editor for computation. In the material thickness direction, five elements with a thickness of 0.6 mm were chosen. Moreover, the pipe was further divided into a weld zone (W) with a width of 2 mm, a zone immediately adjacent to the weld (W + HAZ) with a width of 2 mm and the heat-affected zone (HAZ) with a width of 3 mm. Subsequently, corresponding material cards were assigned to each zone. A detailed view of the finite element mesh used for the computation of the residual stresses can be seen in Figure 13.

The initial state of all components before starting the finite element simulation is shown in Figure 14. For the computation, it was necessary to further define the technological boundary conditions as follows: rigid backing plate, movement of the ram with die in the y-axis direction under a constant speed ($v = 20\ mm{\cdot}s^{-1}$) and friction coefficient between contact surfaces ($f = 0.12$).

When all of these pre-processing steps have been carried out, the numerical simulation of the stamping process (general processing of data) including the springback of the tested pipe (i.e., both the stamping and springback stages) was performed. The area where the residual stresses after the numerical simulation were measured corresponded with the area where the residual stresses were measured through X-ray diffraction. Strictly speaking, it was the transition area between the deformed and undeformed parts of the pipe. A graphical representation of the residual stresses before and after springback can be seen in Figure 15. In this figure is also shown the position of the section plane, which was used to determine the monitored values of the residual stresses before and after springback. Table 5 shows these values of the residual stresses (FEM after springback), which are compared with the values determined via X-ray diffraction (XRD).

From the resulting values in Table 5, it can be seen that in the zone around the weld (position 0°), the residual stress values show a significant gradient. In the other monitored zones (positions 90°, 180 and 270°) this phenomenon does not occur, and differences are negligible. Therefore, for the subsequent data evaluation, a graph just for the residual stresses in the position 0° was made—see Figure 16. There can also be found a comparison of FEM data with the measured values. In the other zones, only the average values were subsequently compared—see Figure 23.

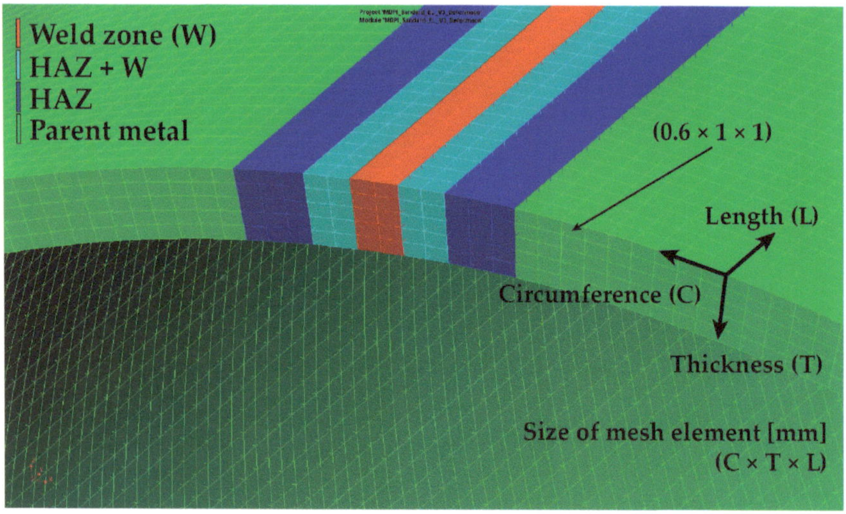

Figure 13. Detail of the applied zones for standard FEM strategy.

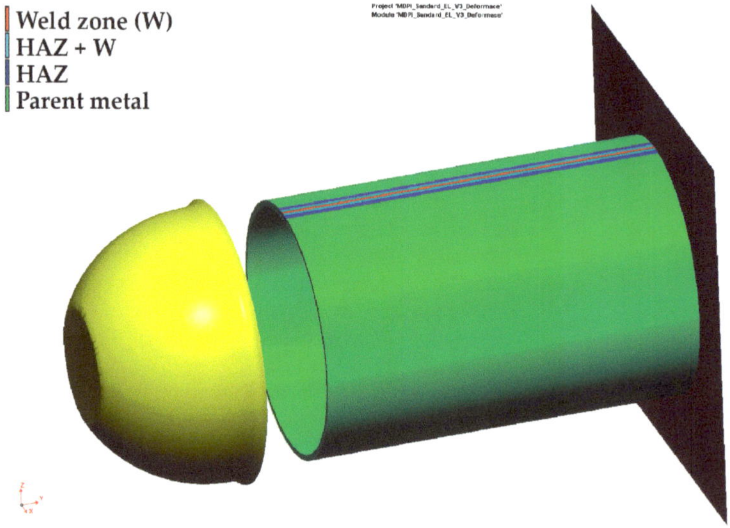

Figure 14. Initial state of all components before the computation.

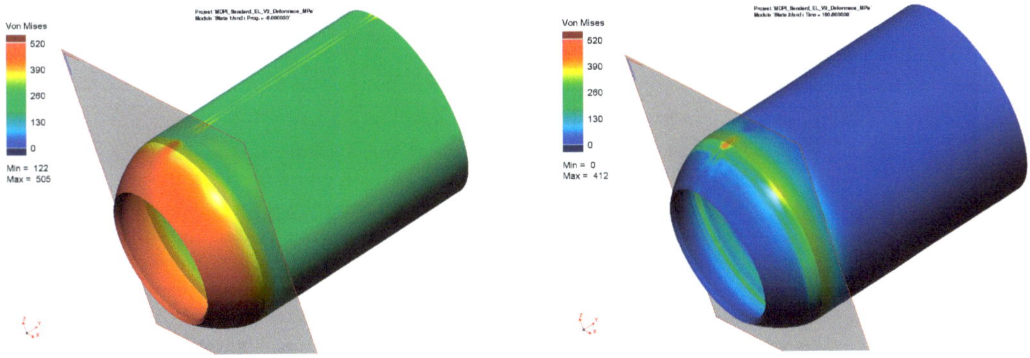

Figure 15. Standard FEM: distribution of residual stresses before (**left**) and after (**right**) springback.

Table 5. Standard FEM: comparison of the residual stress (von Mises) values σ_R (MPa) and XRD.

Distance from the Center of Weld Zone	Measured Position (°)							
	0°		90°		180°		270°	
	XRD	FEM	XRD	FEM	XRD	FEM	XRD	FEM
−10 mm	392	297	378	312	356	316	388	312
−6 mm	412	326	371	311	347	316	371	311
−4 mm	458	344	383	311	341	316	382	312
−2 mm	487	386	369	312	352	317	376	311
−1 mm	506	404	382	312	361	317	374	311
0	492	404	376	312	355	316	367	312
1 mm	502	404	370	312	344	317	369	311
2 mm	496	386	384	312	349	317	382	311
4 mm	442	344	375	311	339	317	372	312
6 mm	403	326	381	311	342	317	379	311
10 mm	386	297	376	312	338	316	370	312

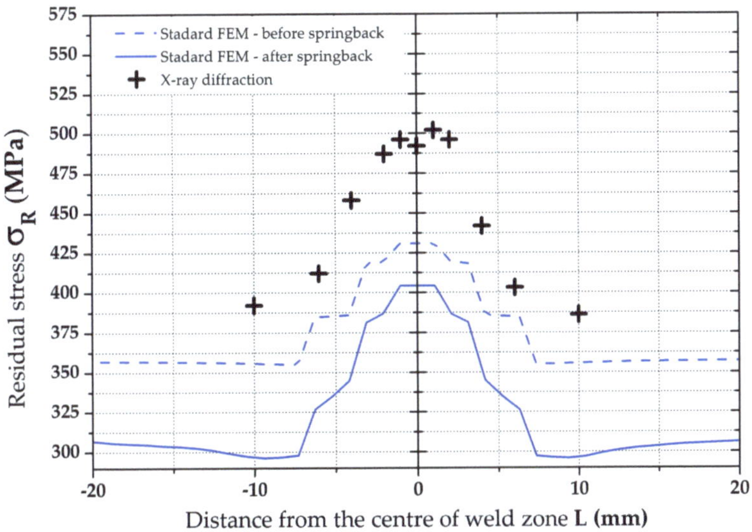

Figure 16. Comparison of residual stresses (von Mises)—XRD vs. standard FEM.

3.2.2. **Modified FEM**—Analysis of the Residual Stresses Considering the Previous Welding

From Table 5, it is also obvious that the results from only the numerical simulation of the metal forming process (without considering the influence of the previous welding technology) do not provide sufficiently accurate results for the residual stress distribution in the formed pipe. For this reason, another FEM of the forming process was defined concerning the initial residual stress values arising from the welding process. A necessary condition for the initial transfer of the residual stresses from the welding process to the forming process is the utilization of an identical finite element mesh. With regard to the welding process, where thermal, deformation and stress phenomena occur in small zones around the weld, it is crucial to use the mesh used in the welding part of the numerical simulation. Thus, for the modified metal forming FEM, with respect to the residual stresses after the welding technology (so results from Sysweld), there was also mesh used that was created in Sysweld (see Figure 7). This mesh, including monitored zones, was applied to the software PAM-Stamp 2G (see Figure 17) and finally was used in the modified FEM strategy.

All further steps in the process definition were identical to the previous approach (material card assignment, technological conditions and tool kinematics). Compared to the previous task, an additional auxiliary stage called "welding stress" was created in a new definition of the forming process (modified FEM). Within this stage, the minimum displacement of the die (in this case, 0.3 mm) without contact with the forming material is defined. Thus, there is no influence on the material to be formed. Once the computation is started, the software generates output files for each stage. After the computation of these "welding stress" stages in the software PAM-Stamp 2G, the resulting values of such computations are saved with zero values of stresses, deformations, displacements, etc. This output file in the format *.erf is then integrated with a file obtained from the Sysweld software. So, the stress values from Sysweld are thus implemented into the "welding stress" stages in the environment of the PAM-Stamp 2G. The next step is to start again the numerical simulation of the metal forming (termed as stamping stage), where the input stresses in the volume elements are already taken into account. The distribution of the residual stresses before the actual forming process is shown in Figure 18. Similarly, as in the previous task, residual stress values were monitored in the relevant section plane (see Figures 19 and 20), and their values (again, only after the springback) are given in Table 6.

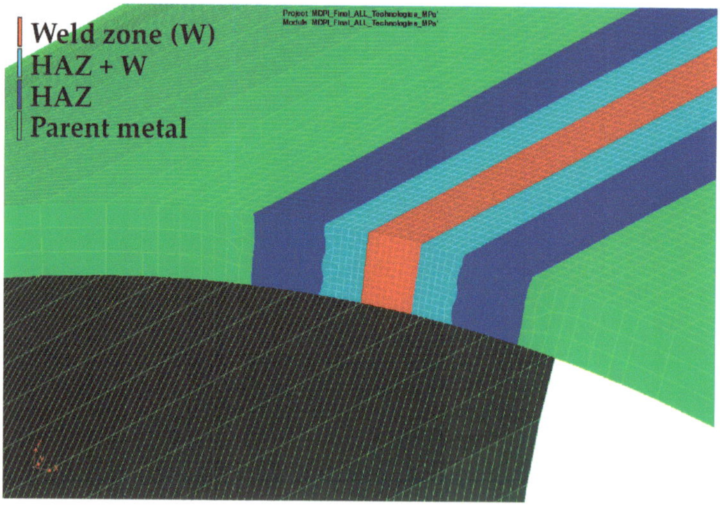

Figure 17. Detail of the applied zones for modified FEM strategy.

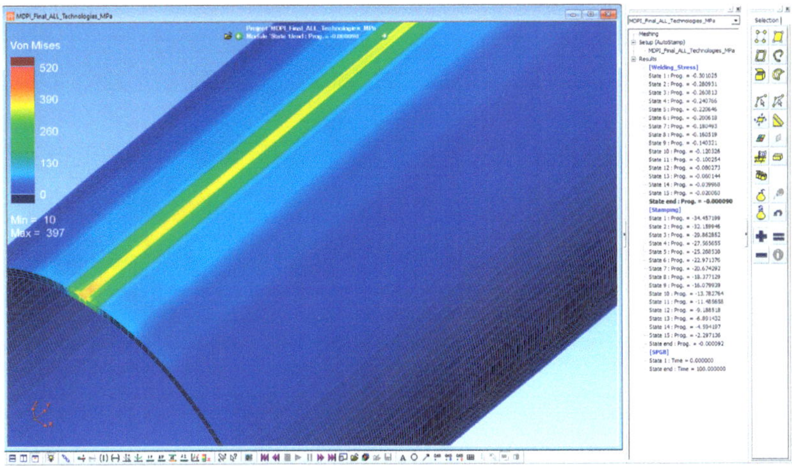

Figure 18. Modified FEM—initial state before the computation of metal forming process.

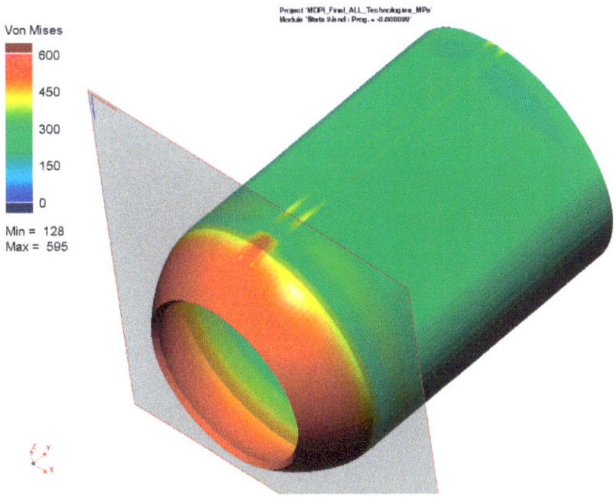

Figure 19. Modified FEM—distribution of residual stresses before springback.

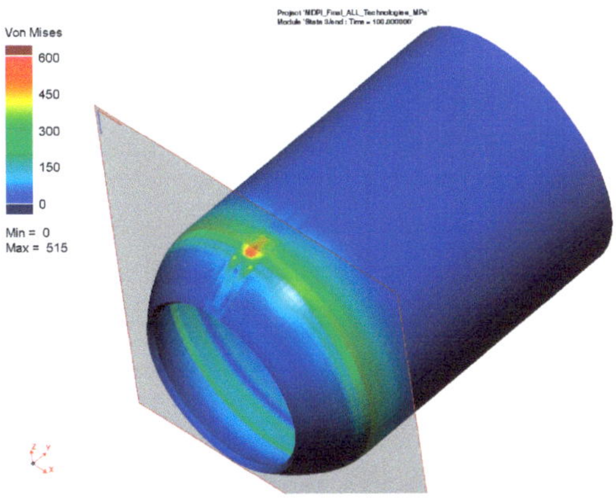

Figure 20. Modified FEM—distribution of residual stresses after springback.

Table 6. Modified FEM—comparison of the residual stress (von Mises) values σ_R (MPa) and XRD.

Distance from the Center of Weld Zone	Measured Position (°)							
	0°		90°		180°		270°	
	XRD	FEM	XRD	FEM	XRD	FEM	XRD	FEM
−10 mm	392	324	378	334	356	336	388	333
−6 mm	412	380	371	333	347	336	371	334
−4 mm	458	403	383	334	341	335	382	334
−2 mm	487	459	369	335	352	335	376	333
−1 mm	506	481	382	334	361	336	374	333
0	492	481	376	335	355	337	367	334
1 mm	502	481	370	334	344	336	369	333
2 mm	496	458	384	335	349	336	382	333
4 mm	442	402	375	334	339	337	372	334
6 mm	403	380	381	333	342	337	379	334
10 mm	386	325	376	334	338	336	370	333

A basic evaluation of this modified FEM was performed as in the previous case. Thus, data in the position 0° were again graphically illustrated—see Figure 21. In the other zones, only the average values were subsequently compared, as shown in Figure 23.

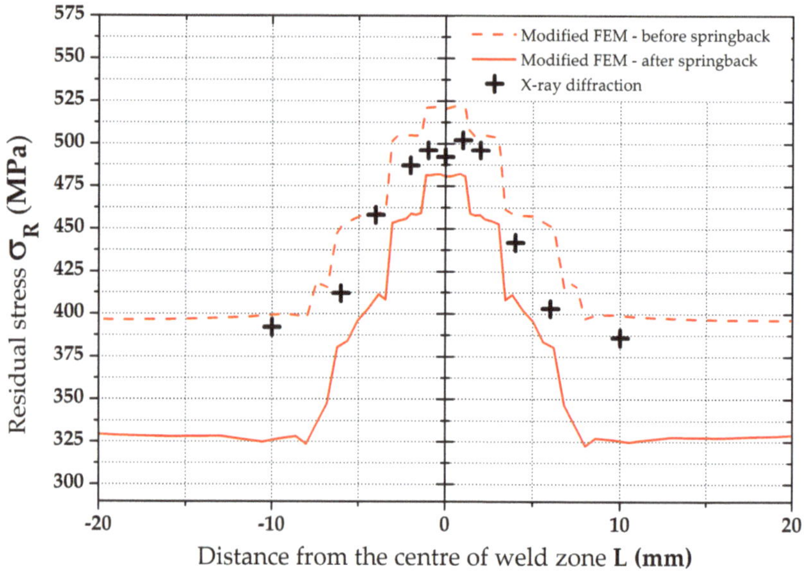

Figure 21. Comparison of residual stresses (von Mises)—XRD vs. modified FEM.

4. Discussion

The development of new hardware and software IT in recent years has enabled acceleration in all sectors of production processes' numerical support. There are still more and more complex mathematical models being introduced, which take into account a considerable number of parameters in the numerical computations. The accuracy of these computations for the individual parts of the technological processes is already at a high level and the obtained results allow us to design complex production processes and also to predict potential problems. The actual trend in the field of industrial IT lies in the creation of a so-called virtual factory, where the chain of all production technologies is numerically simulated. Due to the often completely different requirements for key input parameters of the individual technological processes (e.g., the physical principle of the process, requirements for the finite element mesh, complexity and type of mathematical model, etc.), the challenge of this concept lies primarily in the data compatibility in light of the input and output values required for the individual technological operations. The identification and possible elimination of problems fundamentally affecting the chain of subsequent technological processes are also very needed. Despite the problems mentioned above, the first studies have shown that the concept of a virtual factory, in combination with appropriately adjusted sensors and monitoring of real processes, ensures significant time and cost savings compared to conventional methods, and further development of this concept is expected.

When designing the shape and dimensions of machine components, designers often rely on tabulated values of mechanical properties. However, knowledge of the distribution and magnitude of internal residual stresses, which are, to a large extent, influenced by the history of component production, is also important for the correct functioning and lifetime of the given components. Such manufactured components must therefore be taken as a chain of successive technological procedures. In this paper, the possibility to numerically predict the magnitude of residual stresses in a formed, seamed pipe is presented, taking into account both the welding and forming processes. Both technological processes introduce internal material instability in the form of residual stresses, and, in particular, the welding technology introduces high values of residual stresses that are significantly localized in the weld zone. The uniqueness and novelty of the presented solution for the numerical prediction of residual stresses in a formed, seamed pipe lie in the use of a so-called virtual chain of two consecutive technological operations: welding and forming, realized using specialized software for welding (Sysweld) and sheet metal forming (PAM-Stamp 2G). In contrast to the commonly presented research studies dealing with the numerical simulation of welding or forming processes, the submitted paper presents a comprehensive view of the two successive technological processes, and it also respects the related transfer of results between these two software programs.

To compare the residual stress values determined by the numerical simulations with the real ones, it was also necessary to perform experimental measurements of residual stresses. For this purpose, the X-ray diffraction method was chosen. Four locations at the transition zone between the deformed and undeformed parts of the pipe were chosen for this analysis. The weld zone was taken as the starting position (0°) and three other positions were located at 90° rotations to this reference point. Due to the technical parameters of the X-ray diffraction equipment, it was not possible to measure areas with a diameter lower than 1 mm. Thus, the residual stress analysis was always performed at distances of 1, 2, 4, 6 and 10 mm from the center of the given position. Results from such residual stress analysis revealed a high-stress gradient in the weld zone and almost equal residual stress values in the parent metal in positions 90° and 270°. Average residual stress values measured at these locations (376.9 MPa and 376.0 MPa, respectively) can be considered practically identical from a technical point of view. The position 180° (opposite to the weld) shows that residual stress values are lower by approx. 30 MPa, which is about a 7% decrease compared to positions 90° and 270°. This difference can be explained by the fact that in the zone opposite to the weld is a processed sheet that is fixed during the roll bending process, and thus the residual stresses are partially suppressed. The experimentally determined values of the residual stress values were then compared with the values obtained from the numerical simulations.

To prove the possible positive effect of using a combination of two software to determine residual stresses, two calculation strategies were chosen. In the first case, the standard approach was used in practice, where the numerical simulation of the seamed pipe forming process was performed using only the PAM-Stamp 2G software. In this case, the seamed welded pipe was taken as an inhomogeneous material with different mechanical properties around the weld zone, regardless of any internal stresses introduced into the pipe by the welding process. In this study, this strategy is termed "Standard FEM".

Compared to the common practice, where seamed pipe forming problems are calculated with a homogeneous deformation model neglecting the effect of welding, this approach certainly gives more accurate results. Thus, in order to achieve a higher degree of accuracy in the results computed by the numerical simulation of the forming process, the seamed pipe was divided into several zones (PM, HAZ, HAZ + W and W), where different mechanical properties were taken into account. Thus, to properly define the material cards for the numerical simulations of the forming process in the software PAM-Stamp 2G, it was necessary to measure the mechanical properties of these individual zones. Although the widths of the samples for the static tensile test corresponding to the individual weld zones were chosen to be only 2 mm (see Figure 2), it is clear that it is not possible to strictly separate the individual weld zones when measuring the mechanical properties. The resulting mechanical values in the individual zones thus represent the average properties of the given volume. However, by the nature of the performed experiment and numerical simulation, it is never possible to measure or numerically compute a continuous change in the mechanical properties around the weld.

The determined change in mechanical properties measured by static tensile testing in the weld zone is evident from Table 1 and Figure 3, which graphically illustrates that there was a 44% increase in yield strength, a 22% increase in ultimate tensile strength and that the ductility was lowered by about 6 times compared to the parent metal. As already mentioned above, in the case of the first strategy of residual stress numerical calculation, only a numerical simulation of the forming process was performed without considering the previous welding process. For the finite element computation, a constant size of all elements was chosen, namely 0.6 × 1 × 1 mm (circumferential direction × thickness direction × length of pipe), as is shown in Figures 7 and 13. The measured mechanical properties were assigned to the relevant zones of the seamed pipe, also defined were the technological conditions of the forming process and finally, the numerical computation of forming a well by springback was performed. Based on the results of the numerical simulation "Standard FEM", the magnitudes of the residual stresses after springback were determined in the relevant zones.

In the case of the second strategy, the numerical prediction of residual stresses was implemented as a chain of two consecutive numerical simulations: the welding and forming processes. In this second strategy, termed in this study as "Modified FEM", the transfer of residual stresses results between Sysweld (influence of the welding process) and PAM-Stamp 2G (influence of the forming process) is thus taken into account. At first, a numerical simulation of the welding process was performed, and as a partial result, there was a determination of the stress distribution around the weld. These values were subsequently transferred to the PAM-Stamp 2G. Thus, before the numerical simulation of the forming process, the seamed pipe had already been "loaded" by the residual stresses arising from the welding process. This resulted in the data transfer in the following chain: the numerical simulation of the welding in the Sysweld environment, with the subsequent numerical simulation in the PAM-Stamp 2G environment. The basic condition for the possible data transfer between the software mentioned above was the use of the same finite element mesh for the analyzed pipe. Considering the specific requirements for creating the finite element mesh for the numerical simulation of the welding in Sysweld presented in Chapter 3 (Figure 7, in particular, the high density of the mesh in the weld zone), the same finite element mesh as the one used in Sysweld was chosen for the forming process. Compared to the first numerical simulation strategy (Standard FEM), where only the numerical simulation of the forming process with a constant size of mesh elements (0.6 × 1 × 1 mm) was performed, in the second strategy (Modified FEM), a deformation mesh from Sysweld with variable element size was used for the forming process. Such deformation is significantly different compared to the originally used deformation mesh ("uniform" one). When comparing the two applied strategies of the deformation mesh (Standard and Modified FEM) for numerical simulations of the forming process, it can be seen that in the weld zone, the size of the deformation mesh elements was significantly reduced from the original 1 mm up to 0.3 mm in the circumferential direction from the weld, and in the thickness direction it was from 0.6 mm up to 0.3 mm in the Modified FEM strategy. In the length direction, the element size was again 1 mm. In the area of the base material, mesh elements with a size of 3 × 1 × 1 mm were chosen in the case of the Modified FEM to lower the computational time. Elements of this size were located at a sufficient distance from the welding line (approx. 11 mm) and did not significantly affect the calculation of the welding process. When the results of the numerical simulation were subsequently transferred from Sysweld to the numerical simulation of the forming process in PAM-Stamp 2G, the selected maximum element size of 3 × 1 × 1 mm ensured a sufficiently stable and accurate calculation of the forming process. In light of the relative compatibility of the two software used, the effort to achieve the most accurate results and at the same time an acceptable computational time, it would be possible to consider refining the mesh in the zone of the parent metal from a value of 3 × 1 × 1 mm to a value of 1.5 × 1 × 1 mm. The refinement of the mesh elements in the weld zone no longer makes sense in terms of results accuracy.

Results of the residual stress distributions in the weld zone from both numerical simulation strategies (both Standard and Modified FEM) can be clearly seen in Figure 22, where measured residual stress values from the XRD method are also shown. To compare these values of the residual stresses determined both by the numerical simulation and experimentally, only the residual stress values after springback were considered because they correspond to the actual condition during the pipe production. From the results presented in the comparison in Figure 22, it is evident that a significantly higher agreement with the numerical simulation with experiment was achieved for the FEM computation with respect to the residual stresses after welding as input data (modified FEM). For better visibility, Table 7, showing percentage differences (for both FEM strategies), was created for the weld zone. Experimentally determined residual stress values were taken as the basis (100%) for the calculation of these differences. From Table 7, it can be seen that, for the standard FEM strategy, the average difference is 20.6%. For the second FEM strategy (modified FEM), the average difference is 7.4%. From the results, it is also evident that the highest differences for both FEM strategies are in the zone of parent metal. On the contrary, the lowest differences were determined in the weld zone.

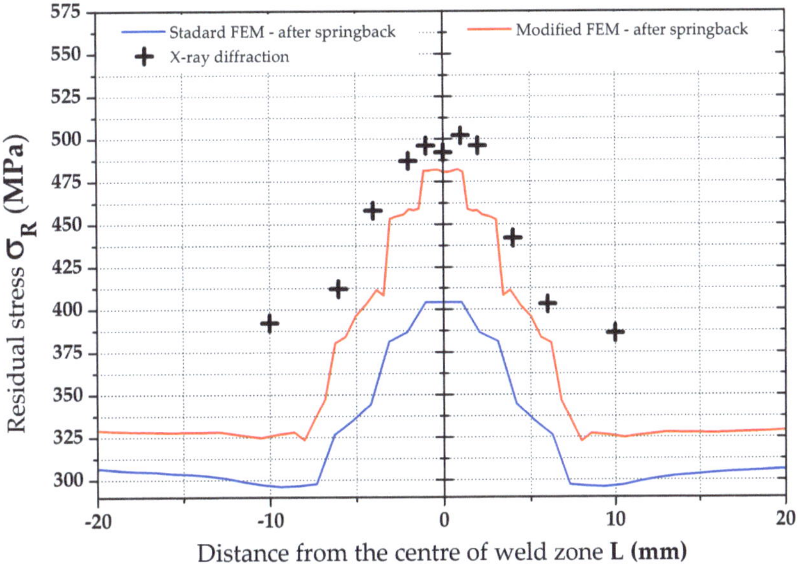

Figure 22. Comparison of residual stresses (von Mises)—XRD vs. standard and modified FEM.

Table 7. Comparison of all determined residual stresses (von Mises)—position 0.

Distance from the Center of Weld Zone	XRD	Standard FEM	Difference	Modified FEM	Difference
	(MPa)	(MPa)	(%)	(MPa)	(%)
−10 mm	392	297	24.2	324	17.3
−6 mm	412	326	20.9	380	7.8
−4 mm	458	344	24.9	403	12.0
−2 mm	487	386	20.7	459	5.7
−1 mm	506	404	20.2	481	4.9
0	492	404	17.9	481	2.2
1 mm	502	404	19.5	481	4.2
2 mm	496	386	22.2	458	7.7
4 mm	442	344	22.2	402	9.0
6 mm	403	326	19.1	380	5.7
10 mm	386	297	23.1	325	15.8

In the positions 90°, 180° and 270°, only the average values of the residual stresses were compared with each other and are graphically shown in Figure 23. An overview of these results is, together with percentage differences, given in Table 8.

Table 8. Comparison of all determined residual stresses (von Mises)—positions 90°, 180° and 270°.

Measured Position	XRD	Standard FEM	Difference	Modified FEM	Difference
	(MPa)	(MPa)	(%)	(MPa)	(%)
90°	377.0	311.7	17.3	334.2	11.3
180°	344.5	316.7	8.1	336.5	2.3
270°	373.2	311.5	16.5	333.5	10.6

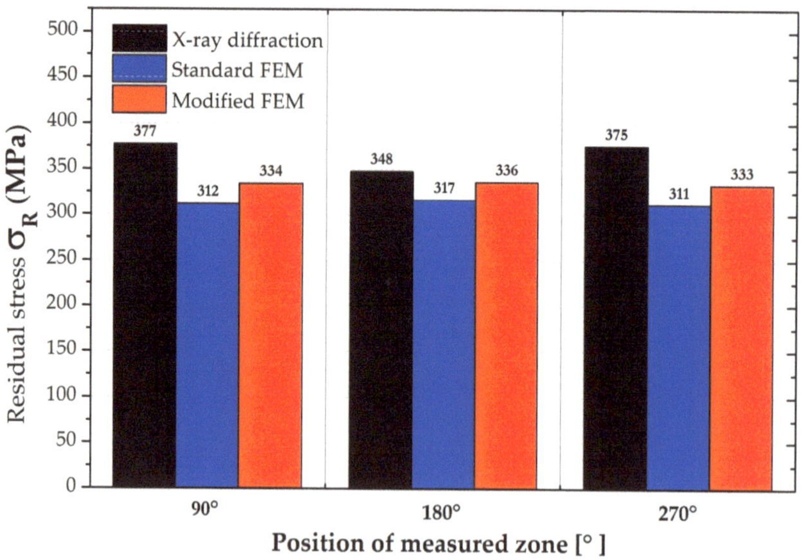

Figure 23. Comparison of residual stresses (von Mises) for measured positions 90°, 180° and 270°.

The results given in Table 8 show that the influence of the computational model is significantly lower in the presented zones than in the weld zone. Both strategies of numerical computation (standard and modified FEM) revealed lower values than the experimentally measured residual stresses. This fact is probably given by the reality that the residual stress values measured via the XRD were already influenced by the roll bending technology as a part of the production chain, where tensile stresses are generated on the pipe surface. The numerical simulation of the welding process can provide information about the stresses around the weld zone but does not substantially affect the remaining parent metal of the tested pipe in terms of the internal stress state. For this reason, in the Sysweld software, zones of the parent metal reveal only minimal residual stresses in the order of ones of MPa, which were subsequently transferred to PAM-Stamp 2G. If the distribution of residual stresses was to be virtually refined, the chain would have to take into account an initial roll bending process, then a welding process and finally the metal forming of the tested pipe.

5. Conclusions

The major aim of the research presented in this paper was to implement the concept of the virtual factory for numerical simulations of two successive production technologies: welding (software: Sysweld) and metal forming (software: PAM-Stamp 2G). In accordance with the main expected outcome, there was an increase in the accuracy of these FEMs. The actual experimental part consisted of welding the seamed pipe and the subsequent formation of its end part. The determined magnitudes of the residual stresses were used as comparison criteria, and their actual values were measured using the XRD. Subsequently, two different FEM strategies were performed, which differed in terms of considering the welding process. In the first case (standard FEM), there was no implementation of the residual stresses after the welding in the computation of the forming process. Only the different values of mechanical properties in the investigated zones (W, W + HAZ, HAZ and PM) were taken into account. The influence of the welding-induced residual stresses was only taken into account in the case of the so-called modified FEM strategy. However, a rather significant modification within the numerical simulation of the forming process was needed in this case. Subsequently, the results of the two strategies were compared, not only with each other but also with the real residual stress values. The following conclusions can be taken from the measured and computed results:

- It was not possible to continuously measure the distribution of residual stresses in local areas of the weld zone and HAZ using the XRD method. Due to the technical parameters of the used equipment, the minimum distance between the measured points was 1 mm;
- In light of the seamed pipe numerical simulation accuracy, it is suitable to divide the pipe into several zones, with respect to the expected change in mechanical properties in the weld zone;
- No good agreement was achieved between the values of residual stresses from FEM and the experimentally measured ones by using the standard FEM definition without taking into account the previous history of the pipe production (i.e., welding in this case). Around the whole circumference of the pipe, the FEM residual stress values were approximately 20% lower than the experimentally measured values;
- More accurate agreement with the experimental results was achieved using a modified FEM method with respect to the residual stresses from the previous welding technology. In the weld zone and HAZ, the difference was only about 5%, and in the parent metal, the difference was about 16%;
- The submitted solution revealed both the possibility and suitability of using such a complex FEM approach in the production process, in this case, welding and forming;
- Both of the used FEM approaches do not provide sufficiently accurate results for the residual stress distribution in the parent material zone. This is probably because it was not taken into account the complete history of the pipe production, so the effect of roll bending was not considered;

- Despite the fact that, for numerical simulations, software programs from the same manufacturer were used, it was very difficult to ensure data compatibility between them (chain Sysweld and PAM-Stamp 2G). Considering the effect of roll bending would require the use of the following chain: PAM-Stamp 2G–Sysweld–PAM-Stamp 2G, which is currently practically impossible to achieve.

Author Contributions: Conceptualization, P.S. and J.S.; methodology P.S.; investigation, J.S. and Š.B.; resources, P.S., J.S. and Š.B.; data curation, P.S., Š.B. and J.B.; writing—original draft preparation, P.S. and J.S.; writing—review and editing, J.S., Š.B. and J.B.; visualization, P.S. and J.S.; supervision, P.S.; funding acquisition, J.S. All authors have read and agreed to the published version of the manuscript.

Funding: This research was funded by the Institutional Endowment for the Long Term Conceptual Development of Research Institutes, as provided by the Ministry of Education, Youth and Sports of the Czech Republic.

Institutional Review Board Statement: Not applicable.

Informed Consent Statement: Not applicable.

Data Availability Statement: Not applicable.

Conflicts of Interest: The authors declare no conflict of interest.

References

1. Kim, N.-H.; Sankar, B.V.; Kumar, A.V. *Introduction to Finite Element Analysis and Design*; John Wiley & Sons: Hoboken, NJ, USA, 2018; ISBN 978-1-119-07872-2.
2. Reddy, J. *Introduction to the Finite Element Method*; McGraw-Hill Education: New York, NY, USA, 2019; ISBN 978-1-259-86190-1.
3. Szabó, B.; Babuška, I. *Finite Element Analysis: Method, Verification and Validation*; John Wiley & Sons: Hoboken, NJ, USA, 2021; ISBN 978-1-119-42642-4.
4. Nielsen, C.V.; Martins, P.A.F. *Metal Forming: Formability, Simulation, and Tool Design*; Academic Press: Cambridge, MA, USA, 2021; ISBN 978-0-323-85256-2.
5. Comini, G. *Finite Element Anylysis in Heat Transfer: Basic Formulation & Linear Problems*; CRC Press: Boca Raton, FL, USA, 2018; ISBN 978-1-315-27510-9.
6. Sravan, K.K.; Jose, M.J. *Application of Sysweld Software in Finding Weld Residual Stress: A Review*; SSRN: Rochester, NY, USA, 2022.
7. Pandey, C.; Mahapatra, M.M.; Kumar, P. A Comparative Study of Transverse Shrinkage Stresses and Residual Stresses in P91 Welded Pipe Including Plasticity Error. *Arch. Civ. Mech. Eng.* **2018**, *18*, 1000–1011. [CrossRef]
8. Wu, G.; Luo, J.; Li, L.; Long, Y.; Zhang, S.; Wang, Y.; Zhang, Y.; Xie, S. Control of Welding Residual Stress in Large Storage Tank by Finite Element Method. *Metals* **2022**, *12*, 1502. [CrossRef]
9. Dai, P.; Wang, Y.; Li, S.; Lu, S.; Feng, G.; Deng, D. FEM Analysis of Residual Stress Induced by Repair Welding in SUS304 Stainless Steel Pipe Butt-Welded Joint. *J. Manuf. Process.* **2020**, *58*, 975–983. [CrossRef]
10. Taraphdar, P.K.; Kumar, R.; Pandey, C.; Mahapatra, M.M. Significance of Finite Element Models and Solid-State Phase Transformation on the Evaluation of Weld Induced Residual Stresses. *Met. Mater. Int.* **2021**, *27*, 3478–3492. [CrossRef]
11. Giudice, F.; Sili, A. A Theoretical Approach to the Residual Stress Assessment Based on Thermal Field Evaluation in Laser Beam Welding. *Int. J. Adv. Manuf. Technol.* **2022**, *123*, 2793–2808. [CrossRef]
12. Chen, L.; Mi, G.; Zhang, X.; Wang, C. Numerical and Experimental Investigation on Microstructure and Residual Stress of Multi-Pass Hybrid Laser-Arc Welded 316L Steel. *Mater. Des.* **2019**, *168*, 107653. [CrossRef]
13. Ramos, H.M.E.; Tavares, S.M.O.; de Castro, P.M.S.T. Numerical Modelling of Welded T-Joint Configurations Using SYSWELD. *Sci. Technol. Mater.* **2018**, *30*, 6–15. [CrossRef]
14. Feng, G.; Wang, Y.; Luo, W.; Hu, L.; Deng, D. Comparison of Welding Residual Stress and Deformation Induced by Local Vacuum Electron Beam Welding and Metal Active Gas Arc Welding in a Stainless Steel Thick-Plate Joint. *J. Mater. Res. Technol.* **2021**, *13*, 1967–1979. [CrossRef]
15. Pavan, A.R.; Arivazhagan, B.; Vasudevan, M.; Sharma, G.K. Numerical Simulation and Validation of Residual Stresses and Distortion in Type 316L(N) Stainless Steel Weld Joints Fabricated by Advanced Welding Techniques. *CIRP J. Manuf. Sci. Technol.* **2022**, *39*, 294–307. [CrossRef]
16. Mokhtarishirazabad, M.; McMillan, M.; Vijayanand, V.D.; Simpson, C.; Agius, D.; Truman, C.; Knowles, D.; Mostafavi, M. Predicting Residual Stress in a 316L Electron Beam Weld Joint Incorporating Plastic Properties Derived from a Crystal Plasticity Finite Element Model. *Int. J. Press. Vessel. Pip.* **2023**, *201*, 104868. [CrossRef]
17. Gkatzogiannis, S. *Finite Element Simulation of Residual Stresses from Welding and High Frequency Hammer Peening*; KIT Scientific Publishing: Karlsruhe, Germany, 2022.
18. Li, S.; Hu, L.; Dai, P.; Bi, T.; Deng, D. Influence of the Groove Shape on Welding Residual Stresses in P92/SUS304 Dissimilar Metal Butt-Welded Joints. *J. Manuf. Process.* **2021**, *66*, 376–386. [CrossRef]

19. Van Puymbroeck, E.; Nagy, W.; Schotte, K.; Ul-Abdin, Z.; De Backer, H. Determination of Residual Welding Stresses in a Steel Bridge Component by Finite Element Modeling of the Incremental Hole-Drilling Method. *Appl. Sci.* **2019**, *9*, 536. [CrossRef]
20. Zhang, H.; Li, L.; Ma, W.; Luo, Y.; Li, Z.; Kuai, H. Effects of Welding Residual Stresses on Fatigue Reliability Assessment of a PC Beam Bridge with Corrugated Steel Webs under Dynamic Vehicle Loading. *Structures* **2022**, *45*, 1561–1572. [CrossRef]
21. Chiocca, A.; Frendo, F.; Aiello, F.; Bertini, L. Influence of Residual Stresses on the Fatigue Life of Welded Joints. Numerical Simulation and Experimental Tests. *Int. J. Fatigue* **2022**, *162*, 106901. [CrossRef]
22. Hemmesi, K.; Mallet, P.; Farajian, M. Numerical Evaluation of Surface Welding Residual Stress Behavior under Multiaxial Mechanical Loading and Experimental Validations. *Int. J. Mech. Sci.* **2020**, *168*, 105127. [CrossRef]
23. Ren, S.; Li, S.; Wang, Y.; Deng, D.; Ma, N. Finite Element Analysis of Residual Stress in 2.25Cr-1Mo Steel Pipe during Welding and Heat Treatment Process. *J. Manuf. Process.* **2019**, *47*, 110–118. [CrossRef]
24. Smith, M.C.; Muránsky, O.; Xiong, Q.; Bouchard, P.J.; Mathew, J.; Austin, C. Validated Prediction of Weld Residual Stresses in Austenitic Steel Pipe Girth Welds before and after Thermal Ageing, Part 1: Mock-up Manufacture, Residual Stress Measurements, and Materials Characterisation. *Int. J. Press. Vessel. Pip.* **2019**, *172*, 233–250. [CrossRef]
25. Ghafouri, M.; Ahola, A.; Ahn, J.; Björk, T. Welding-Induced Stresses and Distortion in High-Strength Steel T-Joints: Numerical and Experimental Study. *J. Constr. Steel Res.* **2022**, *189*, 107088. [CrossRef]
26. Moslemi, N.; Abdi, B.; Gohari, S.; Sudin, I.; Redzuan, N.; Ayob, A.; Ahmed, M.; Rhee, S.; Burvill, C. Influence of Welding Sequences on Induced Residual Stress and Distortion in Pipes. *Constr. Build. Mater.* **2022**, *342*, 127995. [CrossRef]
27. Lü, N.C.; Wang, M.H.; Hao, G.D.; Wang, Y.T. Finite Element Analysis of Residual Welding Stresses and Deformation for a 5A06 Aluminum Alloy Plate. *Strength Mater.* **2020**, *52*, 532–538. [CrossRef]
28. García-García, V.; Mejía, I.; Reyes-Calderón, F.; Benito, J.A.; Cabrera, J.M. FE Thermo-Mechanical Simulation of Welding Residual Stresses and Distortion in Ti-Containing TWIP Steel through GTAW Process. *J. Manuf. Process.* **2020**, *59*, 801–815. [CrossRef]
29. Pavan, A.R.; Arivazhagan, B.; Sharma, G.K.; Kumar, S.A.; Mahadevan, S.; Vasudevan, M. Influence of Hardening Models on the Estimation of Residual Stresses by Finite Element Modeling in Type 316LN Stainless Steel Weld Joints. *J. Mater. Eng. Perform.* **2022**, *31*, 6988–6997. [CrossRef]
30. Baruah, S.; Sarkar, S.; Singh, I.V.; Mishra, B.K. A Computational Framework Based on FEA, ML and GA for Estimation of Welding Residual Stresses. *Finite Elem. Anal. Des.* **2022**, *205*, 103753. [CrossRef]
31. Yaghi, A.H.; Hyde, T.H.; Becker, A.A.; Sun, W.; Wen, W.; Hilson, G.; Simandjuntak, S.; Flewitt, P.E.J.; Pavier, M.J.; Smith, D.J. Comparison of Measured and Modelled Residual Stresses in a Welded P91 Steel Pipe Undergoing Post Weld Heat Treatment. *Int. J. Press. Vessel. Pip.* **2020**, *181*, 104076. [CrossRef]
32. Hashemzadeh, M.; Chen, B.-Q.; Guedes Soares, C. Evaluation of Multi-Pass Welding-Induced Residual Stress Using Numerical and Experimental Approaches. *Ships Offshore Struct.* **2018**, *13*, 847–856. [CrossRef]
33. Zhang, K.; Dong, W.; Lu, S. Finite Element and Experiment Analysis of Welding Residual Stress in S355J2 Steel Considering the Bainite Transformation. *J. Manuf. Process.* **2021**, *62*, 80–89. [CrossRef]
34. Wang, X.; Guo, E. Numerical Analysis and Verification of Residual Stress in T Joint of S355 Steel. *Frat. Ed Integrità Strutt.* **2020**, *14*, 25–32. [CrossRef]
35. Perić, M.; Nižetić, S.; Garašić, I.; Gubeljak, N.; Vuherer, T.; Tonković, Z. Numerical Calculation and Experimental Measurement of Temperatures and Welding Residual Stresses in a Thick-Walled T-Joint Structure. *J. Therm. Anal. Calorim.* **2020**, *141*, 313–322. [CrossRef]
36. Derakhshan, E.D.; Yazdian, N.; Craft, B.; Smith, S.; Kovacevic, R. Numerical Simulation and Experimental Validation of Residual Stress and Welding Distortion Induced by Laser-Based Welding Processes of Thin Structural Steel Plates in Butt Joint Configuration. *Opt. Laser Technol.* **2018**, *104*, 170–182. [CrossRef]
37. Chen, B.-Q.; Hashemzadeh, M.; Guedes Soares, C. Validation of Numerical Simulations with X-ray Diffraction Measurements of Residual Stress in Butt-Welded Steel Plates. *Ships Offshore Struct.* **2018**, *13*, 273–282. [CrossRef]
38. Yaping, L.; Weixin, G. Research on X-ray Welding Image Defect Detection Based on Convolution Neural Network. *J. Phys. Conf. Ser.* **2019**, *1237*, 032005. [CrossRef]
39. Kumar, A.; Kumar, D.R.; Gautam, V. Prediction of Residual Stresses in Biaxial Stretching of Tailor Welded Blanks by Finite Element Analysis. *IOP Conf. Ser. Mater. Sci. Eng.* **2019**, *651*, 012039. [CrossRef]
40. Svetlík, J.; Baron, P.; Dobránsky, J.; Kočiško, M. Implementation of Computer System for Support of Technological Preparation of Production for Technologies of Surface Processing. *Appl. Mech. Mater.* **2014**, *613*, 418–425. [CrossRef]
41. Xiao, M.; Hu, Y.-F.; Jin, H.; Chung, K.-F.; Nethercot, D.A. Prediction of Residual Stresses in High-Strength S690 Cold-Formed Square Hollow Sections Using Integrated Numerical Simulations. *Eng. Struct.* **2022**, *253*, 113682. [CrossRef]
42. Yao, Y.; Quach, W.-M.; Young, B. Simplified Models for Residual Stresses and Equivalent Plastic Strains in Cold-Formed Steel Elliptical Hollow Sections. *Thin-Walled Struct.* **2020**, *154*, 106835. [CrossRef]
43. Kumar, A.; Digavalli, R.K.; Gautam, V.; Krishnaswamy, H. Characterization of Residual Stresses in Conventional Forming and Hydroforming of Tailor Welded Blanks. *J. Mater. Eng. Perform.* **2022**, *31*, 10171–10186. [CrossRef]
44. Hu, Y.-F.; Chung, K.-F.; Ban, H.; Nethercot, D.A. Investigations into Residual Stresses in S690 Cold-Formed Circular Hollow Sections Due to Transverse Bending and Longitudinal Welding. *Eng. Struct.* **2020**, *219*, 110911. [CrossRef]
45. Dobránsky, J.; Pollák, M. Structural Design and Material Cutting Using a Laser End Effector on a Robot Arm. *TEM J.* **2020**, *9*, 1455–1459.

46. Zhao, J.; Tang, J.; Zhou, W.; Jiang, T.; Liu, H.; Xing, B. Numerical Modeling and Experimental Verification of Residual Stress Distribution Evolution of 12Cr2Ni4A Steel Generated by Shot Peening. *Surf. Coat. Technol.* **2022**, *430*, 127993. [CrossRef]
47. Liu, B.; Villavicencio, R.; Guedes Soares, C. Experimental and Numerical Analysis of Residual Stresses and Strains Induced during Cold Bending of Thick Steel Plates. *Mar. Struct.* **2018**, *57*, 121–132. [CrossRef]
48. Díaz, A.; Cuesta, I.I.; Alegre, J.M.; de Jesus, A.M.P.; Manso, J.M. Residual Stresses in Cold-Formed Steel Members: Review of Measurement Methods and Numerical Modelling. *Thin-Walled Struct.* **2021**, *159*, 107335. [CrossRef]
49. Lee, W.; Lee, H.; Bae, J.; Kim, D. Optimization of Manufacturing Process for Exterior Quality of Commercial Vehicle by Multi-Physics Simulations. *Multiscale Sci. Eng.* **2020**, *2*, 114–126. [CrossRef]
50. Li, Y.; Shi, Z.; Lin, J.; Yang, Y.-L.; Saillard, P.; Said, R. Effect of Machining-Induced Residual Stress on Springback of Creep Age Formed AA2050 Plates with Asymmetric Creep-Ageing Behaviour. *Int. J. Mach. Tools Manuf.* **2018**, *132*, 113–122. [CrossRef]
51. Wang, X.; Meng, Q.; Hu, W. Fatigue Life Prediction for Butt-Welded Joints Considering Weld-Induced Residual Stresses and Initial Damage, Relaxation of Residual Stress, and Elasto-Plastic Fatigue Damage. *Fatigue Fract. Eng. Mater. Struct.* **2019**, *42*, 1373–1386. [CrossRef]
52. Monin, V.I.; Lopes, R.T.; Turibus, S.N.; Payão Filho, J.C.; Assis, J.T. de X-ray Diffraction Technique Applied to Study of Residual Stresses after Welding of Duplex Stainless Steel Plates. *Mat. Res.* **2014**, *17*, 64–69. [CrossRef]
53. Ahmed, I.I.; Adebisi, J.A.; Abdulkareem, S.; Sherry, A.H. Investigation of Surface Residual Stress Profile on Martensitic Stainless Steel Weldment with X-ray Diffraction. *J. King Saud Univ.-Eng. Sci.* **2018**, *30*, 183–187. [CrossRef]

materials MDPI

Article

Eutectic In Situ Modification of Polyamide 12 Processed through Laser-Based Powder Bed Fusion

Samuel Schlicht [1,2,*] and Dietmar Drummer [1,2]

1 Institute of Polymer Technology, Friedrich-Alexander-Universität Erlangen-Nürnberg, Am Weichselgarten 10, 91058 Erlangen, Germany
2 Collaborative Research Center 814, Friedrich-Alexander-Universität Erlangen-Nürnberg, Am Weichselgarten 10, 91058 Erlangen, Germany
* Correspondence: samuel.schlicht@fau.de

Abstract: Laser-based powder bed fusion (LPBF) of polymers allows for the additive manufacturing of dense components with high mechanical properties. Due to inherent limitations of present material systems suitable for LPBF of polymers and required high processing temperatures, the present paper investigates the in situ modification of material systems using powder blending of p-aminobenzoic acid and aliphatic polyamide 12, followed by subsequent laser-based additive manufacturing. Prepared powder blends exhibit a considerable reduction of required processing temperatures dependent on the fraction of p-aminobenzoic acid, allowing for the processing of polyamide 12 at a build chamber temperature of 141.5 °C. An elevated fraction of 20 wt% of p-aminobenzoic acid allows for obtaining a considerably increased elongation at break of 24.65% ± 2.87 while exhibiting a reduced ultimate tensile strength. Thermal investigations demonstrate the influence of the thermal material history on thermal properties, associated with the suppression of low-melting crystalline fractions, yielding amorphous material properties of the previously semi-crystalline polymer. Based on complementary infrared spectroscopic analysis, the increased presence of secondary amides can be observed, indicating the influence of both covalently bound aromatic groups and hydrogen-bound supramolecular structures on emerging material properties. The presented approach represents a novel methodology for the energy-efficient in situ preparation of eutectic polyamides, potentially allowing for the manufacturing of tailored material systems with adapted thermal, chemical, and mechanical properties.

Keywords: semi-aromatic polyamides; aramides; powder bed fusion; selective laser sintering; in situ modification; eutectic; eutectic polymers

Citation: Schlicht, S.; Drummer, D. Eutectic In Situ Modification of Polyamide 12 Processed through Laser-Based Powder Bed Fusion. *Materials* **2023**, *16*, 2050. https://doi.org/10.3390/ma16052050

Academic Editors: Jozef Dobránsky and Luboš Běhálek

Received: 8 February 2023
Revised: 26 February 2023
Accepted: 27 February 2023
Published: 1 March 2023

1. Introduction

Laser-based powder bed fusion (LPBF) of polymers represents a well-established additive manufacturing process, allowing for the additive processing of semi-crystalline and amorphous polymers. Considering the predominant application of isothermal processing strategies, the range of materials suitable for LPBF is limited based on the structural requirement of material-specific thermal processing windows. Therefore, the processing of polymers exhibiting tailored mechanical and thermal properties is inherently restricted by technological and economic considerations. To overcome existing limitations in the range of materials suitable for LPBF, the in situ formation of modified material systems has been described for both solid-based and solid–liquid systems, enabling the adaptation of mechanical and thermal properties of manufactured parts. However, the in situ modification in LPBF remains subject to material-specific, complex processing technology, limiting the present in situ enhancement of material properties. Concerning the overall field of polymer engineering, considerable enhancements of thermal and mechanical properties have been obtained by the introduction of liquid crystalline semi-aromatic material systems, being predominantly used in injection molding of thin-walled components due to reduced melt viscosities and increased operating temperatures. Besides well-described synthesis routes for the preparation of aromatically modified polyamides, produced by melt condensation of monomers [1–4] and by solution-based [5,6], complex-mediated condensation of various modified aromatic monomers, additive manufacturing applications of aromatically modified polyamides are scarce. Based on technological restrictions and a lacking economic viability of the additive processing of semi-aromatic polyamides, the direct processing of aforementioned semi-aromatic material systems is inherently limited. To overcome limitations in the processing of tailored material systems and to allow for the considerable extension of material systems suitable for LPBF, the transfer of traditional synthesis routes, described in the early development of semi-aromatic and fully aromatic polyamides, as well as novel sustainable synthesis routes on the in situ modification of polyamides, represents a promising approach.

2. State of the Art

2.1. *Thermal Material Properties in Quasi-Isothermal LPBF of Polymers*

Quasi-isothermal laser-based powder bed fusion of polymers is characterized by the layer-wise manufacturing of semi-crystalline polymers within a material-specific thermal processing window [7]. To allow for minimized geometric deflections of manufactured geometries, the ambient processing temperature is specified to minimize isothermal material crystallization, allowing for the definition of an isothermal processing window [8], defined based on the crystallization and melting onset of the underlying material. However, the occurrence of isothermal crystallization has been confirmed based on experimental [8] and analytical investigations [9,10]. Considering the significance of the isothermal processing window in the LPBF of polymers, the range of materials applicable for LPBF is inherently restricted. A major limitation is represented by the processing of high-melting, technical polyamides that are subject to material aging [11] such as oxidative degradation [12], leading to altered mechanical properties [13], restricting the recyclability and ecologic viability of polyamides processed at high isothermal temperatures. Furthermore, isothermal crystallization kinetics exhibit a substantial influence on the processability of filled material systems [14,15], implicitly influencing the applicability of modified material systems.

2.2. *Properties of Eutectic Polymer-Based Compounds*

The targeted modification of thermal properties of polymers by means of additives of low molecular mass has been extensively described with regard to deep eutectic solvents (DES) [16–18], providing promising, tunable thermal and chemical properties for a variety of applications. Deep eutectic solvents that show structural similarities to ionic liquids rely on the combination of a hydrogen bond acceptor and a hydrogen bond donor, with the hydrogen donor considerably influencing the extent of the melting point reduction [19,20]. Established deep eutectic solvents have been predominantly described based on choline chloride and other quaternary ammonium compounds, with the obtained eutectic systems exhibiting melting point reductions exceeding 100 K relative to the melting point of the reactants [21].

As a sub-class of DES, supramolecular deep eutectic solvents (SUPRADES) [17] based on cyclodextrins [22,23] have been proposed, applying cyclic polysaccharides as hydrogen bond acceptors. By heating prepared dry powder blends, viscous compounds exhibiting good adhesive properties can be obtained from the combination of cyclodextrins and dicarboxylic as well as tricarboxylic acids. Relying on the formation of hydrogen-bonded supramolecular systems [17,22], obtained thermal properties are governed by the formation of hydrogen-bonded supramolecular structures, allowing for the extensive adaption of thermal material properties by adapting the composition of binary material systems. Applying aromatic monomers, de Lacalle et al. [24] described the intermediate formation of deep eutectic monomers, followed by photo-induced polymerization, allowing for tailored material properties with a considerable material toughness.

Based on binary polymer–monomer mixtures, eutectic properties of poly (ε-caprolactone) and trioxane have been described as early as 1977 by Wittmann et al. [25], obtaining a melting peak temperature of 40 °C, demonstrating the targeted modification of thermal properties of polymers using corresponding monomers.

2.3. Synthesis Routes of Semi-Aromatic Polymers

Synthesis routes, described for the preparation of semi- and fully-aromatic polyamides, can be predominantly distinguished between the condensation of diamines and dicarboxylic acids and the auto condensation of aromatic aminocarboxylic acids. Commercially available semi-aromatic polymers are prepared based on the condensation of diamines and dicarboxylic acids, being used in the form of carboxylic chlorides and anhydrides. Formed polymers exhibit considerably altered mechanical and thermal properties compared with aliphatic polyamides, being widely applied in high-temperature environments and challenging tribological and mechanical applications. In contrast, synthesis routes based on auto condensation of aminocarboxylic acids have been described based on melt-condensation and solution-based synthesis. Existing synthesis routes applying solution-based auto condensation of aromatic aminocarboxylic acids are predominantly based on applying phenyl phosphines as condensation agents in the presence of tertiary amines and metal salts, initially described by Yamazaki et al. [5]. Obtained polymers exhibit high molecular weights and melting points, depending on applied synthesis conditions, condensing agents, and solvents. Similar synthesis routes, applied by Shibasaki et al. [6] demonstrate the applicability of modified aromatic monomers. By applying the auto condensation of N-alkylated p-aminobenzoic acid with a varying aliphatic chain length, varying thermal properties could be obtained, hence significantly expanding the range of thermal properties accessible through polymers exhibiting a fully aromatic main chain. Given the complexity of solution-based synthesis of fully and semi-aromatic polyamides, various approaches for obtaining semi-aromatic polymers from melt condensation have been proposed. Based on the bulk polymerization of monomers, the formation of non-isomorphic copolymers yields low-melting products while maintaining similar thermal decomposition properties [1]. Similar approaches have been proposed based on solid-state polycondensation [26–29], obtaining a wide range of accessible thermal properties and molecular weights. However, to date, aforementioned synthesis routes are inherently limited to bulk processing, posing the inherent requirement of further processing steps at elevated processing temperatures.

2.4. Thermal Properties of Semi-Aromatic Polyamides

The length of aliphatic monomers and the occurrence of block copolymers in the chain significantly influence the thermal properties of semi-aromatic polyamides. With regard to semi-aromatic polyamides obtained from diamines and dicarboxylic acids, leading to alternating copolymers, considerably elevated melting points and glass transition temperature values can be observed. A decreasing length of the aliphatic monomer, correlated with an increasing mass fraction of aromatic monomers, is correlated with increasing melting points [30]. In contrast, block copolymers, predominantly obtained from melt condensation and solid state condensation [1,2,4,27,28,30,31], exhibit a non-linear thermal dependency of the fraction of aromatic monomers [2,28,32]. An increasing mass fraction of aromatic monomers added to aliphatic polymers results in the formation of block copolymers, implicitly affecting inter-polymer hydrogen bonding [33]. Therefore, comparatively low contents of aromatic monomers contained in block copolymers induce a reduction of the melting point [1]. Endo et al. [34] demonstrated the influence of varying monomers as well as the formation of block copolymers on resulting thermal properties, achieving a significant reduction in observed melting points, hence facilitating the melt processability of the resulting polymers.

3. Materials and Methods

3.1. Material Preparation and Processing

The in situ modification of Polyamide 12 is based on binary powder blends obtained from p-aminobenzoic acid (pABA), Carl Roth GmbH + Co. KG, Karlsruhe, Germany, and Polyamide 12 powder of type PA 2200 (EOS GmbH, Krailling, Germany). To improve the flowability of prepared powder blends and to facilitate the dispersion of pABA particles, 0.5% of fumed silica (Aerosil R 8200, Evonik AG, Duisburg, Germany), corresponding to the mass of pABA, are added. All educts are processed as received. Blending is conducted using a laboratory mixer, applying a batch size of 250 g and a blending time of 15 min at a controlled ambient temperature of 25 °C. A rotational speed of 5000 rpm is kept constant during the blending process. Applied educts exhibit melting peaks of 186 °C (PA12) and 189 °C (pABA) and crystallization peaks ($dT/dt = -10\ K\ min^{-1}$) of 152 °C (PA12) and 155°C (pABA), respectively, indicating a thermal compatibility concerning the isothermal processing window, displayed in Figure 1. The thermal processing window, defined by the model of isothermal laser sintering [8], indicates the presence of a two-phase state of fused, molten, and unexposed, solid material, requiring a considerable thermal difference between the melting temperature and the crystallization onset. The crystallization of pABA exhibits an influence of the previous thermal history, depicting a dependency of the previous heating run regarding both the crystallization enthalpy and the crystallization temperature.

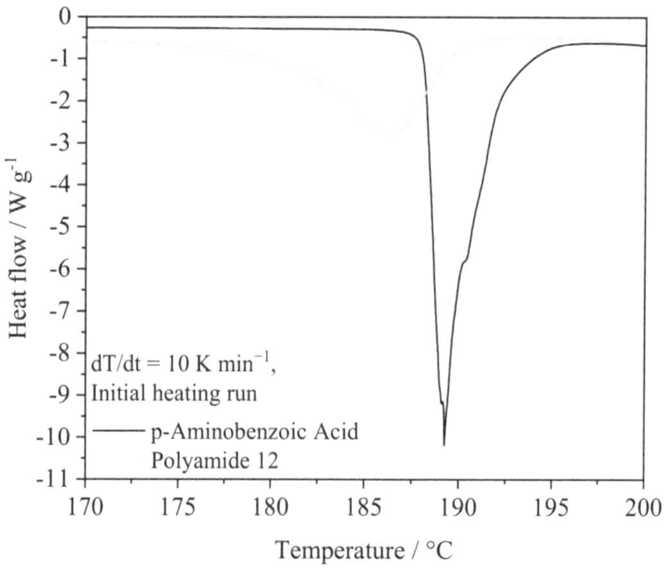

Figure 1. Depiction of differential scanning calorimetric characteristics of PA12 and pABA.

Manufactured specimens comprise tensile bars (DIN EN ISO 3167:2014-11, type 1A) scaled by a factor of 0.5. All processing is conducted using an LPBF machine of the type EOS P 396 (EOS GmbH, Krailling, Germany) equipped with a 70 W CO_2 laser with a wavelength of $\lambda = 10.6\ \mu m$. A build chamber insert exhibiting an area of $100 \times 100\ mm^2$ is applied for limiting the required amount of powder blend. Underlying exposure parameters are chosen according to near-optimal parameters for the processing of unmodified PA12, corresponding to a laser power of $P = 16.8$ W, a hatch distance of $d_{hatch} = 0.2$ mm, an exposure speed of $v = 2000\ mm\ s^{-1}$, and a layer height of $h_{Layer} = 0.12$ mm. Alternating meander scanning is specified as the underlying exposure strategy. The underlying process layout is displayed in Figure 2.

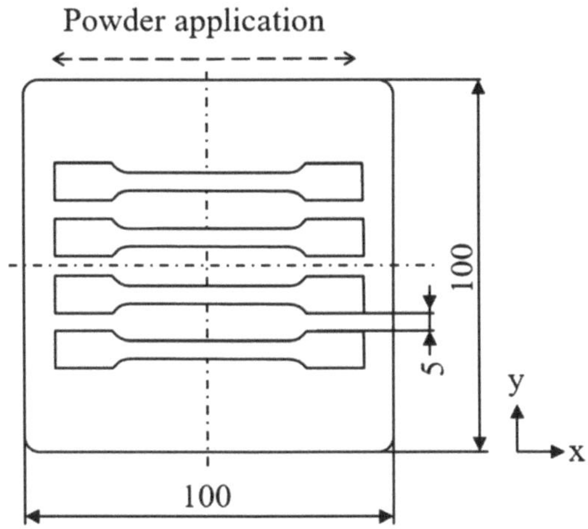

Figure 2. Schematic depiction of the build process layout relative to the powder coating orientation, units in mm.

The isothermal processing temperature is chosen according to preceding differential scanning calorimetric measurements depending on the isothermal processing window of a specific prepared material system. Following the processing, the powder bed surface is tempered at isothermal conditions for 60 min to allow for the isothermal crystallization of manufactured parts. All materials were processed applying a platform temperature of 100 °C. Employed isothermal powder bed temperatures are chosen based on the preceding differential calorimetric characterization of prepared powder blends, displayed in Table 1. Shown molar fractions correspond to the relation of aliphatic and aromatic monomers.

Table 1. Overview of applied mass fractions, corresponding molar fractions of pABA, and applied material-specific isothermal build chamber temperatures.

pABA Mass Fraction/%	pABA Molar Fraction/mol.-%	Isothermal Build Chamber Temperature/°C
0.0	0.00	172.0
1.0	1.56	170.0
2.5	3.87	167.5
5.0	7.63	155.0
10	14.86	144.5
20	28.19	141.5

3.2. Powder Blend Characterization

Characterization of prepared powder blends is conducted using thermal and rheological characterization. Thermal properties are assessed using differential scanning calorimetry (DSC), applying varying heating rates and isothermal tempering at varying temperatures for characterizing time- and temperature-dependent reaction kinetics of prepared material systems. The underlying temperature profile conducted for thermal analysis by means of DSC is displayed in Figure 3, depicting a constant temperature ramp followed by an isothermal reaction phase and a subsequent cooling phase. Two consecutive heating runs are conducted for quantifying the influence of the previous thermal history on resulting thermal material properties.

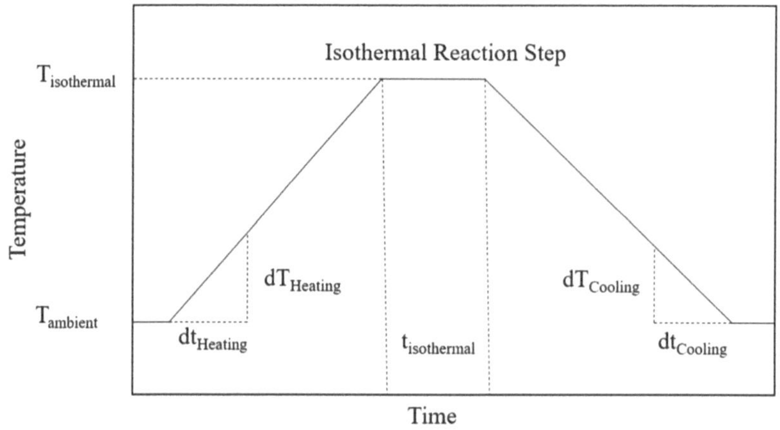

Figure 3. Illustration of the temperature profile and underlying metrices applied for DSC measurements.

DSC measurements conducted for quantifying processing parameters of prepared powder blends are based on a fixed heating and cooling rate of dT/dt = 20 K min^{-1} and a subsequent isothermal phase of $t_{isothermal}$ = 60 s at an isothermal temperature of $T_{isothermal}$ = 240 °C. Further characterizations of temperature-dependent reaction kinetics of a modified material system (ω_{pABA} = 20%), influencing the formation of copolymers and supramolecular structures, are conducted by applying varying isothermal temperatures. An underlying heating rate of $dT_{Heating}/dt_{Heating}$ = 200 K min^{-1} is applied for minimizing the influence of the non-isothermal heating process on the thermal history. Subsequent cooling is conducted by applying a cooling rate of $dT_{Cooling}/dt_{Cooling}$ = -10 K min^{-1}, followed by repeated heating at $dT_{Heating}/dt_{Heating}$ = 10 K min^{-1} for characterizing thermal process properties in dependence of the thermal material history. Varied corresponding isothermal temperatures are displayed in Table 2. All differential calorimetric measurements were conducted using a device of type DSC 2500 (TA Instruments, Inc., New Castle, DE, USA).

Table 2. Isothermal temperature values and corresponding isothermal holding times applied for characterizing high-temperature reaction kinetics of aromatically modified Polyamide 12.

Isothermal Temperature/°C	Isothermal Reaction Time/s
300	600
320	600
340	600
350	600
360	600
370	600

To allow for quantifying the processing properties of aromatically modified material systems with regard to powder coating and the topology of the resulting powder bed surface, the compressibility of prepared powder blends and the superficial powder bed roughness are assessed. Compressibility measurements, proposed by Hesse et al. (2021) for determining rheological powder properties [35], are conducted using a rheometer of type Discovery HR-20 (TA Instruments, New Castle, DE, USA), applying a maximum axial force of $F_{Compression}$ = 25 N, corresponding to a compression stress of $p_{Compression}$ = 0.08 MPa. Complementary characterizations of the powder bed roughness, depicting a negative correlation with the powder bed density [36], are based on applying optical in situ measurements of coated layers using stripe light projection, applying a recoating speed of v = 200 mm s^{-1}.

3.3. Characterization of Manufactured Components

Mechanical testing is conducted according to DIN EN ISO 527-2:2012, applying a strain rate of 25 mm min^{-1}, using an Instron universal testing system (Illinois Tool Works, Glenview, IL, USA) in combination with optical strain measurement. Complementary morphological characterizations include the preparation of polarization micrographs of prepared specimens using a Zeiss AxioImager 2 (Carl Zeiss Microscopy Deutschland GmbH, Jena, Germany) and complementary scanning electron microscopy (SEM) of fractured tensile specimens, applying an SEM of type Zeiss Gemini (Carl Zeiss Microscopy Deutschland GmbH, Jena, Germany). Furthermore, thermal DSC characterizations of manufactured specimens are conducted by applying a constant heating and cooling rate of dT/dt = 10 K min^{-1}. Complementary infrared spectroscopic measurements are based on samples extracted from the center of distinct tensile specimens, allowing for determining the presence of molecular structures characteristic of in situ post-condensation reactions. ATR infrared spectroscopy is conducted using an FT-IR device of the type Invenio FT-IR (Bruker Corporation, Billerica, MA, USA) applying a wavelength resolution of 0.5 cm^{-1}.

4. Results and Discussion

4.1. Powder Blend Characteristics

4.1.1. Thermal Powder Blend Properties

Thermal powder blend properties represent a constituting characteristic in LPBF, determining the applicability and material-specific temperature control with regard to applied isothermal processing conditions. Thermal characteristics, obtained by means of DSC, depict a significant influence of the fraction of p-aminobenzoic acid on the observed melting point, showing a correlation between the fraction of pABA and a corresponding reduction in both melting and crystallization peaks, displayed in Figure 4. Furthermore, increased fractions of pABA lead to a reduced melt and crystallization enthalpy, being correlated with significantly broadened melting and crystallization ranges.

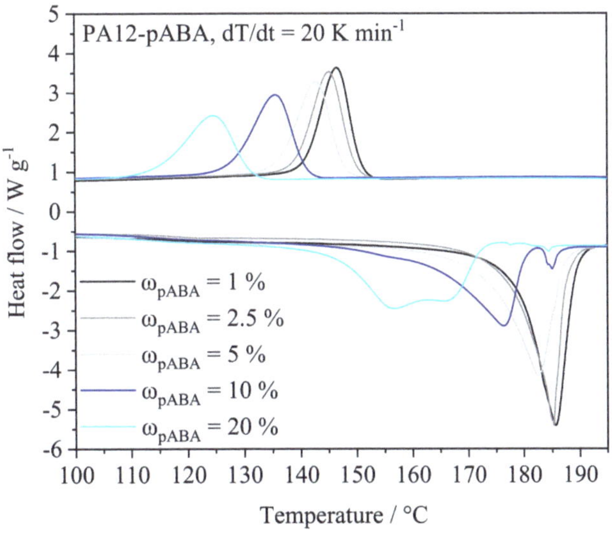

Figure 4. Adapted thermal characteristics of powder blends exhibiting varying mass fractions of pABA, n = 3.

The observed melting point reduction in powder blends is in accordance with observations by Gueche et al. (2021) [37], describing a significant melting point reduction for the application of vinylpyrrolidone–vinyl acetate copolymers modified with aliphatic dicarboxylic acids. Considering time-dependent influences on the resulting thermal material properties, a variation in both the melting process and the crystallization process can be observed. With regard to the second heating run, a decreased melt enthalpy and altered peak temperatures indicate the occurrence of chemical modifications, leading to reduced melting enthalpies. The influence of the thermal history on the melting peak temperature, displayed in Figure 5 can be correlated with experimental findings by Edgar et al. [33], describing reduced melting points of non-isomorphic statistical copolymers.

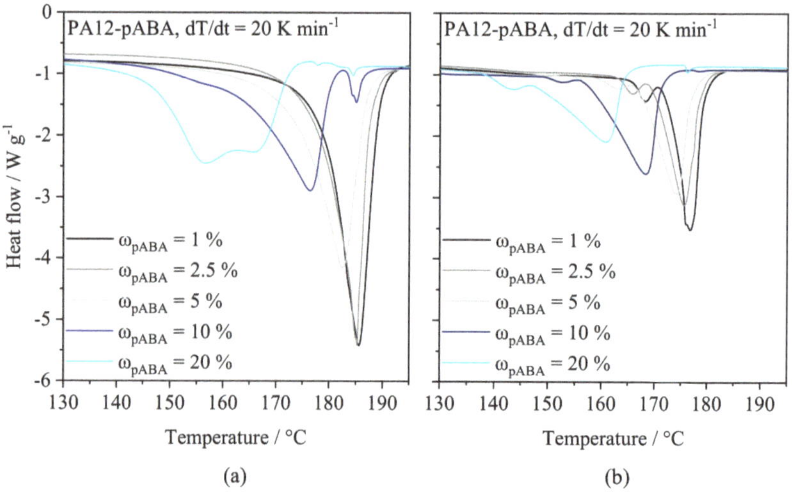

Figure 5. Differential scanning calorimetry of varying fractions of pABA observed in the first (**a**) and second (**b**) heating run, n = 3.

Considering the molecular structures of 12-aminolauric acid and p-aminobenzoic acid, an indication towards the formation of covalently bound aromatic groups, described for the melt condensation of copolymers [1,2], is evident. Therefore, observed altered thermal properties indicate the occurrence of time- and temperature-dependent auto condensation reactions of PA12-pABA blends. The assumption of the formation of condensed copolymers as well as supramolecular structures [22,23] implicitly influencing crystallization kinetics and observed melting range is further supported based on varying the maximum temperature of the initial heating run. Applying a mass fraction of ω_{pABA} = 20%, a significant influence of the thermal history on thermal material properties can be observed, leading to a diminishing melting enthalpy and reduced melting peak temperatures induced by increased peak temperatures applied within the initial heating run, displayed in Figure 6. Considering an applied isothermal reaction temperature of 370 °C, a vanishing melting peak of the prepared compound at a temperature of 109.9 °C is observed, corresponding to a melting point reduction of 76 K relative to virgin polyamide 12.

Observed temperature-dependent thermal properties depict a correlation with findings described by Zhang et al. (2011) [27], describing the formation of oligo (m-phenyleneisophthalamide) from the corresponding monomers at temperatures ranging from 240 °C to 260 °C. In addition to the formation of non-isomorphic copolymers, the influence of intermolecular hydrogen bonds, implicitly affecting thermal material properties, has been shown to considerably influence thermal properties of cyclodextrins, being associated with combined effects of supramolecular deep eutectic solvents and supramolecular polymerization processes [22]. Therefore, emerging thermal material properties are assumed to rely on the simultaneous occurrence of (auto-)condensation reactions and the formation of supramolecular, hydrogen-bonded structures, both influencing intermolecular interactions.

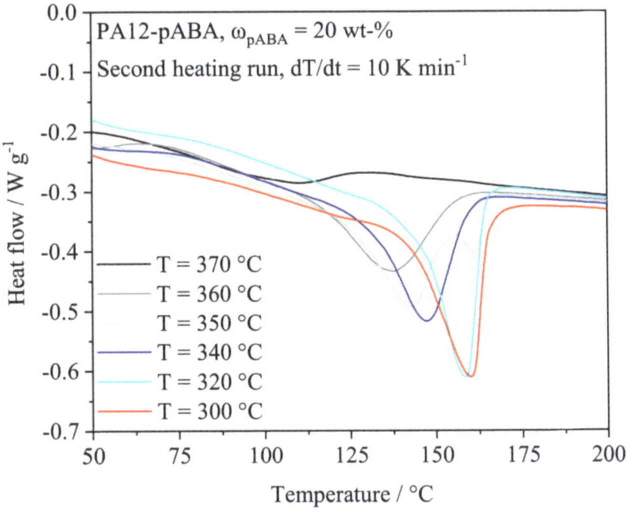

Figure 6. Differential scanning calorimetry of PA12-pABA, ω_{pABA} = 20%, in dependence on the thermal history, $t_{isothermal}$ = 600 s, n = 3.

4.1.2. Rheological Powder Blend Properties

The processability of material systems in the LPBF of polymers considerably depends on the flowability of applied material powders, influencing the powder bed surface roughness [36] and resulting mechanical properties [38] of manufactured components. Based on the determined compressibility of prepared blends, a correlation of the compressibility and the resulting surface roughness can be derived. Increasing fractions of pABA depict a negative correlation with both the compressibility and the optically assessed surface roughness, shown in Figure 7. The observed improvement of rheological processing properties is correlated with an increasing fraction of fumed silica, being added to the fraction of pABA in a constant proportion of 0.5% as a flow agent.

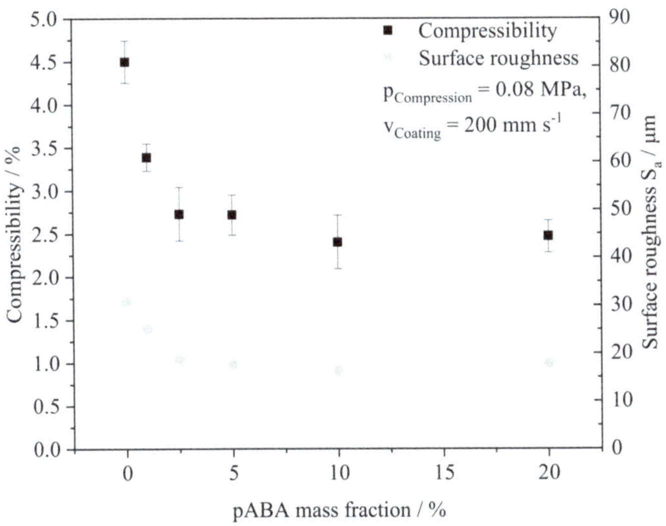

Figure 7. Overview of powder compressibility and powder surface roughness in dependence on contained fractions of pABA, n = 5.

Therefore, increased fractions of pABA allow for improving rheological properties of powder blends, hence facilitating the powder coating process and representing a foundation for the enhanced processability based on an increased powder bed density [36].

4.2. Mechanical and Thermal Part Properties

Thermal properties assessed by means of DSC exhibit significant deviations of thermal powder blend properties and thermal properties of manufactured tensile specimens. In accordance with the aforementioned DSC measurements of corresponding powder blends, a reduction in the melting enthalpy is interlinked with the thermal history of manufactured specimens. Based on the laser-based additive processing followed by an isothermal holding phase, a significantly reduced melting enthalpy relative to the thermal properties of the underlying powder materials is evident, as displayed in Figure 8. In contrast to thermal part properties observed during repetitive heating runs, a bimodal distribution of emerging thermal properties is evident with regard to a fraction of $\omega_{pABA} = 20\%$. Observed melting peaks of $T_1 = 146.6\ °C \pm 0.26$ and $T_2 = 163.8\ °C \pm 0.53$ indicate the formation of a broad range of molecular and supramolecular compositions induced by the laser exposure. Considering the aforementioned DSC measurements, thermal processing conditions in quasi-isothermal LPBF do not initiate the suppression of low-melting crystalline phases. However, in contrast to the ex situ suppression of low-melting crystalline fractions observed in DSC, an occurrence of bimodal melting properties can be observed given a fraction of $\omega_{pABA} = 20\%$. Given the aforementioned time- and temperature-dependency of the emergence of low-melting fractions as well as the considerable vanishing of crystalline fractions, detected low-melting fractions are assumed to be interlinked to a limited thermal exposure in LPBF. The negligible occurrence of melting peaks in the region of the melting point of pure pABA indicates a predominantly homogeneous distribution of pABA, obtained by means of powder blending and subsequent laser-based processing.

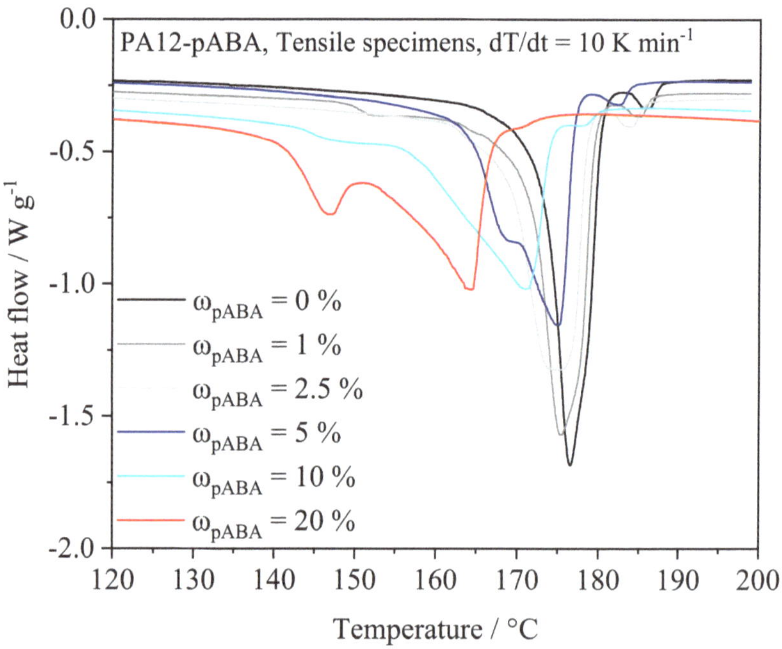

Figure 8. Thermal properties of manufactured tensile specimens assessed using differential scanning calorimetry, n = 3.

The mechanical properties of the prepared specimens depict considerable non-linearities, exhibiting slightly reduced mechanical properties of powder blends containing a proportion less than ω_{pABA} = 20% while exhibiting an inconsistent relation of the fraction of pABA and resulting mechanical properties. In contrast, an elevated fraction of ω_{pABA} = 20% leads to a significant reduction in the ultimate tensile strength and the elastic modulus while depicting a considerably enhanced elongation at break of ε = 24.65% ± 2.87, representing a significant increase compared with an elongation at break of unmodified polyamide 12 of ε = 12.41% ± 1.12, displayed in Figure 9. Furthermore, a fraction of ω_{pABA} = 20% is correlated with a considerably decreased Young's modulus, being interlinked to the aforementioned improvement of the elongation at break.

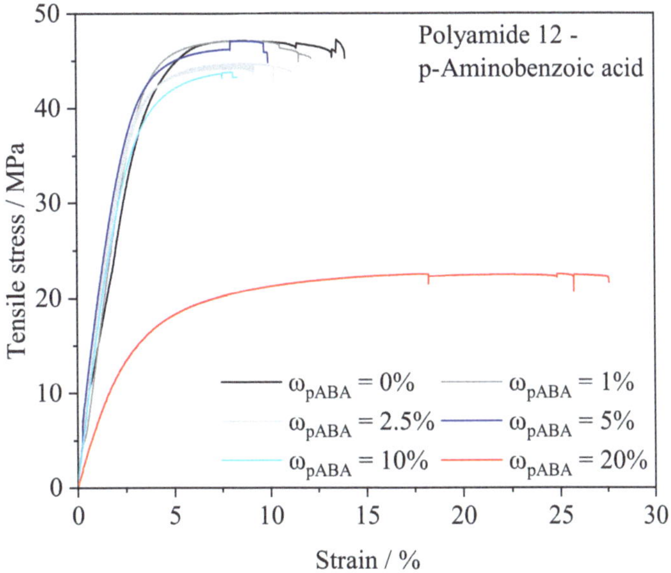

Figure 9. Depiction of the relation of the engineering stress and corresponding engineering strain of manufactured specimens in dependence of the fraction of p-aminobenzoic acid, n = 4, standard deviation indicated as gray shadow.

Considerably altered mechanical properties are assumed to rely on a reduced crystallinity of manufactured specimens, leading to an enhanced elongation at break while limiting the tensile strength [1]. Therefore, observed mechanical characteristics depict a correlation with aforementioned thermal properties. In this regard, a reduced melt enthalpy of manufactured specimens indicates a reduction in the crystallinity of manufactured samples, being attributed to the observed mechanical characteristics, specifically to a decrease in the elastic modulus and the ultimate tensile stress. Therefore, the observed increase in elongation at break when exceeding a particular fraction of pABA is assumed to be associated with the formation of eutectic supramolecular structures and previously described condensation reactions [11,39], potentially leading to the formation of non-isomorphic polymers.

4.3. Part Morphology

Regardless of the underlying fraction of p-aminobenzoic acid, formed part morphologies are characterized by the formation of pores, being associated with occasional thermal degradation. The addition of incremental fractions of pABA leads to the formation of reduced spherulite sizes, observed for all applied fractions of pABA, displayed in Figure 10. Exceeding a proportion of 10% of pABA, local morphological variations, observed due to variations in light scattering, are assumed to rely on residual crystals of pABA.

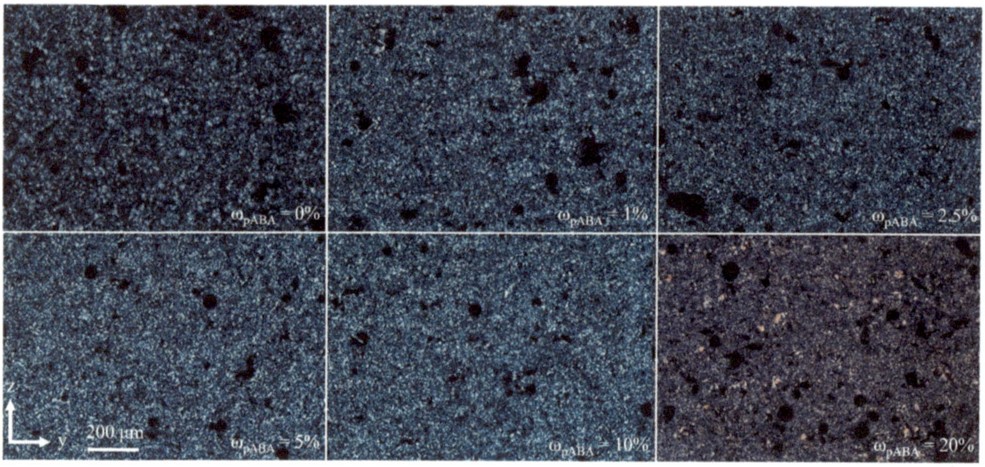

Figure 10. Polarized light micrographs of manufactured tensile specimens in dependence of the fraction of p-aminobenzoic acid.

Complementary scanning electron micrographs, displayed in Figure 11, depict correlations with previously discussed mechanical part properties, showing brittle fracture surfaces at reduced fractions of pABA. In contrast, an elevated elongation at break, evident for a fraction of ω_{pABA} = 20%, is associated with a ductile fracture pattern. Furthermore, the application of elevated fractions of pABA is attributed to the occasional occurrence of recrystallized, undissolved pABA crystals, consistent with polarized light microscopy. Furthermore, the increased formation of pores, observed for a fraction of 20% of pABA, is represented in both polarized light and scanning electron micrographs. Considering previously discussed thermal material properties, the periodic occurrence of pores is assumed to rely on a reduction in the material-specific melt enthalpy. Based on the constant processing parameter, adapted thermal part properties imply a partial thermal degradation due to a material-specific excessive energy input into the melt pool.

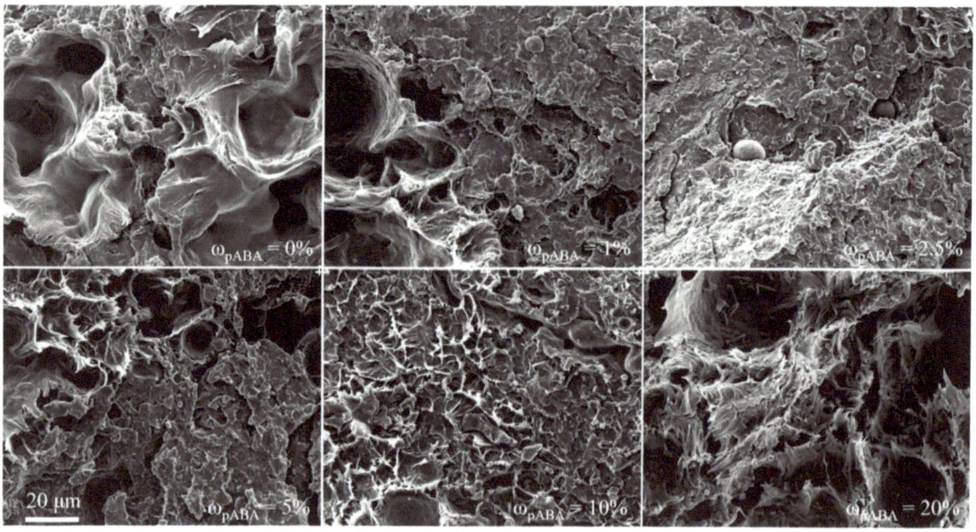

Figure 11. Scanning electron micrographs of fracture surfaces depending on the fraction of pABA.

4.4. Qualitative Characterization of Covalent and Intermolecular Bonds

Previously discussed thermal and mechanical part properties indicate the occurrence of molecular variations induced by the partial thermal condensation of p-aminobenzoic acid and polyamide 12 as well as the assumed formation of supramolecular structures, influencing the eutectic material behavior. Based on infrared spectroscopic measurements, considerable variations, depicting a dependency on the fraction of aromatic additive, can be derived, as displayed in Figure 12.

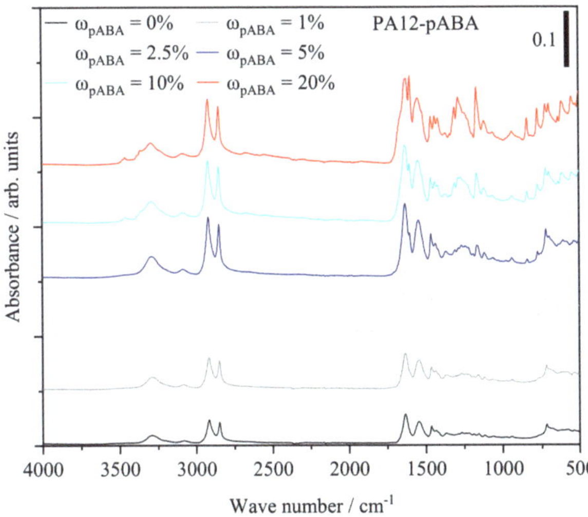

Figure 12. Overview of FTIR spectra of manufactured specimens in dependence on the fraction of pABA; adapted ordinate scaling.

Increasing fractions of pABA are correlated with an increased representation of hydrogen-bonded N-H stretching of secondary amide bonds observed at 3291 cm^{-1}, not observed for the FTIR spectrum of pABA. Similarly, an enhanced absorbance at 1640 cm^{-1}, attributed to the amide I band, depicts a dependency on the extent of aromatic additive, governed by vibrational stretching of C=O and C-N bonds, displayed in Figure 13.

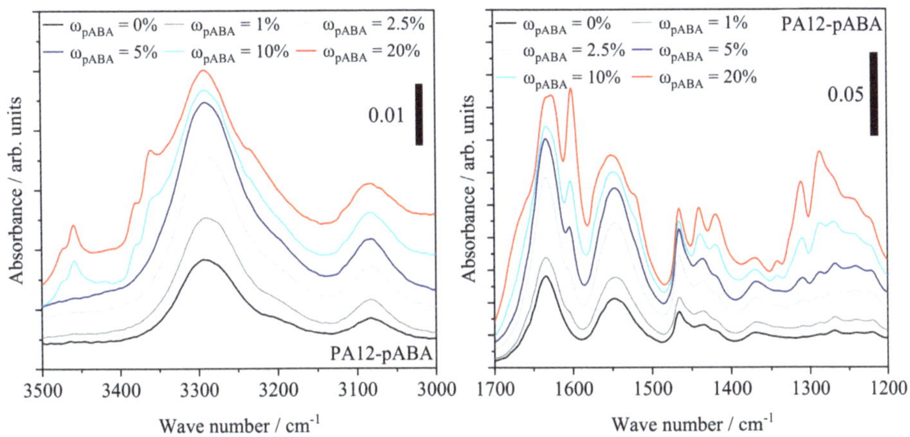

Figure 13. Infrared spectroscopic measurements of manufactured tensile specimens in dependence of the underlying fraction of p-aminobenzoic acid; adapted ordinate scaling.

The amide II band at 1550 cm^{-1}, attributed to N-H bending vibration and C-N stretching vibration, depicts a similarly increased representation. In accordance with FTIR investigations of aromatic polyamides, the emergence of a strong signal at 1285 cm^{-1} depicts a correlation with the underlying fraction of aromatic additive, being associated with C-N stretching of aromatic amide bonds similarly represented in the spectrum of the pure additive, evident in Figure 14.

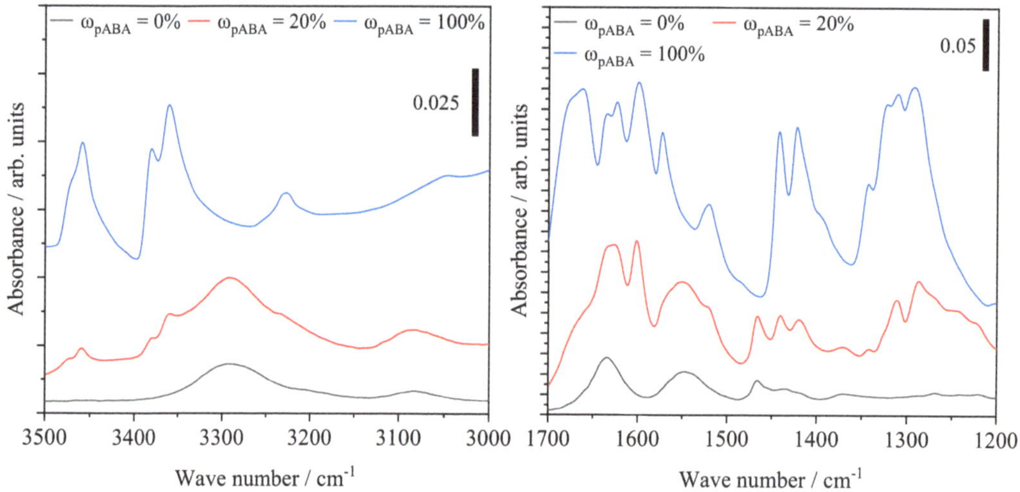

Figure 14. FTIR spectra of exemplary manufactured specimens and the spectrum of pure, unprocessed p-aminobenzoic acid; adapted ordinate scaling.

In addition to the increased occurrence of secondary amides, an emerging peak at 3457 cm^{-1} is correlated with N-H stretching of primary amides, indicating the occurrence of non-covalently bound p-aminobenzoic acid at elevated fractions, evident with regard to the spectrum of pure pABA. However, the signal is limited to pABA fractions of 10 wt% and 20 wt%, respectively. Therefore, infrared spectroscopic investigations indicate the occurrence of post-condensation reactions, previously observed in the form of auto-condensation reactions in powder bed fusion of polyamide 12 [11,39,40], hence increasing the proportion of secondary amide groups based on chain-end modifications, displayed in Figure 15.

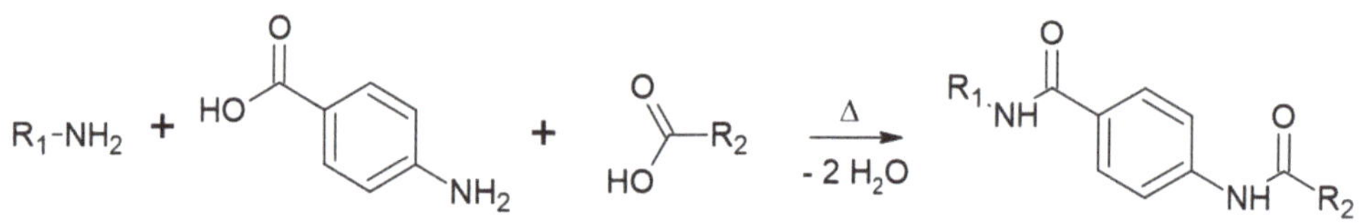

Figure 15. Potential condensation reaction of p-aminobenzoic acid and terminal amino and carboxyl groups of polymeric residues for the formation of secondary amides.

The spectroscopic determination of primary amides indicates the simultaneous formation of covalent bonds and intermolecular hydrogen bonds in accordance with relevant literature on the formation of non-isomorphic copolymers [41] and supramolecular eutectic polymers [22]. The formation of supramolecular structures can potentially be associated with a characteristic peak at 3460 cm^{-1}, interlinked to the intermolecular bonding of OH and C=O-groups [42]. However, given the spectrum of pure pABA, a distinction between peaks originating from intermolecular bonds and amide bands cannot be made, indicating the requirement for further investigations.

For quantifying the proportion of covalent and hydrogen bonds in a filled system containing 20 wt% of pABA, complementary thermogravimetric investigations (dT/dt = 10 K min^{-1}) show a mass loss of 13.51 % in a thermal range from 100 °C to 340 °C, indicating a proportion of intermolecularly bound pABA of 67.5%. Therefore, thermal investigations indicate that quasi-isothermal processing conditions lead to a predominant physical interaction of p-aminobenzoic acid and polyamide 12, in accordance with the spectroscopic identification of primary amines. However, the mechanisms and a potential mutual interaction of covalently bound p-aminobenzoic acid and formed supramolecular structures on emerging mechanical and thermal material properties remain subject to further investigations.

5. Conclusions

To overcome existing limitations in the range of polymer materials accessible for LPBF, a novel approach for the process-integrated eutectic modification of polyamide 12 in the powder bed fusion of polymers is demonstrated within the present paper. Based on the initial powder blending of polyamide 12 and p-aminobenzoic acid, subsequent laser-based additive processing allows for the in situ modification of the applied polymer, associated with significantly reduced processing temperatures. Thermal investigations and infrared spectroscopy indicate the formation of an adapted chemical structure based on the predominant formation of supramolecular hydrogen-bound structures and the formation of covalently bound secondary amides, promoting the formation of a reduced material crystallinity. Mechanical properties of manufactured tensile bars depict a significant, non-linear dependency on the fraction of p-aminobenzoic acid, exhibiting a reduced tensile strength while displaying a considerably enhanced elongation at break when applying a fraction of 20% aromatic additive.

The modification of thermal, chemical, and mechanical part properties based on binary powder blends represents a promising approach for the targeted adaption of part properties. Given the influence of the thermal history on thermal material properties, the laser-induced local adaption of the chemical structure and the application of other polymers are of considerable interest for future research. Furthermore, the underlying mechanisms regarding the interaction of covalently bound aromatic groups and the formation of supramolecular, hydrogen-bound structures are of great interest.

Author Contributions: Conceptualization, validation, methodology, investigations, writing—original draft preparation, S.S.; supervision, funding acquisition, and project administration, D.D. All authors have read and agreed to the published version of the manuscript.

Funding: Funded by the Deutsche Forschungsgemeinschaft (DFG, German Research Foundation) Project-ID 61375930 SFB 814–"Additive Manufacturing", subproject A06. We acknowledge financial support by Deutsche Forschungsgemeinschaft and Friedrich-Alexander-Universität Erlangen-Nürnberg within the funding program "Open Access Publication Funding".

Institutional Review Board Statement: Not applicable.

Informed Consent Statement: Not applicable.

Data Availability Statement: The data presented in this study are not publicly available due to ongoing research in this field.

Conflicts of Interest: The authors declare no conflict of interest.

References

1. Rwei, S.-P.; Ranganathan, P.; Chiang, W.-Y.; Lee, Y.-H. Synthesis of Low Melting Temperature Aliphatic-Aromatic Copolyamides Derived from Novel Bio-Based Semi Aromatic Monomer. *Polymers* **2018**, *10*, 793. [CrossRef] [PubMed]
2. Wolffs, M.; Cotton, L.; Kolkman, A.J.; Rulkens, R. New sustainable alternating semi-aromatic polyamides prepared in bulk by direct solid-state polymerization. *Polym. Int.* **2021**, *70*, 546–554. [CrossRef]
3. Peng, S.; Peng, L.; Yi, C.; Zhang, W.; Wang, X. A novel synthetic strategy for preparing semi-aromatic components modified polyamide 6 polymer. *J. Polym. Sci. Part A Polym. Chem.* **2018**, *56*, 959–967. [CrossRef]

4. Higashihara, T.; Zhang, C.; Tsukuda, A.; Ochi, T.; Ueda, M. Direct synthesis and melt-drawing property of aramids by bulk polycondensation of isophthalic acid with m-phenylenediamine and 3,4′-oxydianiline. *J. Appl. Polym. Sci.* **2012**, *124*, 4398–4402. [CrossRef]
5. Yamazaki, N.; Matsumoto, M.; Higashi, F. Studies on reactions of the N-phosphonium salts of pyridines. XIV. Wholly aromatic polyamides by the direct polycondensation reaction by using phosphites in the presence of metal salts. *J. Polym. Sci. Polym. Chem. Ed.* **1975**, *13*, 1373–1380. [CrossRef]
6. Shibasaki, Y.; Abe, Y.; Sato, N.; Fujimori, A.; Oishi, Y. Direct condensation polymerization of N-alkylated p-aminobenzoic acid and packing of rigid-rod main chains with flexible side chains. *Polym. J.* **2010**, *42*, 72–80. [CrossRef]
7. Greiner, S.; Jaksch, A.; Cholewa, S.; Drummer, D. Development of material-adapted processing strategies for laser sintering of polyamide 12. *Adv. Ind. Eng. Polym. Res.* **2021**, *4*, 251–263. [CrossRef]
8. Drummer, D.; Greiner, S.; Zhao, M.; Wudy, K. A novel approach for understanding laser sintering of polymers. *Addit. Manuf.* **2019**, *27*, 379–388. [CrossRef]
9. Soldner, D.; Greiner, S.; Burkhardt, C.; Drummer, D.; Steinmann, P.; Mergheim, J. Numerical and experimental investigation of the isothermal assumption in selective laser sintering of PA12. *Addit. Manuf.* **2020**, *37*, 101676. [CrossRef]
10. Soldner, D.; Steinmann, P.; Mergheim, J. Modeling crystallization kinetics for selective laser sintering of polyamide 12. *GAMM-Mitteilungen* **2021**, *44*, e202100011. [CrossRef]
11. Wudy, K.; Drummer, D. Aging effects of polyamide 12 in selective laser sintering: Molecular weight distribution and thermal properties. *Addit. Manuf.* **2019**, *25*, 1–9. [CrossRef]
12. El-Mazry, C.; Ben Hassine, M.; Correc, O.; Colin, X. Thermal oxidation kinetics of additive free polyamide 6-6. *Polym. Degrad. Stab.* **2013**, *98*, 22–36. [CrossRef]
13. Pliquet, M.; Rapeaux, M.; Delange, F.; Bussiere, P.O.; Therias, S.; Gardette, J.L. Multiscale analysis of the thermal degradation of polyamide 6,6: Correlating chemical structure to mechanical properties. *Polym. Degrad. Stab.* **2021**, *185*, 109496. [CrossRef]
14. Lanzl, L.; Wudy, K.; Greiner, S.; Drummer, D. Selective laser sintering of copper filled polyamide 12: Characterization of powder properties and process behavior. *Polym. Compos.* **2019**, *40*, 1801–1809. [CrossRef]
15. Lanzl, L.; Wudy, K.; Drummer, D. The effect of short glass fibers on the process behavior of polyamide 12 during selective laser beam melting. *Polym. Test.* **2020**, *83*, 106313. [CrossRef]
16. Roda, A.; Matias, A.A.; Paiva, A.; Duarte, A.R.C. Polymer Science and Engineering Using Deep Eutectic Solvents. *Polymers* **2019**, *11*, 912. [CrossRef] [PubMed]
17. Janicka, P.; Kaykhaii, M.; Płotka-Wasylka, J.; Gębicki, J. Supramolecular deep eutectic solvents and their applications. *Green Chem.* **2022**, *24*, 5035–5045. [CrossRef]
18. Hansen, B.B.; Spittle, S.; Chen, B.; Poe, D.; Zhang, Y.; Klein, J.M.; Horton, A.; Adhikari, L.; Zelovich, T.; Doherty, B.W.; et al. Deep Eutectic Solvents: A Review of Fundamentals and Applications. *Chem. Rev.* **2021**, *121*, 1232–1285. [CrossRef]
19. Stefanovic, R.; Ludwig, M.; Webber, G.B.; Atkin, R.; Page, A.J. Nanostructure, hydrogen bonding and rheology in choline chloride deep eutectic solvents as a function of the hydrogen bond donor. *Phys. Chem. Chem. Phys.* **2017**, *19*, 3297–3306. [CrossRef]
20. Ibrahim, R.K.; Hayyan, M.; AlSaadi, M.A.; Ibrahim, S.; Hayyan, A.; Hashim, M.A. Physical properties of ethylene glycol-based deep eutectic solvents. *J. Mol. Liq.* **2019**, *276*, 794–800. [CrossRef]
21. Abbott, A.P.; Capper, G.; Davies, D.L.; Rasheed, R.K.; Tambyrajah, V. Novel solvent properties of choline chloride/urea mixtures. *Chem. Commun.* **2003**, 70–71. [CrossRef]
22. Wu, S.; Cai, C.; Li, F.; Tan, Z.; Dong, S. Deep Eutectic Supramolecular Polymers: Bulk Supramolecular Materials. *Angew. Chem. Int. Ed.* **2020**, *59*, 11871–11875. [CrossRef] [PubMed]
23. El Achkar, T.; Moufawad, T.; Ruellan, S.; Landy, D.; Greige-Gerges, H.; Fourmentin, S. Cyclodextrins: From solute to solvent. *Chem. Commun.* **2020**, *56*, 3385–3388. [CrossRef] [PubMed]
24. De Lacalle, J.L.; Gallastegui, A.; Olmedo-Martínez, J.L.; Moya, M.; Lopez-Larrea, N.; Picchio, M.L.; Mecerreyes, D. Multifunctional Ionic Polymers from Deep Eutectic Monomers Based on Polyphenols. *ACS Macro Lett.* **2023**, *12*, 125–132. [CrossRef] [PubMed]
25. Wittmann, J.C.; Manley, R.S.J. Polymer–monomer binary mixtures. I. Eutectic and epitaxial crystallization in poly (ε-caprolactone)–trioxane mixtures. *J. Polym. Sci. Polym. Phys. Ed.* **1977**, *15*, 1089–1100. [CrossRef]
26. Papaspyrides, C.D.; Porfyris, A.D.; Vouyiouka, S.; Rulkens, R.; Grolman, E.; Poel, G.V. Solid state polymerization in a micro-reactor: The case of poly (tetramethylene terephthalamide). *J. Appl. Polym. Sci.* **2016**, *133*, 14. [CrossRef]
27. Zhang, C.; Shoji, Y.; Higashihara, T.; Tsukuda, A.; Ochi, T.; Ueda, M. Synthesis of poly (m-phenyleneisophthalamide) by solid-state polycondensation of isophthalic acid with m-phenylenediamine. *J. Polym. Sci. Part A Polym. Chem.* **2011**, *49*, 4725–4728. [CrossRef]
28. Jeyakumar, A.; Goossens, H.; Noordover, B.; Prusty, M.; Scheibitz, M.; Koning, C. Polyamide-6,6-based blocky copolyamides obtained by solid-state modification. *J. Polym. Sci. Part A Polym. Chem.* **2013**, *51*, 5118–5129. [CrossRef]
29. Papaspyrides, C.D.; Vouyiouka, S.N.; Bletsos, I.V. New aspects on the mechanism of the solid state polyamidation of PA 6,6 salt. *Polymer (Guildf).* **2006**, *47*, 1020–1027. [CrossRef]
30. Papaspyrides, C.D.; Porfyris, A.D.; Rulkens, R.; Grolman, E.; Kolkman, A.J. The effect of diamine length on the direct solid state polycondensation of semi-aromatic nylon salts. *J. Polym. Sci. Part A Polym. Chem.* **2016**, *54*, 2493–2506. [CrossRef]
31. Porfyris, A.; Vouyiouka, S.; Papaspyrides, C.; Rulkens, R.; Grolman, E.; Vanden Poel, G. Investigating alternative routes for semi-aromatic polyamide salt preparation: The case of tetramethylenediammonium terephthalate (4T salt). *J. Appl. Polym. Sci.* **2016**, *133*, 13. [CrossRef]

32. Zhang, C. Progress in semicrystalline heat-resistant polyamides. *e-Polymers* **2018**, *18*, 373–408. [CrossRef]
33. Edgar, O.B.; Hill, R. The p-phenylene linkage in linear high polymers: Some structure–property relationships. *J. Polym. Sci.* **1952**, *8*, 1–22. [CrossRef]
34. Endo, T.; Higashihara, T. Direct Synthesis of Thermally Stable Semiaromatic Polyamides by Bulk Polymerization Using Aromatic Diamines and Aliphatic Dicarboxylic Acids. *ACS Omega* **2022**, *7*, 8753–8758. [CrossRef]
35. Hesse, N.; Winzer, B.; Peukert, W.; Schmidt, J. Towards a generally applicable methodology for the characterization of particle properties relevant to processing in powder bed fusion of polymers–From single particle to bulk solid behavior. *Addit. Manuf.* **2021**, *41*, 101957. [CrossRef]
36. Sillani, F.; Schiegg, R.; Schmid, M.; MacDonald, E.; Wegener, K. Powder Surface Roughness as Proxy for Bed Density in Powder Bed Fusion of Polymers. *Polymers* **2022**, *14*, 81. [CrossRef]
37. Gueche, Y.A.; Sanchez-Ballester, N.M.; Bataille, B.; Aubert, A.; Leclercq, L.; Rossi, J.-C.; Soulairol, I. Selective Laser Sintering of Solid Oral Dosage Forms with Copovidone and Paracetamol Using a CO (2) Laser. *Pharmaceutics* **2021**, *13*, 160. [CrossRef]
38. Beitz, S.; Uerlich, R.; Bokelmann, T.; Diener, A.; Vietor, T.; Kwade, A. Influence of Powder Deposition on Powder Bed and Specimen Properties. *Materials* **2019**, *12*, 297. [CrossRef]
39. Benz, J.; Bonten, C. Temperature induced ageing of PA12 powder during selective laser sintering process. *AIP Conf. Proc.* **2019**, *2055*, 140001. [CrossRef]
40. Drummer, D.; Wudy, K.; Drexler, M. Modelling of the aging behavior of polyamide 12 powder during laser melting process. *AIP Conf. Proc.* **2015**, *1664*, 160007. [CrossRef]
41. Jeyakumar, A. Solid-State Modification of Polyamide-6,6. Ph.D. Thesis, Technische Universiteit Eindhoven, Eindhoven, The Netherlands, 2012.
42. Guo, L.; Sato, H.; Hashimoto, T.; Ozaki, Y. FTIR Study on Hydrogen-Bonding Interactions in Biodegradable Polymer Blends of Poly (3-hydroxybutyrate) and Poly (4-vinylphenol). *Macromolecules* **2010**, *43*, 3897–3902. [CrossRef]

materials

Article

The Physical Properties and Crystallization Kinetics of Biocomposite Films Based on PLLA and Spent Coffee Grounds

Jan Novák *, Luboš Běhálek, Martin Borůvka and Petr Lenfeld

Faculty of Mechanical Engineering, Technical University of Liberec, Studentská 1402/2, 46117 Liberec, Czech Republic
* Correspondence: jan.novak@tul.cz

Abstract: In the context of today's needs for environmental sustainability, it is important to develop new materials that are based on renewable resources and biodegrade at the end of their life. Bioplastics reinforced by agricultural waste have the potential to cause a revolution in many industrial applications. This paper reports the physical properties and crystallization kinetics of biocomposite films based on poly(L-lactic acid) (PLLA) and 10 wt.% of spent coffee grounds (SCG). To enhance adhesion between the PLLA matrix and SCG particles, a compatibilizing agent based on itaconic anhydride (IA)-grafted PLLA (PLLA-g-IA) was prepared by reactive extrusion using dicumyl peroxide (DCP). Furthermore, due to the intended application of the film in the packaging industry, the organic plasticizer acetyl tributyl citrate (ATBC) is used to improve processing and increase ductility. The crystallization behavior and thermal properties were observed by differential scanning calorimetry (DSC) and thermogravimetric analysis (TGA). Crystallinity degree increased from 3,5 (neat PLLA) up to 48% (PLLA/PLLA-g-IA/ATBC/SCG) at the highest cooling rate. The physical properties were evaluated by tensile testing and dynamic mechanical analysis (DMA). The combination of the compatibilizer, SCG, and ATBC led to a synergistic effect that positively influenced the supramolecular structure, internal damping, and overall ductility of the composite films.

Keywords: biocomposites; poly(L-lactic acid); spent coffee grounds; itaconic anhydride; plasticizer

Citation: Novák, J.; Běhálek, L.; Borůvka, M.; Lenfeld, P. The Physical Properties and Crystallization Kinetics of Biocomposite Films Based on PLLA and Spent Coffee Grounds. *Materials* **2022**, *15*, 8912. https://doi.org/10.3390/ma15248912

Academic Editor: Gueorgui Gueorguiev

Received: 21 November 2022
Accepted: 12 December 2022
Published: 13 December 2022

1. Introduction

The invention of catalytic polymerization caused massive petroleum-based plastics production that presents a significant challenge for the environmental sustainability [1–3]. The bio-based materials made from renewable resources have the potential to be commercially and environmentally acceptable. Especially in Europe, attention to the production of greenhouse gas emission, renewability, and toxicity is given, and the market with these environmentally friendly materials is growing fast [4,5]. Currently, one of the most commonly used biopolymers is poly(lactic acid) (PLA) [6]. Due to its biocompatibility and bioresorbability, PLA was initially used for medical purposes, e.g., bone fixation [7]. More affordable production and the possibility of biodegradation through the industrial composting process increased PLA application potential also to packaging [8], textiles [9], molding [10], and additive technologies [11]. PLA is a linear aliphatic thermoplastic polyester with high modulus and tensile strength and excellent barrier properties [12]. By contrast, its disadvantages include brittleness [13], a low heat deflection temperature [14], crystallization [15], and a biodegradation rate [16]. Many material modifications could be made to enhance these properties. Blending with other polymers, copolymerization, and the addition of particulate fillers and fibers or plasticizers are commonly used methods [17–21]. For example, adding the organic plasticizers, triethyl citrate (TC) and acetyl-tributyl citrate (ATBC) are used to enhance ductility due to the increase in chain mobility while maintaining biodegradability [22]. Other PLA usage limits, such as low heat deflection temperature and mechanical properties, can be improved through the enhancement of crystallization kinetics [23,24]. One option is to increase the nucleation density by adding heterogeneous nucleation agents [25,26]. The simultaneous addition of the nucleation agents and plasticizers may results in a synergistic effect, which can positively affect the crystallinity degree of PLA [27]. In last decades, many studies have investigated the use of fiber [28] and micro or nano particulate fillers [29,30] and their potential use as nucleating agents. Currently, the natural bio-based fillers, such as hydroxyapatite [31], cellulose nanocrystals [32], biochar [33], rice husks [34], or spent coffee grounds [35], are used. Spent coffee grounds containing cellulose, hemicellulose, lignin, and coffee oil, due to its composition, preferably act as a plasticizer and filler to improve the mixing process [36–38]. For example, Suaduang et al. [38] found a 33% increase in MFR for PLLA containing 10% of spent coffee grounds. Another promising usage of agricultural waste is to obtain new environmental-friendly polymeric material by liquification [39]. Furthermore, due to the lack of PLA-reactive side chain groups, surface and bulk modification is challenging, which can cause poor adhesion in composite structures [40,41]. Among chemical modification possibilities, free radical grafting is arguably one of the most successful and cost-effective treatments for preparing compatibilizers and improving interface adhesion. A number of studies was published for effect assessment of silanes, styrenes, or maleic anhydride (MAH) as compatibilizing agents [41–43]. Using MAH in PLA composites with low filler concentrations shows also significant impact on the degree of crystallinity, especially at slow cooling rates. Běhálek et al. [44] observed an increase in the absolute degree of crystallinity (X_c) from 54.4 to 60.1% for PLA with 10 wt.% of cellulose fibers by the grafting MAH onto a PLA backbone. However, due to the poor activity of MAH toward macro-radicals, the grafting efficiency is low compared to other polyesters [45]. Additionally, the reaction of MAH-grafted polymers with proteins may result in unstable amid bonds that can easily be hydrolyzed [40]. Itaconic anhydride (IA), with its chemical similarity, represents an alternative to MAH, and moreover, it is derived from renewable resources [46]. The high degree of grafting was reached by Verbeek et al. [47] for polyethylene. Crosslinking during the grafting reactions reduced elongation at break while improving the impact strength, creep resistance and not affecting the tensile strength [48]. Ku Marsilla et al. [40] found an increase in elongation at break, tensile strength, and energy to break for PLA grafted by 5 wt.% IA. The main aim of this study was to evaluate the thermal, mechanical, and structural properties of prepared bio-composite structures, where the reactive extrusion of itaconic anhydride with PLA was applied to reduce the interfacial tension between the matrix and spent coffee grounds (SCG) filler. Due to the brittleness of the PLA and in the view of the intended application, an organic plasticizer, ATBC, was incorporated into a bio-composite. A variety of characterization methods were used, such as Fourier transform infrared spectroscopy (FT-IR), scanning electron microscopy (SEM), differential scanning calorimetry (DSC), thermogravimetric analysis (TGA), tensile test, and dynamic mechanical analysis (DMA). Furthermore, the effect of SCG, ATBC, IA grafting onto PLA chain, and cooling rate of the non-isothermal crystallization process on the crystallization kinetics was investigated in detail using a modified Avrami equation.

2. Materials and Methods

Poly(L-lactic acid) (PLLA) matrix under the trade name of Luminy L130 was supplied by Total Corbion PLA (Gorinchem, The Netherlands). It has an approximate molar mass of 170,000 g/mol, a glass transition temperature (T_g) of 60 °C, a melting point of 175 °C, and contains more than 99% of L-lactide. Particulate spent coffee ground filler (SCG) Kimbo 100% Arabica (De'Longhi S.p.A., Treviso, Italy) was ground to an average particle size lower than 20 μm. In addition to that, 95% itaconic anhydride (IA) (Sigma-Aldrich, Darmstadt, Germany) with an approximate molar mass of 112 g/mol in the form of a powder as a compatibilizer agent and 98% dicumyl peroxide (DCP) (Sigma-Aldrich, Darmstadt, Germany) with an approximate molar mass of 270 g/mol as an initiator was used. Dehydrated acetone (ACROS Organics, Geel, Belgium) was applied to dissolve DCP and IA. A liquid form of ATBC, Citroflex A-4 delivered by Vertellus Holding LLC (Indianapolis, IN, USA), was used as a plasticizer.

2.1. Masterbatch Preparation and Film Processing

For the bio-composite film extrusion, at first, it was necessary to prepare masterbatches (MB) (Table 1), which were combined with PLLA pellets to reach the targeted compositions (Table 2). All additives, as well as the PLLA, were dried in vacuum oven VD53 (Binder GmbH, Tuttlingen, Germany) to remove moisture for 24 h at 60 °C. For the MB1 preparation, IA and DCP were dissolved in 50 mL of dehydrated acetone and added to PLLA. Further, the material was dried again and reactive extrusion with a subsequent granulation process was carried out. As a processing device, the laboratory micro-compounder MC 15 HT (Xplore, Sittard, The Netherlands) with conical screws, a constant temperature profile of 180 °C and a screw speed of 60 rpm was used. In this study, the IA and DCP concentration was fixed to 6 and 0.5% according to conclusions of Ku Marsilla et al. [40]. The dispersion and distribution were controlled by the level of torque. Consequently, the dried bio-composite pellets were processed on the twin-screw extruder MC 15 HT with a flat film die (210 °C, 0.2 mm gap size) at a melt temperature of 200 °C and 60 rpm screw speed. The extruded films were drawn by air-cooled rolls.

Table 1. Masterbatch compositions.

Samples	PLLA (g)	DCP (g)	IA (g)	ATBC (g)	SCG (g)
MB1	50.00	0.85	10.20	0	0
MB2	50.00	0.85	10.20	0	17.00
MB3	74.95	0	0	17.00	0
MB4	60.00	0	0	0	17.00

Table 2. Sample compositions.

Samples	PLLA (wt.%)	IA (wt.%)	ATBC (wt.%)	SCG (wt.%)
PLLA	100	0	0	0
PLLA/SCG	90	0	0	10
PLLA/PLLA-g-IA	94	6	0	0
PLLA/PLLA-g-IA/SCG	84	6	0	10
PLLA/ATBC	90	0	10	0
PLLA/ATBC/SCG	80	0	10	10
PLLA/PLLA-g-IA/ATBC	84	6	10	0
PLLA/PLLA-g-IA/ATBC/SCG	74	6	10	10

2.2. Spectral Analysis

To study the grafting of IA onto the PLLA, the Fourier transform infrared spectroscopy (FT-IR) using a Nicolet iS10 (Thermo Fisher Scientific, Waltham, MA, USA) spectrometer was performed. Measurements were taken with single-reflection attenuated total reflectance (ATR) system equipped with a diamond crystal. The specimens were dried in a vacuum oven VD53 at 60 °C for 24 h before the characterization. Each measurement consisted of 32 scans with 4 cm^{-1} resolution in the 400 to 4000 cm^{-1} range.

2.3. Scanning Electron Microscopy (SEM)

The morphology of fractured surfaces was examined by field emission scanning electron microscopy (FE-SEM) using the TESCAN MIRA 3 (Tescan, Brno, Czech Republic) instrument with an accelerated voltage of 5 kV. The film specimens were frozen in liquid nitrogen (−196 °C) and crushed. The fractured surfaces were coated with 5 nm of gold using a sputter coater LEICA EM ACE 200 (Leica Microsystems, Wetzlar, Germany).

2.4. Crystallization Kinetics

The non-isothermal study of crystallization kinetics was conducted by differential scanning calorimetry (DSC) analysis using a DSC 1/700 (Mettler Toledo, Greifensee, Switzerland) at different cooling rates at 5 to 40 °C min^{-1}. The facility was calibrated via zinc and indium standards. For each measurement, it was prepared about 4–8 mg of the film sample, and an empty pan was used as a reference. The samples were heated from 0 to 200 °C with a heating rate of 10 °C min^{-1} and performed under a constant nitrogen flow of 50 mL min^{-1}. The following thermal properties were observed for each composite film: the enthalpy of melting (ΔH_m), the enthalpy of cold crystallization (ΔH_{cc}), and the enthalpy of pre-melt crystallization (ΔH_{pc}). In addition to that, the degree of crystallinity (X_c) was determined according to the following equation:

$$X_c = \frac{\Delta H_m - \Delta H_{cc} - \Delta H_{pc}}{\Delta H_m^0 \cdot w_m} \quad (1)$$

where ($\Delta H_m{}^0$) is the melting enthalpy of 100% crystalline PLLA (106 J g^{-1}) [49] and w_m is the mass content of PLLA in the composites.

Further, the melting temperature (T_m), cold crystallization temperature (T_{cc}), pre-melt crystallization temperature (T_{pc}), melt crystallization temperature (T_c), and enthalpy of melt crystallization (ΔH_c) were tracked. Relative crystallinity (X_T) was calculated according to the following equation:

$$X_T = \frac{\Delta H_T}{\Delta H_C} = \frac{\int_{T_0}^{T} \left(\frac{dH_c}{dT}\right) dT}{\int_{T_0}^{T^\infty} \left(\frac{dH_c}{dT}\right) dT} \quad (2)$$

where (ΔH_T) is the heat of melt crystallization, (T_1) initial temperature, (T) temperature at a given moment, and (T^∞) final temperature. The melt crystallization time (t_c) was determined by the following equation:

$$t_c = \frac{T_0 - T}{CR} \quad (3)$$

where (CR) is the cooling rate.

Jeziorny-modified model of the logarithmic transformation of Avrami Equation (4) was used to consider the temperature changes during the crystallization process [50,51]. The growth rate parameter (Z_t) is corrected for CR, giving a modified growth rate constant (Z_c), as follows:

$$\log(-\ln(1 - X_T)) = \log Z_t - n \log t \quad (4)$$

$$\log Z_c = \frac{\log Z_t}{CR} \quad (5)$$

2.5. Thermal Degradation

The temperature of thermal decomposition of each additive and matrix was set by thermogravimetric analysis (TGA) using TGA2 instrument (Mettler Toledo, Greifensee, Switzerland). The samples were heated from 50 to 600 °C under N_2 atmosphere at the heating ramp of 10 °C min^{-1}. The initial decomposition temperature was determined at 5% mass loss. Prior to the measurements, the temperature sensor was calibrated using the Curie temperature of ferromagnetic materials based on isotherm, trafoperm, and nickel alloy.

2.6. Mechanical Measurement

Measurement of mechanical properties was performed by using a TIRA test 2300 (Tira, Schalke, Germany) testing machine equipped with a load cell of 1 kN and extensometer MFX 500-B (Mess & Feinwerktechnik GmbH, Velbert, Germany). Measurements have been performed according to the ISO 527 standard. Young's modulus measurements were performed at a crosshead speed of 1 mm min^{-1} and tensile strength with elongation at break at a crosshead speed of 5 mm min^{-1}. According to ISO 291 standard, all the specimens were conditioned in a KSP climatic chamber (Teseco, Kostelec nad Orlici, Czech Republic) at 23 °C and 50% relative humidity for one week.

2.7. Dynamic Mechanical Analysis (DMA)

The measurement of dynamic mechanical properties was carried out on the DMA DX04T (RMI s.r.o., Lazne Bohdanec, Czech Republic) equipment with the constant sinusoidal strain under the temperature of 0 to 80 °C, frequency of 1 Hz, and deformation amplitude of 0.2 mm. The square specimens (10 × 10 mm) were obtained from the films and tensile tested. The heating rate was 2 °C min^{-1}, and the storage modulus (E′), loss modulus (E″), complex modulus (E*), and loss factor (tan δ) were evaluated according to Equations (6)–(9). The values of these parameters were determined at temperatures of 10, 23, 40, and 50 °C.

$$E' = E^* \, . \, \cos\delta \tag{6}$$

$$E'' = E^* \, . \, \sin\delta \tag{7}$$

$$\tan\delta = \frac{E''}{E'} \tag{8}$$

$$E^* = E' + i \, . E'' \tag{9}$$

3. Results

3.1. Spectral Analysis

The presence of IA on the PLLA chain was identified through FT-IR measurements, which were compared to neat PLLA and IA (Figure 1). Full spectra of each component are shown in Figure 1a. As shown in Figure 1b,c, new absorption bands appear at 1649 cm^{-1} and 928 cm^{-1} in the spectra of PLLA-g-IA. A new peak can be observed at 1649 cm^{-1} for PLLA-g-IA, which is not noticeable for a neat PLLA sample (Figure 1b). This absorption was ascribed to the stretching vibrations of the cyclic anhydride functional group (C=C), which prove the mechanism of IA grafting onto PLA by Ku Marsilla [40]. Absorption band at around 2920 cm^{-1} can be assigned to CH_2 functional groups, which could be considered as proof of grafted IA since CH_2 groups are included in its molecule [52]. A peak reflecting C=O stretching vibration of an anhydride ring was probably not observed due to the low concentration of grafted IA [46]. Absorption band at 928 cm^{-1} can be ascribed to the symmetric stretching vibration of the C-O-C functional group (Figure 1c). This conclusion confirms successful grafting of IA monomers onto the PLLA backbone.

3.2. Scanning Electron Microscopy (SEM)

Structural analysis of the fractured surfaces of the composite films using electron microscopy was performed to assess the effect of the compatibilizing agent (IA) and plasticizer (ATBC) on the interfacial adhesion between the SCG and the PLLA matrix. The brittle fracture of all studied films could be observed in the microscopic images. In this case, Figure 2b shows that the adhesion between the hydrophobic matrix and the hydrophilic natural filler is sufficient even without the use of a compatibilizing agent. However, with the addition of PLLA-g-IA, SCG particles were coated and bonded to the matrix better than in the PLLA/SCG composite (Figure 2d). The reduction in the interfacial tension between the filler and matrix should be mainly reflected in the mechanical properties of the films. For the sample PLLA/ATBC/SCG, the interfacial adhesion is also at a very good level, which may be caused by higher chain mobility due to the addition of a plasticizer that contributed to better envelopment of the SCG filler (Figure 3b). The same results were observed for PLLA/PLLA-g-IA/ATBC/SCG composite films (Figure 3d).

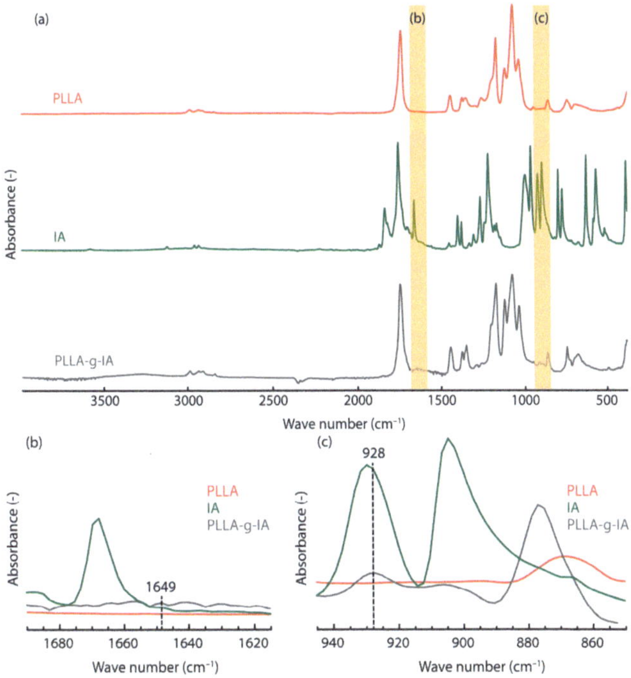

Figure 1. FT-IR spectra of PLLA Luminy L130, IA, and PLLA-g-IA: (**a**) full FT-IR spectrum from 4000 cm^{-1} to 400 cm^{-1}, (**b**) absorbance of the C=C stretching at 1649 cm^{-1}, and (**c**) absorbance of the C-O-C stretching at 928 cm^{-1}.

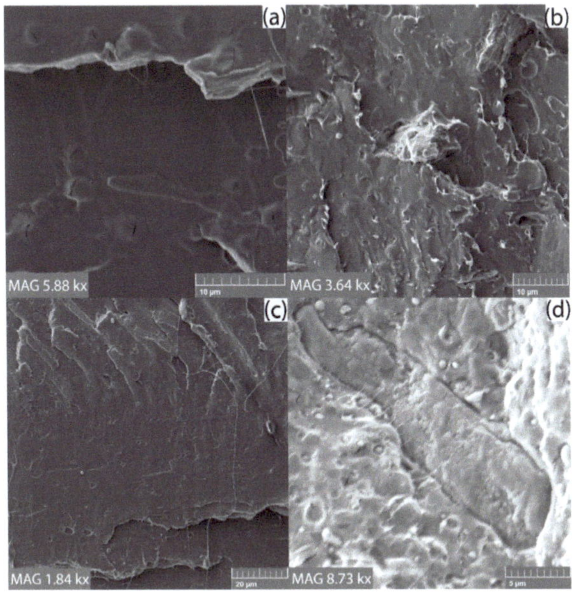

Figure 2. SEM analysis of fractured surfaces: (**a**) PLLA, (**b**) PLLA/SCG, (**c**) PLLA/PLLA-g-IA, and (**d**) PLLA/PLLA-g-IA/SCG.

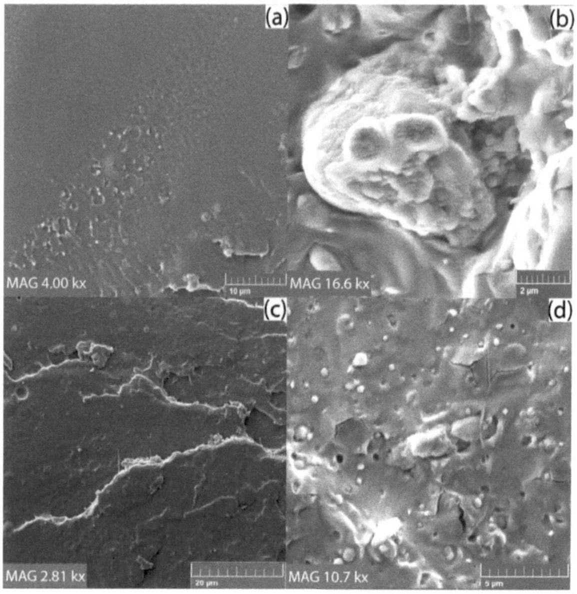

Figure 3. SEM analysis of fractured surfaces: (**a**) PLLA/ATBC, (**b**) PLLA/ATBC/SCG, (**c**) PLLA/PLLA-g-IA/ATBC, and (**d**) PLLA/PLLA-g-IA/ATBC/SCG.

3.3. Crystallization Kinetics

The extruded film samples were studied using DSC analysis according to the ISO 11,357 standard to specify morphology and thermal properties. In this section, the results obtained from the melt cooling run and the second heating run, which determine the crystallization capability of the materials, are discussed. In terms of the transition temperatures, the results of the DSC analysis show that the melting temperature of the composite structures is mainly determined by the polymer matrix in range of 172 ± 2 °C, see Table 3. In the presence of ATBC, the melting temperature slightly decreases to values of 168 ± 2 °C. As expected, the effect of ATBC is particularly pronounced in changing the T_g, which is in range of 45 ± 5 while for PLLA material structures without ATBC in range of 59 ± 2 °C. However, the determination of the T_g by DSC is affected by a methodological error, which is particularly visible at lower cooling rates, when the primary crystallization of the polymer system is greatest, increasing its degree of crystallinity and resulting in difficulty in determining the glass transition region. The study of crystallization kinetics has shown, in addition to well-known facts, such as the effect of CR on the X_c and a positive effect of SCG on the formation of the PLLA crystalline structure, see Figure 4. In contrast, no positive effect of IA on the X_c of the PLLA/SCG composite structure was demonstrated, although grafting IA onto PLLA chain alone resulted in an increase in X_c. Comparing the PLLA and PLLA/PLLA-g-IA matrices, an increase in crystallinity of 2.6 times can be found at CR of 5 °C/min. In contrast, the comparison of the PLLA/SCG and PLLA/PLLA-g-IA/SCG composites shows a 5% decrease in crystallinity, which could be observed at all investigated CRs. Thus, it can be concluded that the grafting of IA onto the PLLA chain does not achieve a synergistic effect in the crystallization of SCG composite structures, whereas in the case of the PLLA/ATBC/SCG, it is noticeable, and an increase in the X_c was observed at all CRs.

Table 3. The results from the second heating run of DSC analysis.

Samples	CR (°C/min)	T_g (°C)	T_m (°C)	T_{pc} (°C)	T_{sc} (°C)	ΔH_m (J/g)	ΔH_{pc} (J/g)	ΔH_{sc} (J/g)	X_c (%)
PLLA	5	60.9	175.5	160.9	113.0	40.2	1.4	25.9	12.2
	10	60.3	174.7	160.4	111.9	39.1	0.5	32.2	6.0
	20	60.9	175.5	160.9	113.0	38.2	0.3	33.3	4.4
	40	60.9	175.7	156.1	113.0	38.4	0.3	34.4	3.5
PLLA/SCG	5	-	173.9	-	-	43.4	0.0	0.0	45.5
	10	-	173.6	160.3	-	41.2	1.8	0.0	41.3
	20	59.0	172.5	155.3	94.1	46.5	6.6	15.8	25.4
	40	58.8	172.4	155.3	97.2	45.9	6.7	26.7	13.1
PLLA/PLLA-g-IA	5	59.6	172.4	156.1	105.1	44.6	4.5	8.6	31.7
	10	58.9	172.7	157.2	108.2	41.6	1.8	28.7	11.1
	20	59.4	173.1	158.6	109.2	41.2	1.0	33.7	6.4
	40	59.3	173.1	158.6	109.4	40.4	1.0	35.0	4.4
PLLA/PLLA-g-IA/SCG	5	-	172.6	159.6	-	39.7	1.3	0.0	43.1
	10	-	172.0	155.5	94.8	43.1	5.0	4.1	38.2
	20	59.1	171.7	154.9	99.2	43.7	5.7	24.9	14.6
	40	59.0	171.7	155.1	99.2	43.3	5.5	28.3	10.6
PLLA/ATBC	5	-	169.2	153.1	-	40.4	1.5	0.0	40.8
	10	-	169.1	147.2	86.6	42.5	5.2	10.2	28.4
	20	40.8	168.9	148.9	92.7	42.3	4.4	24.2	14.3
	40	40.1	168.9	149.3	93.0	42.2	4.1	27.5	11.2
PLLA/ATBC/SCG	5	-	169.8	-	-	38.5	0.0	0.0	45.4
	10	-	169.3	154.9	-	38.5	0.8	0.0	44.5
	20	-	169.0	145.7	77.1	42.0	5.8	5.6	36.1
	40	40.9	168.7	143.9	82.5	42.3	6.5	18.6	20.4
PLLA/PLLA-g-IA/ATBC	5	-	170.0	153.1	-	37.5	1.4	0.0	40.6
	10	-	169.1	146.7	90.1	40.5	4.9	12.0	26.5
	20	43.9	169.5	148.9	94.5	39.2	4.1	23.7	12.8
	40	43.6	169.4	149.1	94.4	39.1	3.7	26.1	11.0
PLLA/PLLA-g-IA/ATBC/SCG	5	-	171.5	-	-	37.7	0.0	0.0	48.0
	10	-	170.2	152.0	91.8	39.9	3.9	0.9	45.2
	20	49.1	170.2	149.3	91.3	42.5	5.9	19.2	22.2
	40	49.1	170.2	148.9	91.4	42.3	5.9	23.8	16.1

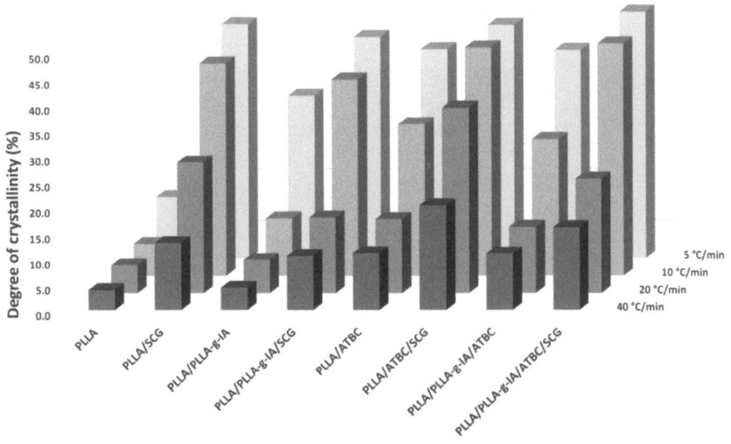

Figure 4. Crystallinity degree of biocomposite structures at different CRs.

A positive finding is that the increase in X_c is particularly evident even at high CRs, which otherwise generally hinder polymer crystallization. At slow melt CR (5 °C/min), the PLLA/PLLA-g-IA/ATBC/SCG achieved the highest value of X_c = 48%, showing an increase of 18% compared to PLLA/PLLA-g-IA/ATBC and a fourfold increase compared to neat PLLA matrix. A similar result can be observed for the PLLA/SCG structure. At higher CRs (40 °C/min), the crystallization ability of PLLA is limited. The values of X_c show that a higher change in the macromolecule arrangement can be achieved at higher CRs due to the SCG addition; for example, a more than threefold X_c increase was observed for the PLLA/SCG structure, while PLLA/PLLA-g-IA/SCG increased 2.4 times (only 1.4-fold at 5 °C/min). The highest values of X_c at CR = 40 °C/min were achieved by the PLLA/ATBC/SCG (X_c = 20%), where an increase of about 2 times was observed compared to the PLLA/ATBC matrix, while at CR = 5 °C/min this increase was only 10%. Overall, the organic plasticizer based on ATBC can be attributed to an increase in the conformational processes of macromolecules promoting the crystallization behavior of the polymer. Its addition to PLLA results in an increase in X_c at all CRs. The positive effect of the SCG can be clearly seen in the increase in enthalpy changes in the primary crystallization and the suppression of the secondary crystallization. The temperature of secondary crystallization (T_{sc}) shifts to significantly lower values (Table 3), indicating that SCG composites require lower kinetic energy for the internal rotations of macromolecules, similar to the PLLA/ATBC structures.

Such as cellulose nanocrystals [26], lignin nanoparticles [53] or cellulose fibers, and rice husks [44], SCG act as a nucleating agent to promote overall crystallization of PLLA. This is due to the higher onset of the crystallization temperature and the peak of the primary crystallization temperature (T_c), respectively, see Table 4. For example, at CR of 5 °C/min, T_c = 101.7 °C for PLLA and 108.2 °C for PLLA/SCG. Similarly, this phenomenon can be observed for the PLLA/ATBC/SCG structure, where at the same CR, the T_c increased from 93.8 to 101.6 °C, see Table 4. The relative crystallization (X_T) determined at T_c and halftime of the crystallization process ($t_{1/2}$) corresponding to X_T = 50% were also monitored to assess the crystallization kinetics. As was mentioned above, the presence of SCG increases the primary crystallization temperature, whereas ATBC decreases it. Thus, the PLLA/SCG crystallinity half-time is smaller than PLLA/ATBC/SCG at all studied CRs. The effect of SCG presence in the PLLA, PLLA/PLLA-g-IA, PLLA/ATBC, and PLLA/PLLA-g-IA/ATBC is shown in Figures 5 and 6.

An increase in the crystallization halftime also occurs when IA is grafted onto the PLLA chain. At a CR = 5 °C/min, in all composite structures, the $t_{1/2}$ decreased due to the presence of SCG. At higher melt CRs, this trend persisted especially for the PLLA/SCG and PLLA/ATBC/SCG. For the PLLA/PLLA-g-IA/SCG and PLLA/PLLA-g-IA/ATBC/SCG, there is a slight increase in the $t_{1/2}$. Thus, it can be assumed that grafting of IA onto the PLLA backbone suppresses the rate of crystallization. The results of the change in X_T clearly indicate a higher rate of the PLLA crystallization process in the presence of SCG and ATBC. Especially at lower CRs, the synergistic effect of SCG with ATBC in the PLLA matrix was confirmed (X_T has grown up to 52% compared to neat PLLA). These conclusions are further supported by the Jeziorny modification of Avrami crystallization kinetics model. The statistical parameters obtained by the linear fitting of the dependence of log[-ln(1-X_T)] and log t are listed in Table 5.

The Avrami exponent n as a measure of uniform crystal growth increases by the addition of SCG from 3.25 to 4.58 at CR = 5 °C/min. The values of n could be concluded to be a mixed-type growth with variable nucleation. With the increase in the CR, the values of Z_c increase and $t_{1/2}$ decreases, which indicates the higher crystallization rate with the increase in CR. Additionally, in the presence of SCG, the values of Z_c and $t_{1/2}$ changed from 0.00069 and 3.23 min at CR = 5 °C/min to 0.037839 and 0.34 min at CR = 40 °C/min. Therefore, based on the results of $t_{1/2}$ and Z_c, we could draw the conclusion that the presence of SCG has a positive effect on PLLA crystallization kinetics, especially at higher CRs. The same conclusion was reached for PLLA/PLLA-g-IA and PLLA/ATBC matrix. The value of n is depending on the nature of the nucleation and growth geometry of crystals [54]. This indicates that the presence of SCG has a great influence on the nucleation and growth geometry of PLLA. The highest effect could be observed for PLLA/PLLA-g-IA/SCG and PLLA/ATBC/SCG, where the value of n was 2.63 and 2.96 at CR = 40 °C/min, while for the other composite structures, there was not a noticeable area of primary crystallization.

Table 4. The results from the cooling run of DSC analysis.

Samples	CR (°C/min)	T_c (°C)	ΔH_c (J/g)	ΔH_{Tc} (J/g)	X_T (%)
PLLA	5	101.7	5.43	2.40	44.2
	10	100.2	1.18	0.48	40.7
	20	-	-	-	-
	40	-	-	-	-
PLLA/SCG	5	108.2	35.45	20.13	56.8
	10	105.7	33.35	17.22	51.6
	20	96.2	12.20	5.97	48.9
	40	93.7	0.81	0.44	54.3
PLLA/PLLA-g-IA	5	100.9	21.94	9.60	43.8
	10	96.6	3.29	1.61	48.9
	20	-	-	-	-
	40	-	-	-	-
PLLA/PLLA-g-IA/SCG	5	105.9	33.37	17.87	53.6
	10	99.3	23.48	10.91	46.5
	20	94.5	2.67	1.46	54.7
	40	91.7	0.10	0.06	60.0
PLLA/ATBC	5	93.8	31.48	17.95	57.0
	10	86.6	16.97	8.22	48.4
	20	87.5	1.37	0.58	42.3
	40	-	-	-	-
PLLA/ATBC/SCG	5	101.6	33.17	22.33	67.3
	10	94.2	29.35	18.14	61.8
	20	86.9	17.34	8.21	47.3
	40	83.7	0.99	0.53	53.5
PLLA/PLLA-g-IA/ATBC	5	95.3	28.86	16.02	55.5
	10	88.6	11.92	5.55	46.6
	20	88.1	1.24	0.50	40.7
	40	-	-	-	-
PLLA/PLLA-g-IA/ATBC/SCG	5	101.9	31.84	18.27	57.4
	10	95.0	25.47	12.36	48.5
	20	90.9	4.21	1.91	45.4
	40	-	-	-	-

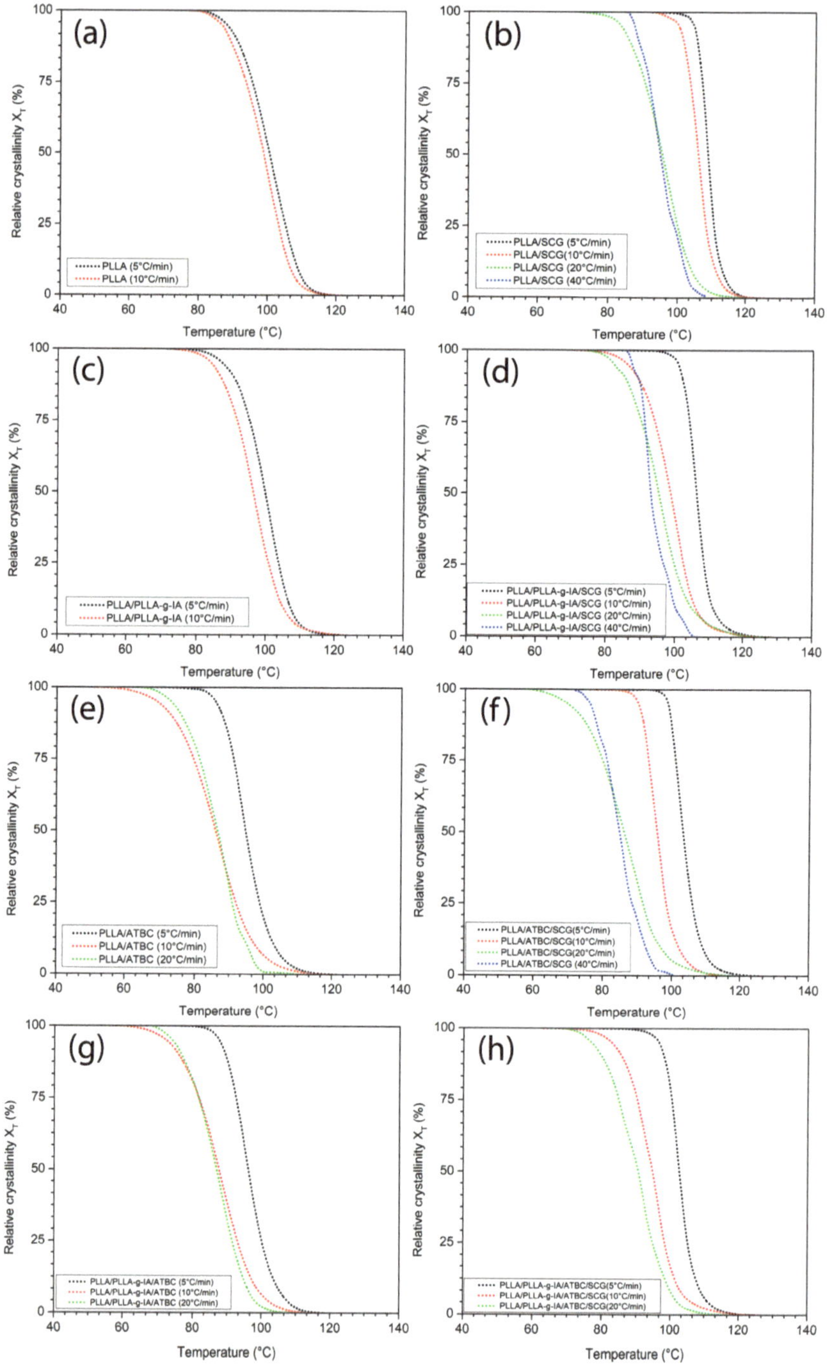

Figure 5. Variation in X_T during the cooling rates 5, 10, 20, and 40 °C/min for (**a**) neat PLLA, (**b**) PLLA/SCG, (**c**) PLLA/PLLA-g-IA, (**d**) PLLA/PLLA-g-IA/SCG, (**e**) PLLA/ATBC, (**f**) PLLA/ATBC/SCG, (**g**) PLLA/PLLA-g-IA/ATBC, and (**h**) PLLA/PLLA-g-IA/ATBC/SCG.

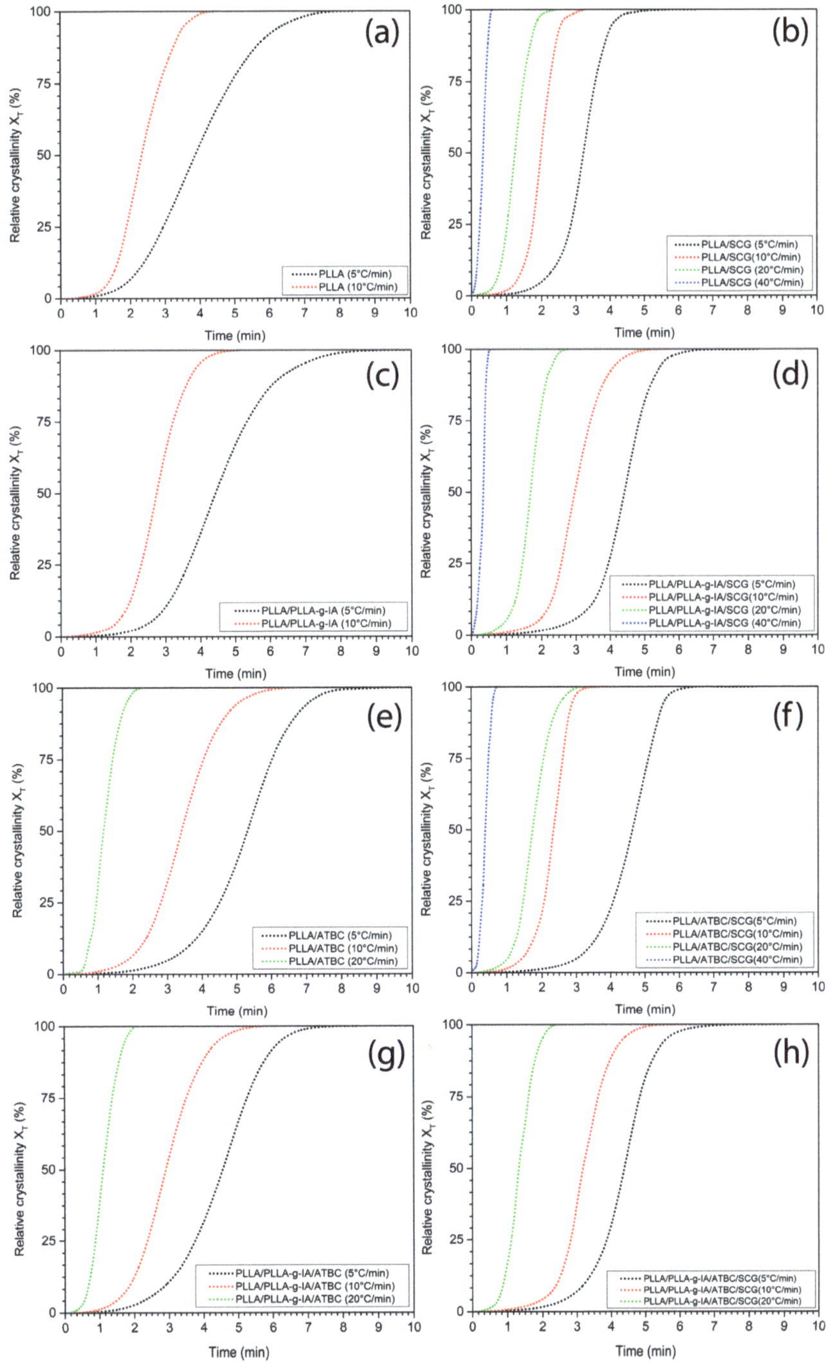

Figure 6. Plots of X_T as a function of time during the cooling rates 5, 10, 20, and 40 °C/min for (**a**) neat PLLA, (**b**) PLLA/SCG, (**c**) PLLA/PLLA-g-IA, (**d**) PLLA/PLLA-g-IA/SCG, (**e**) PLLA/ATBC, (**f**) PLLA/ATBC/SCG, (**g**) PLLA/PLLA-g-IA/ATBC, and (**h**) PLLA/PLLA-g-IA/ATBC/SCG.

Table 5. The modified Avrami parameters for the non-isothermal crystallization of PLLA, PLLA/PLLA-g-IA, PLLA/ATBC, and PLLA/PLLA-g-IA/ATBC composites.

Samples	CR (°C/min)	$t_{1/2}$ (min)	log Z_t	Z_c	n	r^2
PLLA	5	3.87	−2.10	0.00159	3.25	0.99
	10	2.35	−1.70	0.00201	3.98	0.99
	20	-	-	-	-	-
	40	-	-	-	-	-
PLLA/SCG	5	3.23	−2.46	0.00069	4.58	0.98
	10	2.03	−1.57	0.00267	4.75	0.98
	20	1.25	−0.58	0.01315	3.86	0.99
	40	0.34	1.18	0.37839	2.71	0.98
PLLA/PLLA-g-IA	5	4.43	−2.72	0.00038	3.84	0.99
	10	2.50	−1.83	0.00148	3.71	0.97
	20	-	-	-	-	-
	40	-	-	-	-	-
PLLA/PLLA-g-IA/SCG	5	4.42	−3.07	0.00017	4.55	0.96
	10	2.97	−2.03	0.00094	3.81	0.96
	20	1.70	−1.08	0.00416	4.05	0.98
	40	0.35	1.19	0.38720	2.63	0.96
PLLA/ATBC	5	5.32	−2.92	0.00024	3.83	0.99
	10	3.79	−2.06	0.00088	3.50	0.99
	20	1.18	−0.54	0.01442	4.06	0.99
	40	-	-	-	-	-
PLLA/ATBC/SCG	5	4.65	−3.13	0.00015	4.56	0.95
	10	2.54	−1.64	0.00228	4.18	0.96
	20	1.75	−1.05	0.00446	3.22	0.95
	40	0.40	1.05	0.28050	2.96	0.99
PLLA/PLLA-g-IA/ATBC	5	4.54	−2.71	0.00039	3.93	0.99
	10	3.10	−1.86	0.00138	3.57	0.99
	20	1.11	-0.38	0.02084	3.64	0.99
	40	-	-	-	-	-
PLLA/PLLA-g-IA/ATBC/SCG	5	4.37	−2.88	0.00027	4.25	0.97
	10	3.18	−2.23	0.00059	4.01	0.96
	20	1.34	−0.67	0.01069	3.79	0.99
	40	-	-	-	-	-

3.4. Thermal Degradation

The thermodegradation temperatures are summarized in Table 6. Thermal stability of neat PLLA, SCG, and ATBC reaches its limit at 328, 253, and 222 °C, respectively. Thus, the addition of ATBC into PLLA films caused a significant decrease in the initial decomposition temperature. Similar results were observed by Maiza et al. [22] and others [55,56]. Grafting IA onto the PLLA backbone slightly increased thermal stability, which is ascribed to the new C=C bond. The addition of SCG reduced the decomposition temperature to approximately 310 °C, which is related to the degradation of low molecular weight components, such as hemicellulose, that is present in the lignocellulosic residue [57]. A similar reduction in thermal stability was observed by Terroba-Delicado et al. [25] or Mendes et al. [58] when studying PE/SCG thermal properties. Nekhamanurak et al. [59] observed PLLA/$CaCO_3$ fatty acid-treated composites and concluded the reaction of the fatty acids and oils with ester linkage of PLLA decreases its thermal stability. The incorporation of SCG into PLLA/PLLA-g-IA/ATBC matrix points to the synergistic effect between SCG and ATBC in the presence of PLLA-g-IA.

Table 6. Thermodegradation temperatures.

Samples	$T_{d,5}$ (°C)
PLLA	328
PLLA/SCG	310
PLLA/PLLA-g-IA	337
PLLA/PLLA-g-IA/SCG	316
PLLA/ATBC	272
PLLA/ATBC/SCG	272
PLLA/PLLA-g-IA/ATBC	272
PLLA/PLLA-g-IA/ATBC/SCG	279

3.5. Mechanical Measurements

The results from tensile tests of neat PLLA, modified PLLA, and SCG-based composites prepared by extrusion are shown in Table 7. The measured values show that the Young's modulus depending on the presence of SCG does not change in practice, which is positive since the moisture content of the SCG composite film 0.7–0.9% is higher than that of the neat PLLA 0.3–0.4%. The same conclusion was reached for the composite with grafted itaconic anhydride onto the PLLA chain. Significant changes are observed for the PLLA/ATBC/SCG film, where the presence of coffee grounds in the PLLA/ATBC matrix causes a decrease in tensile modulus of about 28% (from 2834 MPa to 2049 MPa) and an increase in the elongation more than five times (from 7 to 36%), although the degree of crystallinity increased by 63%. These results indicate the synergistic effect of ATBC and SCG containing fatty acids and other components reducing the stiffness and increasing the ductility of the material, which was not achieved by PLLA containing SCG and ATBC separately. This synergistic effect was also observed for the composite films PLLA/PLLA-g-IA/ATBC and PLLA/PLLA-g-IA/ATBC/SCG, respectively. The presence of SCG led to a 12% decrease in the average modulus and in an increase in ductility by approximately 90%. A comparison of the tensile test results shows that the addition of SCG to PLLA, PLLA/PLLA-g-IA, and PLLA/ATBC matrix reduces the tensile strength as expected. This effect is also supported by the increase in the elongation at break, which confirms the SCG plasticizing ability. Similar results were observed by Suaduang et al. [35] on the bio-composite PLA/SCG blown films. It was found that a greater content of SCG significantly decreases the tensile strength, and the results of elongation at break showed the reduction to brittleness compared to neat PLA. This brittleness reduction is supported by ATBC content. The ATBC with abundant polar groups easily enters the gap between PLA chains, which weakens the binding and entanglement force between PLA chains and increases their mobility [60]. Tian et al. observed that 20 wt.% of ATBC resulted in the great increase in the elongation at break while the tensile strength and tensile modulus reduce observably when compared with the 15 wt.% ATBC modified PLA, which is attributed to the reduction in intermolecular forces [60,61].

Table 7. Tensile modulus, strength, and elongation at break of PLLA-based films.

Samples	E_t (MPa)			σ_m (MPa)			ε_{tb} (%)		
	$\bar{x}$	s	CI(95)	$\bar{x}$	s	CI(95)	$\bar{x}$	s	CI(95)
PLLA	3530	193	120	50	3	2	8	1	1
PLLA/SCG	3599	222	138	36	2	1	19	4	2
PLLA/PLLA-g-IA	3603	173	107	46	2	1	4	1	1
PLLA/PLLA-g-IA/SCG	3386	221	137	33	2	1	11	4	2
PLLA/ATBC	2834	300	186	39	2	1	7	3	2
PLLA/ATBC/SCG	2049	193	120	25	1	1	36	11	7
PLLA/PLLA-g-IA/ATBC	1475	135	84	25	2	1	127	88	55
PLLA/PLLA-g-IA/ATBC/SCG	1293	267	165	23	3	2	241	35	22

3.6. Dynamic Mechanical Analysis (DMA)

The experimental research has investigated the temperature dependence of the dynamic modulus of elasticity (E*), including its real (E′) and imaginary (E″) components, and the loss factor (tan δ) characterizing the internal damping of the material. The parameters determined by DMA at temperatures of 10, 23, 40, and 50 °C, respectively, are shown in Table 8. The temperature of 10 °C represents cooler conditions, the temperature of 23 °C corresponds to standard environmental conditions, the temperature of 40 °C corresponds to the mean glass transition temperature of PLLA/ATBC-based films, and the temperature of 50 °C corresponds to the glass transition of other studied composite films. With the addition of SCG, the results of the DMA show an increase in the imaginary component of the dynamic modulus, which is characterizing the damping properties of the material. The change in tan δ is insignificant as the imaginary component of the dynamic modulus increases along with its real component. This is not the case for the PLLA/ATBC/SCG film due to the content of a plasticizer, which reduces the elastic modulus. In contrast to the static tensile test, a positive effect of SCG on the elastic modulus and damping properties was observed. An increase in these properties may be promoted by the presence of fatty acids in SCG, which is in line with the conclusions of Coelho et al. [62]. The differences in the values of the elastic modulus found in static and dynamic loading are caused by the nature of the test and the range of deformation regions. The greatest effect of SCG onto damping properties was clearly observed for the PLLA/ATBC/SCG film, which corresponds to the change in the nominal relative elongation at break, see Table 7. The SCG in the given composite system increased the imaginary elastic modulus and the loss factor up to 2.9 and 3.8 times, respectively. The deviations observed for the PLLA/PLLA-g-IA/ATBC/SCG composite structure may be due not only to the frequency of the measurements but also to the frequency of the recorded data and their approximation.

Table 8. Values of storage modulus, loss modulus, complex modulus, and loss factor.

Samples	T (°C)	E′ (MPa)	E″ (MPa)	tan δ (-)	E* (MPa)
PLLA	10	2276	37	0.016	2277
	23	2222	42	0.019	2223
	40	2140	51	0.024	2141
	50	2088	65	0.031	2090
PLLA/SCG	10	2632	43	0.017	2633
	23	2596	44	0.017	2597
	40	2465	56	0.023	2466
	50	2354	76	0.032	2356
PLLA/PLLA-g-IA	10	2449	35	0.015	2449
	23	2374	45	0.019	2375
	40	2260	58	0.026	2261
	50	2147	80	0.038	2150
PLLA/PLLA-g-IA/SCG	10	3074	45	0.015	3074
	23	3012	48	0.015	3013
	40	2833	69	0.024	2834
	50	2617	139	0.047	2623
PLLA/ATBC	10	2282	65	0.028	2284
	23	2251	63	0.028	2253
	40	2017	90	0.044	2021
	50	-	-	-	-
PLLA/ATBC/SCG	10	2009	185	0.093	2027
	23	1907	181	0.096	1925
	40	1412	238	0.169	1453
	50	-	-	-	-

Table 8. *Cont.*

Samples	T (°C)	E′ (MPa)	E″ (MPa)	tan δ (-)	E* (MPa)
PLLA/PLLA-g-IA/ATBC	10	2506	133	0.053	2513
	23	2361	144	0.061	2370
	40	1007	384	0.038	1131
	50	-	-	-	-
PLLA/PLLA-g-IA/ATBC/SCG	10	2733	122	0.045	2738
	23	2516	133	0.053	2523
	40	1131	376	0.033	1241
	50	-	-	-	-

4. Conclusions

In this paper, the itaconic anhydride (IA) was successfully grafted onto PLLA backbone by a reactive extrusion process, and consequently, acetyl tributyl citrate (ATBC) and spent coffee grounds (SCG) were incorporated. To increase the application potential, the effect of the organic plasticizer on the supramolecular structure, thermal, mechanical, and dynamic-mechanical properties of the PLLA/SCG bio-composite were investigated. The presence of SCG and ATBC caused not only an increase in elongation at break of PLLA composites but also enhanced crystallization kinetics. The positive effect of SCG was observed as the nucleation barrier in the primary crystallization region decreased. A synergistic effect occurring especially at higher cooling rates (CR) was achieved for the PLLA/ATBC/SCG composite structure. At lower CRs, PLLA/PLLA-g-IA/ATBC/SCG reached the highest values of the degree of crystallinity (X_c). Thus, the study of crystallization kinetics showed that SCG significantly increased the crystallization rate in all studied materials. In terms of thermal stability, SCG reduced the initial temperature of degradation. However, the chemical reaction of the fatty acids contained in SCG with the ester groups of PLLA was partially prevented by grafting IA onto its backbone. It also had a positive effect on the formation of a supramolecular structure by increasing the X_c. The same effect was achieved by the presence of an ATBC plasticizer. The combination of the abovementioned modifications resulted into highest X_c values while maintaining the desirable properties of the extruded films, such as high elongation at break. A positive effect of SCG on the overall ductility and the damping properties was also demonstrated by DMA analysis and an increase in the imaginary component of the dynamic modulus of elasticity.

Author Contributions: Conceptualization, L.B., J.N., M.B. and P.L.; methodology, L.B.; validation, P.L. and M.B.; formal analysis, L.B., J.N. and M.B.; resources, J.N. and M.B.; writing—original draft preparation, J.N.; writing—review and editing, M.B.; visualization, J.N., M.B. and L.B.; supervision, M.B. and P.L., project administration, J.N.; funding acquisition, J.N. All authors have read and agreed to the published version of the manuscript.

Funding: This research was funded by the Student Grand Competition of the Technical University of Liberec under the project No. SGS-2022-5011 and the Institutional Endowment for the Long Term Conceptual Development of Research Institutes, as provided by the Ministry of Education, Youth and Sports of the Czech Republic.

Institutional Review Board Statement: Not applicable.

Informed Consent Statement: Not applicable.

Data Availability Statement: Not applicable.

Conflicts of Interest: The authors declare no conflict of interest.

References

1. Mülhaupt, R. Catalytic Polymerization and Post Polymerization Catalysis Fifty Years After the Discovery of Ziegler's Catalysts. *Macromol. Chem. Phys.* **2003**, *204*, 289–327. [CrossRef]
2. Philp, J.C.; Ritchie, R.J.; Guy, K. Biobased Plastics in a Bioeconomy. *Trends Biotechnol.* **2013**, *31*, 65–67. [CrossRef] [PubMed]
3. Barker, T.; Bashmakov, I.; Bernstein, L.; Bogner, J.; Bosch, P.; Dave, R.; Davidson, O.; Fisher, B.; Grubb, M.; Gupta, S. Technical Summary. In *Climate Change 2007: Mitigation. In Contribution of Working Group III to the Fourth Assessment Report of the Intergovernmental Panel on Climate Change*; Cambridge University Press: Cambridge, UK, 2007.
4. La Mantia, F.; Morreale, M. Green Composites: A Brief Review. *Compos. Part A Appl. Sci. Manuf.* **2011**, *42*, 579–588. [CrossRef]
5. Moshood, T.D.; Nawanir, G.; Mahmud, F.; Mohamad, F.; Ahmad, M.H.; Abdul Ghani, A. Expanding Policy for Biodegradable Plastic Products and Market Dynamics of Bio-Based Plastics: Challenges and Opportunities. *Sustainability* **2021**, *13*, 6170. [CrossRef]
6. EUBIO_Admin Market. *European Bioplastics e.V.*
7. Farah, S.; Anderson, D.G.; Langer, R. Physical and Mechanical Properties of PLA, and Their Functions in Widespread Applications—A Comprehensive Review. *Adv. Drug Deliv. Rev.* **2016**, *107*, 367–392. [CrossRef]
8. Auras, R.; Harte, B.; Selke, S. An Overview of Polylactides as Packaging Materials. *Macromol. Biosci.* **2004**, *4*, 835–864. [CrossRef]
9. Mochizuki, M. *Textile Applications. Poly (Lactic Acid) Synthesis, Structures, Properties, Processing, Applications, and End of Life*; John Wiley & Sons, Inc.: Hoboken, NJ, USA, 2022; pp. 619–629.
10. Muller, J.; González-Martínez, C.; Chiralt, A. Poly (Lactic) Acid (PLA) and Starch Bilayer Films, Containing Cinnamaldehyde, Obtained by Compression Moulding. *Eur. Polym. J.* **2017**, *95*, 56–70. [CrossRef]
11. Rasselet, D.; Caro-Bretelle, A.-S.; Taguet, A.; Lopez-Cuesta, J.-M. Reactive Compatibilization of PLA/PA11 Blends and Their Application in Additive Manufacturing. *Materials* **2019**, *12*, 485. [CrossRef]
12. Sin, L.T. *Polylactic Acid: PLA Biopolymer Technology and Applications*; William Andrew: Norwich, NY, USA, 2012; ISBN 1-4377-4460-5.
13. Zengwen, C.; Pan, H.; Chen, Y.; Bian, J.; Han, L.; Zhang, H.; Dong, L.; Yang, Y. Transform Poly (Lactic Acid) Packaging Film from Brittleness to Toughness Using Traditional Industrial Equipments. *Polymer* **2019**, *180*, 121728. [CrossRef]
14. Tábi, T.; Ageyeva, T.; Kovács, J.G. Improving the Ductility and Heat Deflection Temperature of Injection Molded Poly (Lactic Acid) Products: A Comprehensive Review. *Polym. Test.* **2021**, *101*, 107282. [CrossRef]
15. Salmerón Sánchez, M.; Mathot, V.B.F.; Vanden Poel, G.; Gómez Ribelles, J.L. Effect of the Cooling Rate on the Nucleation Kinetics of Poly(l-Lactic Acid) and Its Influence on Morphology. *Macromolecules* **2007**, *40*, 7989–7997. [CrossRef]
16. Aitor, L.; Erlantz, L. Review on the Thermomechanical Properties and Biodegradation Behaviour of Polyester. *Eur. Polym. J* **2019**, *121*, 1–31.
17. Yang, Y.; Zhang, L.; Xiong, Z.; Tang, Z.; Zhang, R.; Zhu, J. Research Progress in the Heat Resistance, Toughening and Filling Modification of PLA. *Sci. China Chem.* **2016**, *59*, 1355–1368. [CrossRef]
18. Ramot, Y.; Haim-Zada, M.; Domb, A.J.; Nyska, A. Biocompatibility and Safety of PLA and Its Copolymers. *Adv. Drug Deliv. Rev.* **2016**, *107*, 153–162. [CrossRef]
19. Fortelny, I.; Ujcic, A.; Fambri, L.; Slouf, M. Phase Structure, Compatibility, and Toughness of PLA/PCL Blends: A Review. *Front. Mater.* **2019**, *6*, 206. [CrossRef]
20. Choi, K.; Choi, M.-C.; Han, D.-H.; Park, T.-S.; Ha, C.-S. Plasticization of Poly (Lactic Acid)(PLA) through Chemical Grafting of Poly (Ethylene Glycol)(PEG) via in Situ Reactive Blending. *Eur. Polym. J.* **2013**, *49*, 2356–2364. [CrossRef]
21. Murariu, M.; Dubois, P. PLA Composites: From Production to Properties. *Adv. Drug Deliv. Rev.* **2016**, *107*, 17–46. [CrossRef]
22. Maiza, M.; Benaniba, M.T.; Quintard, G.; Massardier-Nageotte, V. Biobased Additive Plasticizing Polylactic Acid (PLA). *Polimeros* **2015**, *25*, 581–590. [CrossRef]
23. Dartora, P.C.; da Rosa Loureiro, M.; de Camargo Forte, M.M. Crystallization Kinetics and Morphology of Poly (Lactic Acid) with Polysaccharide as Nucleating Agent. *J. Therm. Anal. Calorim.* **2018**, *134*, 1705–1713. [CrossRef]
24. Battegazzore, D.; Bocchini, S.; Frache, A. Crystallization Kinetics of Poly (Lactic Acid)-Talc Composites. *Express Polym. Lett.* **2011**, *5*, 849–858. [CrossRef]
25. Terroba-Delicado, E.; Fiori, S.; Torres-Giner, S.; Gomez-Caturla, J.; Montanes, N.; Sanchez-Nacher, L. Improving the Mechanical Ductility and Toughness of Injection-Molded Polylactide Pieces by the Dual Incorporation of Liquor Waste Derived Spent Coffee Grounds and Oligomers of Lactic Acid. 2 November 2021. Preprint (Version 1). Available online: https://assets.researchsquare.com/files/rs-1035447/v1/97b0a683-ae91-4208-92fc-0b971337b616.pdf?c=1648436444 (accessed on 20 November 2022). [CrossRef]
26. Gupta, A.; Simmons, W.; Schueneman, G.T.; Mintz, E.A. Lignin-Coated Cellulose Nanocrystals as Promising Nucleating Agent for Poly (Lactic Acid). *J. Therm. Anal. Calorim.* **2016**, *126*, 1243–1251. [CrossRef]
27. Saeidlou, S.; Huneault, M.A.; Li, H.; Park, C.B. Poly (Lactic Acid) Crystallization. *Prog. Polym. Sci.* **2012**, *37*, 1657–1677. [CrossRef]
28. Běhálek, L.; Seidl, M.; Dobránsky, J. Crystallization of Polylactic Acid Composites with Banana and Hemp Fibres by Means of DSC and XRD Methods. *Appl. Mech. Mater.* **2014**, *616*, 325–332. [CrossRef]
29. Brzeziński, M.; Biela, T. Polylactide Nanocomposites with Functionalized Carbon Nanotubes and Their Stereocomplexes: A Focused Review. *Mater. Lett.* **2014**, *121*, 244–250. [CrossRef]

30. Murariu, M.; Laoutid, F.; Dubois, P.; Fontaine, G.; Bourbigot, S.; Devaux, E.; Campagne, C.; Ferreira, M.; Solarski, S. Pathways to Biodegradable Flame Retardant Polymer (Nano) Composites. In *Polymer Green Flame Retardants*; Elsevier: Amsterdam, The Netherlands, 2014; pp. 709–773.
31. Akindoyo, J.O.; Beg, M.D.; Ghazali, S.; Heim, H.P.; Feldmann, M. Impact Modified PLA-Hydroxyapatite Composites–Thermo-Mechanical Properties. *Compos. Part A Appl. Sci. Manuf.* **2018**, *107*, 326–333. [CrossRef]
32. Herrera, N.; Salaberria, A.M.; Mathew, A.P.; Oksman, K. Plasticized Polylactic Acid Nanocomposite Films with Cellulose and Chitin Nanocrystals Prepared Using Extrusion and Compression Molding with Two Cooling Rates: Effects on Mechanical, Thermal and Optical Properties. *Compos. Part A Appl. Sci. Manuf.* **2016**, *83*, 89–97. [CrossRef]
33. Arrigo, R.; Bartoli, M.; Malucelli, G. Poly (Lactic Acid)–Biochar Biocomposites: Effect of Processing and Filler Content on Rheological, Thermal, and Mechanical Properties. *Polymers* **2020**, *12*, 892. [CrossRef]
34. Battegazzore, D.; Bocchini, S.; Alongi, J.; Frache, A. Rice Husk as Bio-Source of Silica: Preparation and Characterization of PLA–Silica Bio-Composites. *RSC Adv.* **2014**, *4*, 54703–54712. [CrossRef]
35. Suaduang, N.; Ross, S.; Ross, G.; Pratumshat, S.; Mahasaranon, S. Effect of Spent Coffee Grounds Filler on the Physical and Mechanical Properties of Poly (Lactic Acid) Bio-Composite Films. *Mater. Today Proc.* **2019**, *17*, 2104–2110. [CrossRef]
36. Ballesteros, L.F.; Teixeira, J.A.; Mussatto, S.I. Chemical, Functional, and Structural Properties of Spent Coffee Grounds and Coffee Silverskin. *Food Bioprocess Technol.* **2014**, *7*, 3493–3503. [CrossRef]
37. Mussatto, S.I.; Machado, E.; Martins, S.; Teixeira, J.A. Production, Composition, and Application of Coffee and Its Industrial Residues. *Food Bioprocess Technol.* **2011**, *4*, 661–672. [CrossRef]
38. Suaduang, N.; Ross, S.; Ross, G.M.; Wangsoub, S.; Mahasaranon, S. *The Physical and Mechanical Properties of Biocomposite Films Composed of Poly (Lactic Acid) with Spent Coffee Grounds.*; Trans Tech Publications Ltd.: Bäch, Switzerland, 2019; Volume 824, pp. 87–93.
39. Naguib, H.M.; Hou, G. Exploitation of Natural and Recycled Biomass Resources to Get Eco-Friendly Polymer. *J. Polym. Environ.* **2022**. [CrossRef]
40. Ku Marsilla, K.I.; Verbeek, C.J.R. Modification of Poly(Lactic Acid) Using Itaconic Anhydride by Reactive Extrusion. *Eur. Polym. J.* **2015**, *67*, 213–223. [CrossRef]
41. Essabir, H.; Raji, M.; Laaziz, S.A.; Rodrique, D.; Bouhfid, R. Thermo-Mechanical Performances of Polypropylene Biocomposites Based on Untreated, Treated and Compatibilized Spent Coffee Grounds. *Compos. Part B Eng.* **2018**, *149*, 1–11. [CrossRef]
42. Jo, M.Y.; Ryu, Y.J.; Ko, J.H.; Yoon, J. Effects of Compatibilizers on the Mechanical Properties of ABS/PLA Composites. *J. Appl. Polym. Sci.* **2012**, *125*, E231–E238. [CrossRef]
43. Du, J.; Wang, Y.; Xie, X.; Xu, M.; Song, Y. Styrene-Assisted Maleic Anhydride Grafted Poly (Lactic Acid) as an Effective Compatibilizer for Wood Flour/Poly (Lactic Acid) Bio-Composites. *Polymers* **2017**, *9*, 623. [CrossRef]
44. Běhálek, L.; Borůvka, M.; Brdlík, P.; Habr, J.; Lenfeld, P.; Kroisová, D.; Veselka, F.; Novák, J. Thermal Properties and Non-Isothermal Crystallization Kinetics of Biocomposites Based on Poly(Lactic Acid), Rice Husks and Cellulose Fibres. *J. Therm. Anal. Calorim.* **2020**, *142*, 629–649. [CrossRef]
45. Ma, P.; Jiang, L.; Ye, T.; Dong, W.; Chen, M. Melt Free-Radical Grafting of Maleic Anhydride onto Biodegradable Poly(Lactic Acid) by Using Styrene as A Comonomer. *Polymers* **2014**, *6*, 1528–1543. [CrossRef]
46. Petruš, J.; Kučera, F.; Petrůj, J. Post-Polymerization Modification of Poly (Lactic Acid) via Radical Grafting with Itaconic Anhydride. *Eur. Polym. J.* **2016**, *77*, 16–30. [CrossRef]
47. Verbeek, C.J.R.; Hanipah, S.H. Grafting Itaconic Anhydride onto Polyethylene Using Extrusion. *J. Appl. Polym. Sci.* **2010**, *116*, 3118–3126. [CrossRef]
48. Tamboli, S.M.; Mhaske, S.T.; Kale, D.D. Crosslinked Polyethylene. *Indian J. Chem. Technol.* **2004**, *11*, 853–864.
49. Sarasua, J.-R.; Prud'homme, R.E.; Wisniewski, M.; Le Borgne, A.; Spassky, N. Crystallization and Melting Behavior of Polylactides. *Macromolecules* **1998**, *31*, 3895–3905. [CrossRef]
50. Jeziorny, A. Parameters Characterizing the Kinetics of the Non-Isothermal Crystallization of Poly (Ethylene Terephthalate) Determined by DSC. *Polymer* **1978**, *19*, 1142–1144. [CrossRef]
51. Coburn, N.; Douglas, P.; Kaya, D.; Gupta, J.; McNally, T. Isothermal and Non-Isothermal Crystallization Kinetics of Composites of Poly (Propylene) and MWCNTs. *Adv. Ind. Eng. Polym. Res.* **2018**, *1*, 99–110. [CrossRef]
52. Muenprasat, D.; Suttireungwong, S.; Tongpin, C. Functionalization of Poly (Lactic Acid) with Maleic Anhydride for Biomedical Application. *J. Met. Mater. Miner.* **2010**, *20*, 189–192.
53. Yang, W.; Fortunati, E.; Dominici, F.; Giovanale, G.; Mazzaglia, A.; Balestra, G.M.; Kenny, J.; Puglia, D. Synergic Effect of Cellulose and Lignin Nanostructures in PLA Based Systems for Food Antibacterial Packaging. *Eur. Polym. J.* **2016**, *79*, 1–12. [CrossRef]
54. Qiu, Z.; Fujinami, S.; Komura, M.; Nakajima, K.; Ikehara, T.; Nishi, T. Nonisothermal Crystallization Kinetics of Poly (Butylene Succinate) and Poly (Ethylene Succinate). *Polym. J.* **2004**, *36*, 642–646. [CrossRef]
55. Brdlík, P.; Borůvka, M.; Běhálek, L.; Lenfeld, P. Biodegradation of Poly (Lactic Acid) Biocomposites under Controlled Composting Conditions and Freshwater Biotope. *Polymers* **2021**, *13*, 594. [CrossRef]
56. Arrieta, M.P.; Peponi, L.; López, D.; Fernández-García, M. Recovery of Yerba Mate (Ilex Paraguariensis) Residue for the Development of PLA-Based Bionanocomposite Films. *Ind. Crops Prod.* **2018**, *111*, 317–328. [CrossRef]
57. Morrell, J.J. Degradation of Lignocellulosic Materials and Its Prevention. *Jom* **2014**, *66*, 580–587. [CrossRef]

58. Mendes, J.F.; Martins, J.T.; Manrich, A.; Luchesi, B.R.; Dantas, A.P.S.; Vanderlei, R.M.; Claro, P.C.; Neto, A.R.D.S.; Mattoso, L.H.C.; Martins, M.A. Thermo-Physical and Mechanical Characteristics of Composites Based on High-Density Polyethylene (HDPE) e Spent Coffee Grounds (SCG). *J. Polym. Environ.* **2021**, *29*, 2888–2900. [CrossRef]
59. Nekhamanurak, B.; Patanathabutr, P.; Hongsriphan, N. The Influence of Micro-/Nano-CaCO3 on Thermal Stability and Melt Rheology Behavior of Poly (Lactic Acid). *Energy Procedia* **2014**, *56*, 118–128. [CrossRef]
60. Tian, J.; Cao, Z.; Qian, S.; Xia, Y.; Zhang, J.; Kong, Y.; Sheng, K.; Zhang, Y.; Wan, Y.; Takahashi, J. Improving Tensile Strength and Impact Toughness of Plasticized Poly(Lactic Acid) Biocomposites by Incorporating Nanofibrillated Cellulose. *Nanotechnology* **2022**, *11*, 2469–2482. [CrossRef]
61. Gálvez, J.; Correa Aguirre, J.P.; Hidalgo Salazar, M.A.; Vera Mondragón, B.; Wagner, E.; Caicedo, C. Effect of Extrusion Screw Speed and Plasticizer Proportions on the Rheological, Thermal, Mechanical, Morphological and Superficial Properties of PLA. *Polymers* **2020**, *12*, 2111. [CrossRef]
62. Coelho, G.O.; Batista, M.J.; Ávila, A.F.; Franca, A.S.; Oliveira, L.S. Development and Characterization of Biopolymeric Films of Galactomannans Recovered from Spent Coffee Grounds. *J. Food Eng.* **2021**, *289*, 110083. [CrossRef]

materials

MDPI

Article

The Influence of the Use of Technological Waste and the Simulation of Material Lifetime on the Unnotched Impact Strength of Two Different Polymer Composites

Jozef Dobránsky [1,*], **Miroslav Gombár** [2] and **Tomáš Stejskal** [3]

[1] Faculty of Manufacturing Technologies with a Seat in Presov, Technical University of Kosice, Sturova 31, 080 01 Presov, Slovakia
[2] Faculty of Management and Business, University of Presov, Namestie Legionarov 3, 080 01 Presov, Slovakia
[3] Faculty of Mechanical Engineering, Technical University of Kosice, Letna 9, 042 00 Kosice, Slovakia
* Correspondence: jozef.dobransky@tuke.sk; Tel.: +421-55-602-6350

Abstract: The article deals with the assessment of the impact of technological polymer waste resulting from plastic injection technology and the subsequent simulation of the lifetime of polymer products on the impact strength of two different polymer composites. Two different types of polymer composites used to produce plastic parts in the automotive industry were chosen for the research. Based on the evaluation of the PBT composite before exposing the specimens to elevated temperature, it can be concluded that the concentration of the recycled material in the virgin material only affects the values of the unnotched impact strength of the PBT composite when the recycled material accounts for at least 50 wt.%. The results of the unnotched impact strength of the test specimens exposed at 150 °C/500 h make it evident that the addition of recycled material to the virgin material significantly reduces the components' service lives. The same trend could be observed with the second tested composite material, PPA.

Keywords: technological waste; simulation; material lifetime; impact strength; polymer composites

check for updates

Citation: Dobránsky, J.; Gombár, M.; Stejskal, T. The Influence of the Use of Technological Waste and the Simulation of Material Lifetime on the Unnotched Impact Strength of Two Different Polymer Composites. *Materials* **2022**, *15*, 8516. https://doi.org/10.3390/ma15238516

Academic Editor: Jānis Andersons

Received: 16 November 2022
Accepted: 27 November 2022
Published: 29 November 2022

Publisher's Note: MDPI stays neutral with regard to jurisdictional claims in published maps and institutional affiliations.

1. Introduction

In their study, the authors Tomar and Maiti looked at the unnotched impact strength of a polymeric PBT blend containing up to 39% ABAS. By mixing the materials, the tensile properties were reduced, while the unnotched impact strength increased. Morphological studies showed good ABAS dispersion into the PBT matrix. The strength properties were dependent on PBT crystallinity and interphase adhesion between phases. The unnotched impact strength depended on the concentration of admixed rubber and the interparticle spacing of the discrete phase [1].

In his study, the author Szostak analyzed the effect of the method of PET and PBT mixture preparation. He found that the method of preparation had the greatest effect on the mechanical properties, especially the impact strength. Materials must first be mixed in the extrusion machine before they are processed in the injection machine. This improves the properties of the injected PET/PBT mixtures [2].

In their work, the authors Sharma et al. used two types of impact modifiers to monitor their effect on the unnotched impact strength of the modified materials. They compared ultra low density polyethylene grafted glycidyl methacrylate (ULDPE-g-GMA, IM1) and ethylene-methyl acrylate-glycidyl methacrylate (E-MA-GMA, IM2). The application of 2% ULDPE-g-GMA weight produced a better result compared to 2% E-MA-GMA weight in terms of unnotched impact strength. Therefore, 2% ULDPE-g-GMA weight is considered an optimized percentage for nanocomposite preparation. As an unnotched impact strength modifier, ULDPE-g-GMA reduces the yield strength and tensile modulus of pure PBT [3].

In their research, the authors Ambrosio et al. focused on the effect of the parameters of mixtures processed on a twin-screw extruder. In processing mixtures of polybutylene terephthalate acrylonitrile-butadiene-styrene with and without a reactive compatibilizer, they monitored the morphology and impact strength. In their study, they found that increasing the feed rate greatly reduced the ductile brittle transition temperature (DBTT) and slightly increased the impact strength at room temperature (RTIS) of compatible mixtures [4].

The main goal of the study by the authors Ajorloo et al. was to investigate the feasibility of using waste materials. They used different ratios of virgin and recycled PP. The authors found that increasing the concentration of recycled PP in the composite led to decreases in the mechanical properties, especially the ductility and impact strength. Due to the lower viscosity in the form of recycled PP, a better dispersion of the filler particles was observed by SEM analysis, leading to an improvement in tensile strength. In addition, the use of recycled material caused decreases in the percentage of crystallinity and the melting point. During the study, it was found that the use of a 20/80 to 40/60 ratio of recycled PP/virgin PP in composites adequately met the requirements of automotive parts, with advantages in environmental and economic aspects. Compared to composites filled with talc, the ductility of composites filled with fly ash is higher. The simultaneous use of talc and fly ash as a hybrid system results in higher ductility but lowers the impact strength compared to samples filled with talc. The use of recycled PP mixed with fly ash provides advantages in cost reduction and sustainable and ecological production [5].

Guo et al. observed the damage of polymer composites reinforced with wood fibers. They used waste WF and recycled HDPE (Re-HDPE) to understand the failure mechanism of wood-fiber-reinforced (WF) high-density polyethylene (HDPE) composites. Highly filled WF/Re-HDPE composites were prepared by extrusion. The results showed that acoustic emission (AE) technology can better understand the damage progression and fracture process of WF/Re-HDPE composites and determine the damage degree, damage accumulation, and damage mode. Matrix deformation, fiber breakage, interfacial delamination, fiber-to-matrix debonding, fiber pull-out, and matrix cracking were the dominant damage modes of highly filled WF/Re-HDPE composites under flexural loading, and the AE signal varied with different degrees of damage and modes of damage [6].

Plastics are cheap, easy to shape, and light. Plastics have substantially supplanted traditional materials in many areas. The problem of recycling is still a big challenge. There are both technological and economic challenges that hinder progress in this area. In their study, the authors Ignatyev et al. provided an overview of recycling, along with a future outlook using popular polymers such as polyolefins, poly(vinyl chloride), polyurethane, and poly(ethylene terephthalate). The different types of recycling, primary, secondary, tertiary, quaternary, and biological recycling, were discussed, together with related issues such as compatibility and cross-linking [7,8].

In addition to the positive effects on environmental sustainability, polymer recycling is also beneficial in terms of reducing production costs and material resources. The aim of the study by Jiun et al. was to find out how the recycling cycle affects the physical properties and mechanical properties of the thermoplastic polymer and thermoplastic elastomer polymer. The obtained results showed that the tensile strength and density decreased for the thermoplastic polymer, and the appearance changes were significant when the number of recycling cycles increased [9,10].

An article by Hugo et al. dealt with the formulation of composite materials for structural or semistructural applications using thermoplastic polymer waste. The mechanical and thermal properties of a proprietary blend of recycled polymers with a range of different fillers were investigated. The authors found that the addition of small amounts of mica to glass-fiber-reinforced composites showed significant synergy in the tensile strength and modulus [11,12].

The article was created based on a request from practice. The company that processes the composite materials in question wanted to verify their properties. At the same time, there is also an ecological side. Reusing waste material in the manufacturing process reduces the amount of waste in the automotive supply chain.

2. Experiment Preparation

For the analyses, two types of materials were selected from the group of polymer composites that are used for the production of moldings (connectors) intended for the automotive industry. The materials that form the matrices of the composites belong to the group of engineering (structural) or highly durable (high-tech) polymer composites. From the group of structural polymers, the PBT material (polybutylene terephthalate) and PPA material (polyphthalamide) were selected [13,14].

The first composite of polybutylene terephthalate (PBT), marketed as Ultradur® B 4406 G6, consists of 30 wt.% glass fibers. It is designed for products requiring increased resistance to burning. An overview of the basic material properties of this material is given in Table 1.

Table 1. Typical properties of ULTRADUR® B 4406 G6.

Properties	Value	Unit
Density	1650	kg/m^3
Melt volume-flow rate (MVR) (275 °C; 2.16 kg)	8	$cm^3/10$ min
Tensile modulus	11,300	MPa
Tensile stress at break	145	MPa
Tensile strain at break	2.3	%
Charpy unnotched impact strength (+23 °C)	30	kJ/m^2
Charpy unnotched impact strength (−30 °C)	25	kJ/m^2

The second composite, polyphthalamide (PPA), with the trade name ZYTEL® HTN FR52G30NH NC010, is filled with 30% glass fibers and nonhalogenated additives: flame retardants, lubricants, and separation agents. Injection molds for the use of this type of material must be made of corrosion-resistant and wear-resistant steels. When processing the material, it is important to make sure that it is not in the injection unit for a long time because it can degrade. An overview of the basic material properties of the composite is given in Table 2.

Table 2. Typical properties of ZYTEL® HTN FR52G30NH NC010.

Properties	Value	Unit
Density	1.44	g/cm^3
Tensile modulus	10,500	MPa
Tensile stress at break	150	MPa
Tensile strain at break	2.2	%
Charpy unnotched impact strength (+23 °C)	45	kJ/m^2
Charpy unnotched impact strength (−30 °C)	40	kJ/m^2

Before the experimental measurement, it was necessary to prepare the test specimens, which consisted of mixing recycled material in the form of pulp with the virgin granulate. Mixing was carried out manually in specified ratios, corresponding to the recycled content in the virgin material, of 0, 10, 20, 30, 40, 50, 60, 80, and 100 wt.%. From the mixture (batches) prepared in this way, multipurpose type A test specimens were injected according to STS EN ISO 3167 or 1A specimens were injected according to STS EN ISO 527-2. The production of multipurpose test specimens in the form of double-sided blades and rectangular prisms for the Charpy unnotched impact strength test was carried out in accordance with STS EN ISO 294-1 [15].

The injection process took place on a standard Demag D 60–182 injection machine with a column structure. Both materials were dried before injection under the conditions indicated in their respective material data sheets. At the end of the production cycle, all specimens were conditioned under standard conditions of 23/50. To simulate the assessment of the lives of the moldings according to the recyclate content, half of the test specimens were exposed to an elevated temperature of 150 °C (PBT) or 230 °C (PPA) in a forced air circulation oven for 500 h prior to studying their performance.

The determination of the unnotched impact strength by the Charpy method was carried out under the conditions specified in the STS EN ISO 179-1 standard. Before the test, the span of the instrument brackets was set, and the instrument was calibrated. Afterwards, a hammer was let down loosely to determine the friction losses. At the same time, it was checked whether the Ceast Resil 5.5 (Figure 1) instrument was able to perform the test at the prescribed impact velocity and whether it was in the correct range of the absorbed energy. For the purposes of the experimental measurement, we used an impact hammer with a nominal energy of 5 J for the PBT material before exposure to elevated temperature. A hammer with a nominal energy of 2 J was used for the PBT after exposure and the PPA before and after exposure [16–18].

Figure 1. CEAST Resil 5.5 test device with sample location.

Half of the test specimens intended to determine the unnotched impact strength at low temperatures were conditioned in the ProfiMaster PMU 0450 laboratory freezer at –30 °C for 24 h. A thorough type C puncture was recorded in all test specimens.

3. Evaluation of Results and Discussion

This chapter deals with the assessment of unnotched impact strength by the Charpy method. The unnotched impact strength was first measured at 23 °C on test specimens before and after exposure to elevated temperature. Subsequently, the bodies were exposed to a temperature of −30 °C for 24 h, and the unnotched impact strength at low temperatures was measured. The influence of the presence of the recycled material on the lives of the samples is also evaluated, where the lives were simulated by exposing the materials to an elevated temperature of 150 °C for 500 h (PBT) or 230 °C for 500 h (PPA), according to the internal regulations of the VW Group [19,20].

3.1. Evaluation of Unnotched Impact Strength by the Charpy Method for Composite PBT

Figure 2 shows the impact of the amount of recyclate on the unnotched impact strength. As shown in the figure, the amount of recycled material in the virgin material affects the value of the unnotched impact strength. The unnotched impact strength value decreased from 30.7 kJ/m^2 (virgin material) to 24.8 kJ/m^2 when the recyclate accounted for 100% of the amount, representing an overall drop of 19.2%.

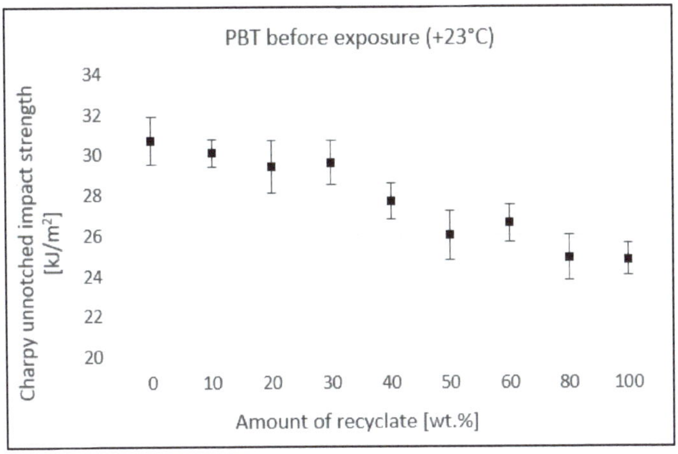

Figure 2. Dependence of the unnotched impact strength of PBT on the amount of recycled material before the samples were exposed to elevated temperature (+23 °C).

Up to 40 wt.% recycled material, the unnotched impact strength decreased slightly. With 50 wt.% recycled material, the value recorded a greater drop, falling from 30.7 kJ/m^2 (virgin material) to 26 kJ/m^2, a decrease of 15.3%.

Figure 3 shows a comparison of the PBT composite unnotched impact strength values before and after exposure to elevated temperature. The value of unnotched impact strength after the specimens were exposed to elevated temperature decreased gradually from the value of 23.4 kJ/m^2 (virgin material) to the value of 17.1 kJ/m^2 (100% recyclate), which represented a reduction of 26.9%.

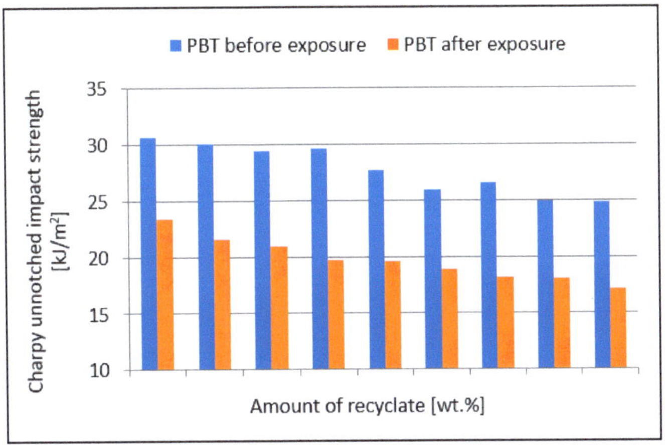

Figure 3. Dependence of the unnotched impact strength of PBT on the amount of recycled material before and after the samples were exposed to elevated temperature (+23 °C).

The value of the unnotched impact strength, shown in Figure 2, after the specimens were exposed to elevated temperature decreased, on average, by between 23.8 and 31% compared to unexposed specimens, which represented a significant reduction. The results also confirm the conclusions of Kim et al. [21].

Figure 4 shows the effect of added recycled material on unnotched impact strength measured in the specimens exposed to −30 °C. Up to a concentration of 40 wt.% recycled material, the value of the unnotched impact strength fluctuated around 20.5 kJ/m^2. At a 50 wt.% concentration of recycled material, it decreased to 16.8 kJ/m^2, a drop of 19.2%. The further addition of recyclate caused only a marginal reduction to a final value of 15.8 kJ/m^2, which represented an overall reduction of 24% compared to the virgin material.

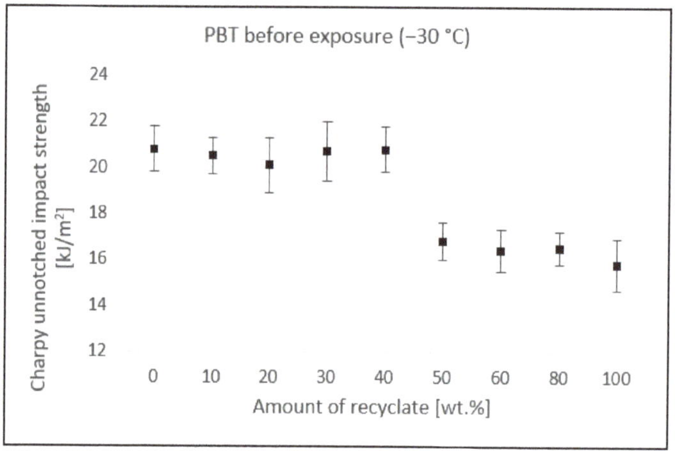

Figure 4. Dependence of the unnotched impact strength of PBT on the amount of recycled material before the samples were exposed to elevated temperature (−30 °C).

Figure 5 shows a comparison of the unnotched impact strength values before and after exposure to elevated temperature, measured at the specimen temperature of −30 °C. The unnotched impact strength value was lower in specimens composed of up to 40 wt.% recycled material compared to specimens before exposure, where it decreased by 6.2% in the virgin material and by as much as 17.3% with the recyclate accounting for 40 wt.%. At concentrations between 50 and 100 wt.% of the recycled material, the unnotched impact strength value varied only marginally compared to the specimens before they were exposed to elevated temperature. The same finding was observed by Behalek et al. [22].

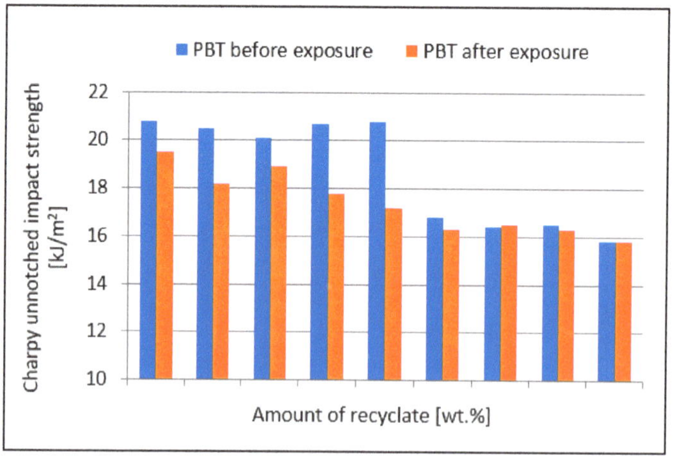

Figure 5. Dependence of the unnotched impact strength of PBT on the amount of the recycled material before and after the samples were exposed to elevated temperature (−30 °C).

Based on the evaluation of the PBT composite before the specimens were exposed to elevated temperature, it can be concluded that the concentration of the recycled material in the virgin material only affects the values of the unnotched impact strength of the PBT composite when the recycled material accounts for at least 50 wt.%.

The results of the unnotched impact strength of the test specimens exposed to 150 °C/500 h make it evident that the addition of recycled material to the virgin material significantly reduces the components' service lives.

3.2. *Evaluation of Unnotched Impact Strength by the Charpy Method for Composite PPA*

Figure 6 shows the effect of the recyclate concentration on the unnotched impact strength of the PPA composite. The virgin material value of unnotched impact strength decreased slightly up to a 40 wt.% concentration of the recycled material, from 30.5 kJ/m^2 to 26.7 kJ/m^2, a reduction of 12.5%. At a 50 wt.% concentration of the recycled material, the unnotched impact strength dropped rapidly to 16.9 kJ/m^2, a reduction of 44.6% compared to virgin material. Adding further recycled material did not result in a further reduction in the unnotched impact strength.

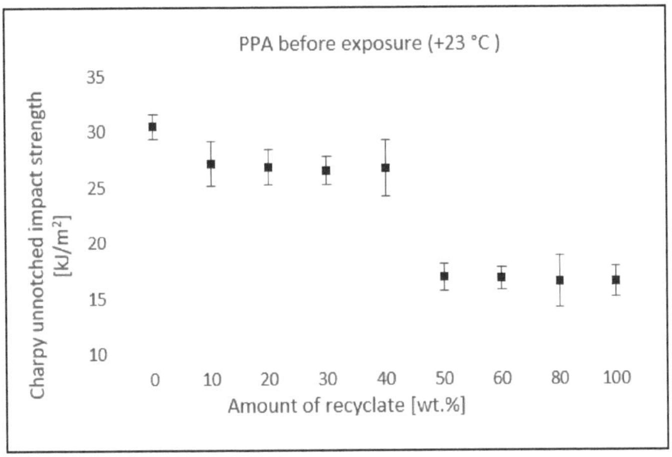

Figure 6. Dependence of the unnotched impact strength of PPA on the amount of recycled material before the samples were exposed to elevated temperature (+23 °C).

Figure 7 shows a comparison of the unnotched impact strength of the specimens before and after exposure to elevated temperature. The value of the unnotched impact strength of specimens exposed to elevated temperature in virgin material decreased marginally from the value of 21.3 kJ/m^2 to 19.7 kJ/m^2 at a 40 wt.% concentration of the recycled material, a reduction of 7.5%. At the recycled material concentration of 50 wt.%, the value dropped to 15.7 kJ/m^2, a reduction of 26.3% compared to the virgin material. By adding more recycled material, the value of the unnotched impact strength decreased marginally.

The unnotched impact strength differed between specimens exposed to elevated temperature and those that were not yet exposed to elevated temperature up to the recycled material concentration of 40 wt.%. The difference in values ranged from 27.6 to 30.2%. At a recyclate concentration of 50 wt.% or more, no significant differences in unnotched impact strength values were recorded. The results also confirm the conclusions of Wang et al. [23] and Svetlik et al. [24].

Figure 8 shows the effect of the addition of recycled material on the unnotched impact strength measured in specimens exposed to a temperature of −30 °C. The value of the unnotched impact strength up to 40 wt.% recycled material decreased from at least 23.3 kJ/m^2 in virgin material to 20.7 kJ/m^2, a reduction of 11.2%. At the recyclate concentration of

50 wt.%, it dropped to the value of 14.1 kJ/m^2, a reduction of 39.5%. With the further addition of recyclate, it dropped only marginally, to a final value of 11.7 kJ/m^2, which represented an overall reduction of 49.8% compared to the virgin material.

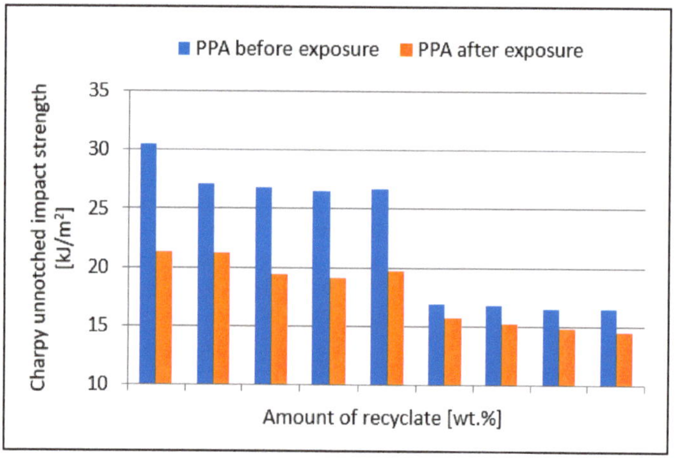

Figure 7. Dependence of the unnotched impact strength of PPA on the amount of recycled material before and after the samples were exposed to elevated temperature (+23 °C).

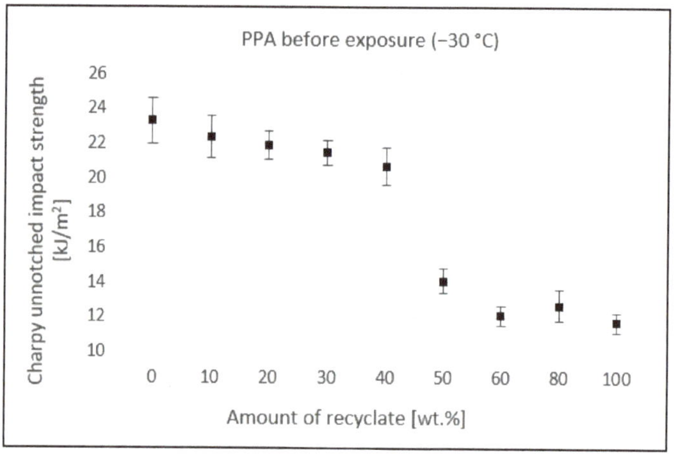

Figure 8. Dependence of the unnotched impact strength of PPA on the amount of recycled material before the samples were exposed to elevated temperature (−30 °C).

Figure 9 shows a comparison of the unnotched impact strength values before and after the exposure to elevated temperature, measured on specimens at a low temperature.

After the specimens were exposed to elevated temperature, they exhibited lower values of unnotched impact strength up to the recycled material concentration of 40 wt.% compared to the specimens before exposure, ranging from 11.9 to 21.3%. At a recyclate concentration of 50 wt.% or more, the unnotched impact strength of the specimens after exposure was slightly greater than it was before exposure. The results also confirm the conclusions of Khalili et al. [25] and Milosevic [26].

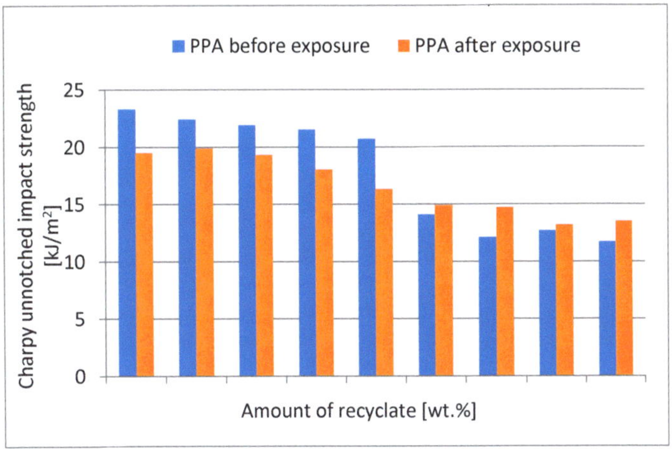

Figure 9. Dependence of the unnotched impact strength of PPA on the amount of recycled material before and after the samples were exposed to elevated temperature (−30 °C).

Based on the evaluation of the PPA composite before the specimens were exposed to elevated temperature, it can be concluded that the concentration of the recycled material in the virgin material only affects the values of the unnotched impact strength of the PPA composite when the recycled material accounts for at least 50 wt.%.

The unnotched impact strength results measured on the test specimens exposed to 230 °C/500 h make it evident that the addition of recycled material to the virgin material significantly reduces the components' service lives. The results also confirm the conclusions of Hernandez-Diaz et al. [27].

The same trend could be observed in both cases involving specimens where measurements were performed at a low temperature. The values of the unnotched impact strength decreased in specimens both before and after exposure to elevated temperature.

4. Conclusions

The basic statistical analysis of the general model (1) that was used to predict the examined response y depending on the change in the studied independent variables, xi, in the nominal (state) and ordinal scale (the recycled material amount) was performed through an analysis of variance. The analysis of variance of the studied parameter, y, was a basic statistical analysis of the suitability of the general model that was used (1). An analysis of variance can, on one hand, analyze whether the variability due to random errors is significantly smaller than the variability in the measured values explained by the model. On the other hand, ANOVA results can be viewed in terms of the basic nature of ANOVA, testing the zero statistical hypothesis stating that neither of the effects used in the models (the state and the recycled material amount) had a significant effect on the change in the studied variable (y). The statistical analysis of the experimental results utilized a factorial analysis of variance, where the influence of the main effects of the independent variables and their mutual interaction were considered. The basic general ANOVA table is shown in Table 3.

$$\widehat{y} = b_0 \cdot x_0 + \sum_{j=1}^{N} b_j \cdot x_j + \sum_{\substack{u,j=1 \\ u \neq j}}^{N} b_{uj} \cdot x_u \cdot x_j \quad (1)$$

In order to better understand the interaction of the independent input variables on the studied value of the unnotched impact strength expressed by the a_{cU} value ($kJ \cdot m^{-2}$) in two independent test procedures, we analyzed both tests conducted for PPA and PBT separately. In the first unnotched impact strength test performed on the PPA material, based on the applied analytical method (ANOVA), we concluded that the considered independent input variables (the recycled material amount, the specimen temperature, and the exposure) at the selected level of significance (α = 0.05) significantly affected the conditioned value of the unnotched impact strength. In the sense of the general model (1), where the effect of the mutual interactions of the input factors was also considered, the mutual interaction of the amount of recycled material and the temperature of the specimen (p = 0.000) contributed 1.720% of the change in the conditional value of the studied variable. Another significant interaction defined by the mutual relationship between the amount of recycled material and the exposure (p = 0.000) accounted for 6.880% of the change in the conditional value of the studied variable. The interaction of the specimen temperature and the exposure (p = 0.000) accounted for 1.130% of the change in the conditional value of the studied variable, and finally the interaction of all three considered input variables (p = 0.001) accounted for 1.130% of the change in the conditional value of the studied variable. At the same time, it has to be said that the quantity of the recycled material, accounting for 41.950% of the change in the conditional value of the studied variable, had the greatest impact on the change in the value of the energy of the unnotched impact strength in the PPA test. For the sake of completeness, the model error accounted for 13.170% of the change, i.e., the model (1) defined for the experimentally obtained values can explain 86.830% of the variability in the studied a_{cU} (kJ·m^{-2}) variable. In view of the fact that the previous analysis confirmed the statistical significance of the interactions, it is necessary and needed to also consider the simultaneous effects of the independent input variables. As documented in Figure 10, there were significant differences between the specimen temperatures of −30 °C and +23 °C containing individual quantities of recycled material (from 0 to 100%). First, it can be observed that the average value of the unnotched impact strength for the specimen temperature of +23 °C (20.178 ± 0.802 kJ·m^{-2}) was higher than the average value of the unnotched impact strength at the specimen temperature of −30 °C (16.610 ± 0.556 kJ·m^{-2}) over the entire range of the amount of recyclate that was used. This average deviation was 4.185 kJ·m^{-2}. However, under the effect of the amount of recycled material used, we can see that the difference in the value of the unnotched impact strength between the specimen temperature was the smallest for +23 °C and −30 °C at a 100 wt.% amount of recycled material, and this difference was 2.143 ± 0.821 kJ·m^{-2} (p = 0.001). On the contrary, the highest value of the studied difference was observed when the recycled material accounted for 80 wt.%, with a value of 6.259 ± 1.984 kJ·m^{-2} (p = 0.000). It should be mentioned that all differences in the values of the unnotched impact strength were significant at the selected level of significance (α = 0.05).

Table 3. Basic analysis of an ANOVA.

Source	DF	Sum of Squares	Mean Square	F Ratio	Prob > F
Model	$DF_{Model} = a - 1$	S_{Model}	$MS_{Model} = S_{Model}/DF_{Model}$	$F = MS_{Model}/MS_{Error}$	p_M
Error	$DF_{Error} = N - a$	S_{Error}	$MS_{Error} = S_{Error}/DF_{Error}$		
C. Total	$DF_{C.Total} = N - 1$	$S_{C.Total}$	$MS_{C.Total} = S_{C.Total}/DF_{C.Total}$		

When examining the mutual interaction between the amount of recycled material and the exposure (Figure 11), it can be said that, with the exception of the recycled material accounting for 60 wt.% and 100 wt.%, the value of the unnotched impact strength was higher in the state of the specimen before exposure. The average value of the difference within the entire range of the recycled material amount was 3.046 kJ·m^{-2}. Just as in the previous case, it was also important to monitor the amount of recycled material. The maximum difference in the values of the studied variable was achieved at 40 wt.%, with this difference being significant at the selected level of significance (p = 0.000) and reaching the value of 6.430 ± 1.697 kJ·m^{-2}. The next values of the a_{cU} variable difference were found when the recycled material accounted for 20 wt.% (5.833 ± 1.331 kJ·m^{-2}, p = 0.000) and when the recycled amount was 0 wt.%, with a value of 5.826 ± 1.719 kJ·m^{-2} (p = 0.000).

On the other hand, two negative differences in the unnotched impact strength value, namely with the amount of the recyclate accounting for 60 wt.% (-0.231 ± 1.108 kJ·m^{-2}, $p = 0.708$) and 100 wt.% (-0.188 ± 0.945 kJ·m^{-2}, $p = 0.761$) were statistically insignificant. We observed other statistically insignificant differences in the evaluated unnotched impact strength at 10 wt.% (0.610 ± 1.646 kJ·m^{-2}, $p = 0.166$) and 50 wt.% (0.666 ± 1.026 kJ·m^{-2}, $p = 0.281$).

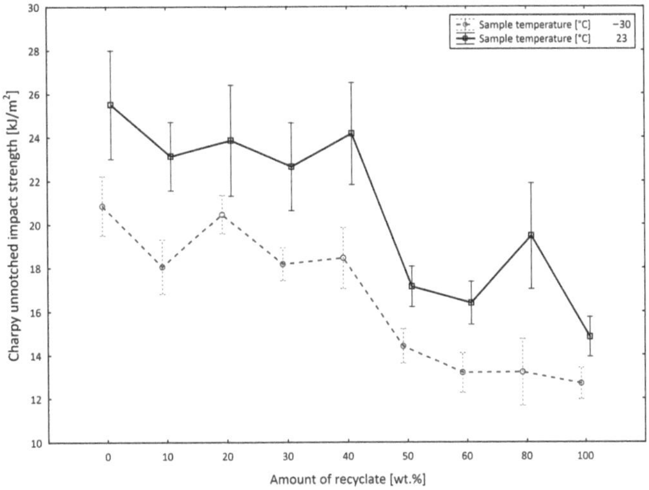

Figure 10. Dependence of the mutual interaction of the amount of recyclate and the temperature of the sample on the value of the unnotched impact strength of PPA.

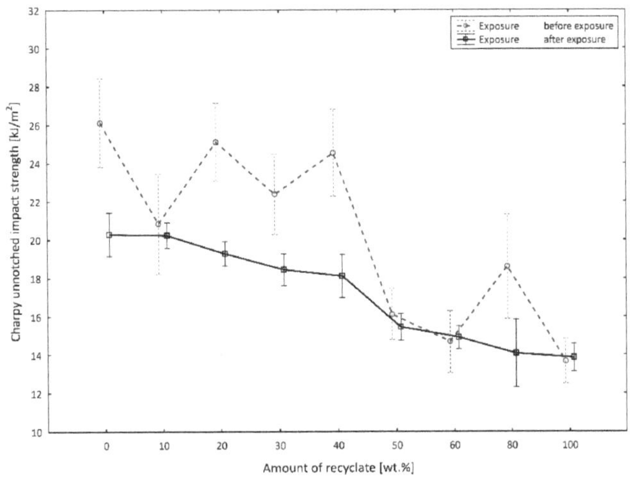

Figure 11. Dependence of the mutual interaction of the amount of recyclate and the exposure of the sample on the value of the unnotched impact strength of PPA.

When examining the mutual interaction of the specimen temperature and exposure (Figure 12), without considering the effect of the recycled material amount, we can see that at the specimen temperature of −30 °C there was no significant difference in the difference between the studied value of the unnotched impact strength, a_{cU} (0.045 ± 0.799 kJ·m^{-2}, $p = 0.878$).

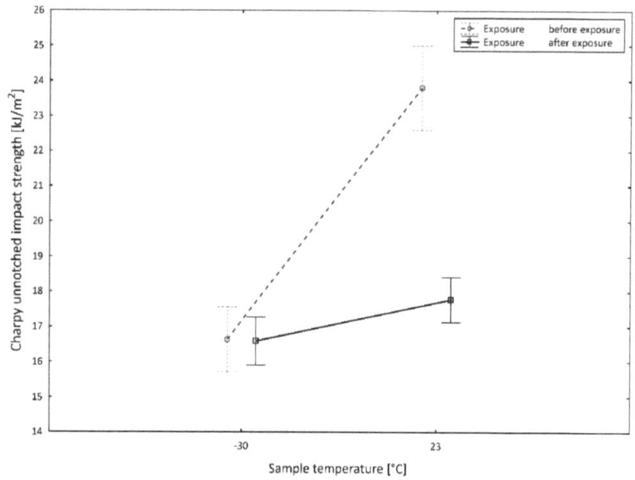

Figure 12. Dependence of the mutual interaction of the sample temperature and the exposure of the sample on the value of the unnotched impact strength of PPA.

Furthermore, Figure 13 shows that at the specimen temperature of +23 °C the value of the unnotched impact strength before exposure (23.802 ± 1.195 kJ·m^{-2}) was significantly higher than the value of the unnotched impact strength after exposure (17.771 ± 0.644 kJ·m^{-2}). This difference was also significant (p = 0.000) at the selected level of significance of p = 0.05.

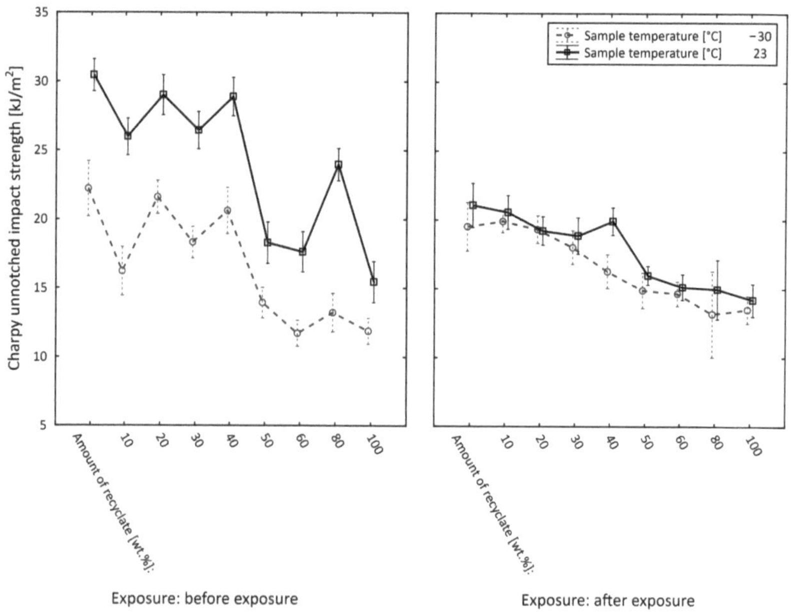

Figure 13. Dependence of the mutual interaction of the sample temperature and the exposure of the sample on the value of the unnotched impact strength of PPA.

When analyzing all three input variables at the same time, as the last significant interaction (Figure 13), we observed the highest value of the difference in the studied value of unnotched impact strength with the recyclate accounting for 20 wt.%, a specimen temperature of +23 °C, and between the states before and after the exposure. This difference reached 9.784 ± 0.644 kJ·m^{-2} (p = 0.000). The second most significant difference, with a value of 9.393 ± 1.384 kJ·m^{-2} (p = 0.000) was between the states before and after the exposure of the specimens with 0 wt.% consisting of the recyclate and a temperature of +23 °C. Significant values for the differences in the unnotched impact strength could be observed between the states before and after the exposure in the specimens with the recyclate accounting for 80 wt.% with the specimen temperature of +23 °C (9.005 ± 1.668 kJ·m^{-2}, p = 0.000) and with the recyclate accounting for 40 wt.% with the specimen temperature of +23 °C (8.975 ± 1.188 kJ·m^{-2}, p = 0.000). For all the above relationships with a specimen temperature of +23 °C, the unnotched impact strength values were higher in the state before the exposure. The opposite trend, i.e., greater values of unnotched impact strength in the state after the exposure compared to the state before the exposure, was observed only under the specimen temperature of −30 °C, namely with the recyclate accounting for 10 wt.% (3.688 ± 1.284 kJ·m^{-2}, p = 0.000), with the recyclate accounting for 60 wt.% (2.953 ± 0.911 kJ·m^{-2}, p = 0.001), and with the recyclate accounting for 100 wt.% (1.618 ± 0.966 kJ·m^{-2}, p = 0.044).

The initial analysis of the PBT unnotched impact strength test indicated that the change in the studied variable, a_{cU} (kJ·m^{-2}), was significantly influenced by the amount of recycled material (p = 0.000), with a 9.650% effect in model (1); specimen temperature (p = 0.000), with a 23.450% effect, as well as the exposure (p = 0.000), with a 22.720% effect. Compared to the previous analysis of PPA, the most striking difference was the reduced effect of the recyclate amount. When we considered the interactions of mutually independent input variables within the studied model (1), the change in the value of the unnotched impact strength was significantly influenced by the interaction between the amount of recycled material and the exposure (p = 0.005), with a 2.260% effect; the interaction of the specimen temperature and the exposure (p = 0.000), with a 6.560% effect; and the interaction of the recycled material amount, the specimen temperature, and the exposure (p = 0.018), with a 1.870% effect. In contrast to PPA, in the case of PBT, no significant effect resulted from the interaction between the amount of recycled material and the temperature of the specimen (p = 0.183), and at the same time there was a higher share of model error (32.350%) in PBT than in the PPA models. When we considered the basic input variables as separate effects, the average value of the unnotched impact strength energy for all specimens was 21.198 ± 0.568 kJ·m^{-2}. The highest value of unnotched impact strength was achieved by a specimen with 20 wt.% recycled material, namely 23.877 ± 1.651 kJ·m^{-2}, and the lowest value (18.839 ± 1.506 kJ·m^{-2}) was achieved by specimens with 60 wt.% recycled material. However, the unnotched impact strength values were relatively homogeneous in terms of the amount of recycled material, and the difference between the minimum and the maximum was 26.743%. In terms of the specimen temperature as a separate effect at −30 °C, the unnotched impact strength value reached 18.546 ± 0.511 kJ·m^{-2}, and at +23 °C the values reached 23.849 ± 0.859 kJ·m^{-2}. The difference, with a value of 5.304 ± 0.685 kJ·m^{-2}, was significant at the level of significance of ± = 0.05 (p = 0.000). In terms of the exposure, as the last main effect that was studied, we achieved higher values of unnotched impact strength (23.808 ± 0.852 kJ·m^{-2}) in specimens in the state before the exposure in contrast to specimens after the exposure (18.588 ± 0.532 kJ·m^{-2}), while the difference was significant at the selected level of significance (p = 0.000). The first significant interaction between the amount of recycled material and the exposure points to the fact that higher unnotched impact strength values were always achieved in specimens prior to the exposure. Differences between the states before and after the exposure were highest when the recyclate accounted for 10 wt.% (7.629 ± 1.983 kJ·m^{-2}), followed by the difference at 0 wt.% recyclate (6.845 ± 2.034 kJ·m^{-2}) and at 50 wt.% recyclate (6.720 ± 1.976 kJ·m^{-2}). All these differences were significant at the chosen level of significance. On the other hand, the minimum difference in the unnotched impact strength value between the states before and after the exposure was achieved when the recyclate accounted for 80 wt.% (2.025 ± 2.017 kJ·m^{-2}). However, this difference was not statistically significant (p = 0.052).

The second smallest difference was the difference when the recyclate accounted for 20 wt.%, with a value of 3.694 ± 2.249 kJ·m^{-2} (p = 0.000), followed by the difference when the recyclate accounted for 30 wt.%, with a value of 4.469 ± 2.328 kJ·m^{-2} (p = 0.000). The second significant interaction, namely between the exposure and the specimen temperature indicates that specimens before the exposure achieved higher mean unnotched impact strength values than after the exposure. Specifically, for a specimen temperature of −30 °C, this difference was 2.416 ± 0.678 kJ·m^{-2} (p = 0.000), and for a specimen temperature of +23 °C this difference was 8.024 ± 0.886 kJ·m^{-2} (p = 0.000).

In the mutual interaction of all input variables in PBT, as in the previous cases, the unnotched impact strength values for specimens before the exposure were higher compared to specimens after the exposure. In terms of their size, we observed the maximum difference in unnotched impact strength at 10 wt.% of abrasive with a specimen temperature of +23 °C, with a value of 12.930 ± 1.283 kJ·m^{-2} (p = 0.000). In order, the second highest value relating to the unnotched impact strength was achieved when the recyclate accounted for 50 wt.% with a temperature of +23 °C (11.170 ± 2.163 kJ·m^{-2}; p = 0.000), followed by the difference in the unnotched impact strength values at 0 wt.% of the abrasive and, as in previous cases, the specimen temperature of +23 °C (9.848 ± 2.628 kJ·m^{-2}; p = 0.000). On the contrary, the smallest values of the differences in the unnotched impact strength of specimens in the states before and after the exposure were achieved when the recyclate accounted for 100 wt.% with a specimen temperature of −30 °C (1.358 ± 2.935 kJ·m^{-2}; p = 0.356), when the recyclate accounted for 80 wt.% with a specimen temperature of −30 °C (1.500 ± 1.917 kJ·m^{-2}; p = 0.308), when the recyclate accounted for 30 wt.% with a specimen temperature of −30 °C (1.803 ± 1.779 kJ·m^{-2}; p = 0.220), and when the recyclate accounted for 40 wt.% with a specimen temperature of −30 °C (1.993 ± 1.286 kJ·m^{-2}; p = 0.176), all of which, however, were statistically insignificant. Interestingly, the differences in the value of the unnotched impact strength at −30 °C reached an average value of 2.456 kJ·m^{-2}, while for the specimen temperature of +23 °C the average value of the unnotched impact strength difference was 8.024 kJ·m^{-2}, from which it was possible to draw one of the partial conclusions that at the specimen temperature of −30 °C neither the exposure nor the amount of recycled material had a significant effect on the change in the value of the unnotched impact strength. A graphical representation of the complex effect of the input variables (the recycled material amount, the specimen temperature, and the exposure) on the change in the value of the unnotched impact strength is given in Figure 14.

Finally, we analyzed the effect of the material (PPA or PBT) on the impact test value of the unnotched impact strength, a_{cU} (kJ·m^{-2}), depending on the amount of recycled material, the specimen temperature, and the exposure, with the exposure being the dividing factor (Figures 15 and 16).

If, in the first step, we focused on the value of the parameter studied in the specimens before the exposure, we arrived at the conclusion that the average value of the impact test of the unnotched impact strength (regardless of the specimen temperature and the amount of recycled material) in the case of PPA reached the value of 20.136 ± 0.915 kJ·m^{-2} and in the case of PBT reached the value of 23.808 ± 0.852 kJ·m^{-2}. In the unnotched impact strength test in specimens after the exposure, the mean value in the case of PPA was 17.179 ± 0.472 kJ·m^{-2}, and in the case of PBT it was 18.588 ± 0.532 kJ·m^{-2}. If we also started to consider the effect of the amount of the recycled material and the specimen temperature, the average difference between PPA and PBT in the specimens before the exposure reached −1.409 kJ·m^{-2}, which means that on average we achieved higher values of the studied a_{cU} (kJ·m^{-2}) variable in PBT.

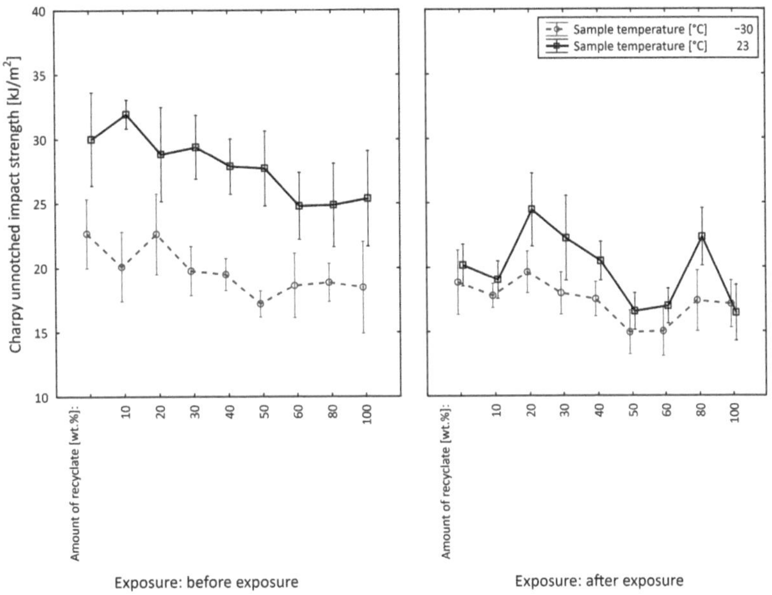

Figure 14. Dependence of the mutual interaction of the sample temperature and the exposure of the sample on the value of the unnotched impact strength of PBT.

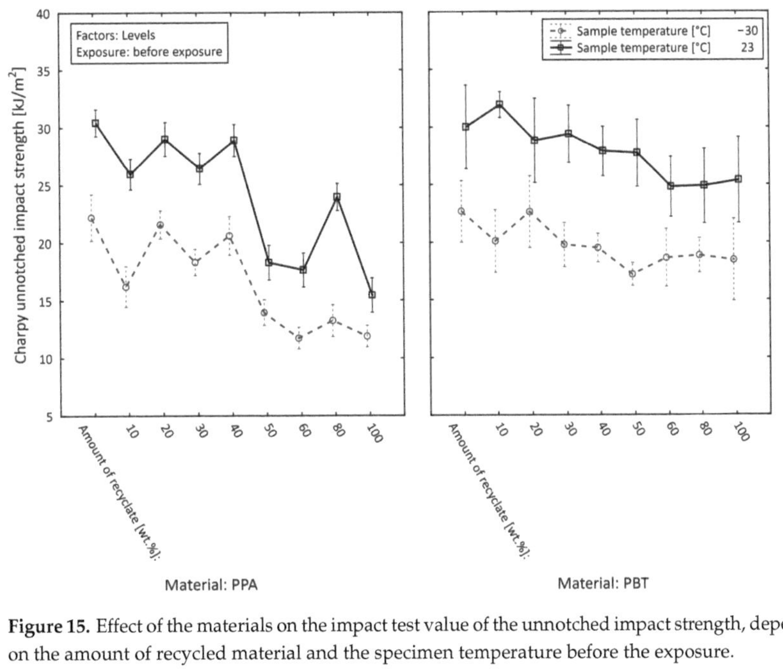

Figure 15. Effect of the materials on the impact test value of the unnotched impact strength, depending on the amount of recycled material and the specimen temperature before the exposure.

On closer analysis, we arrived at the conclusion that there were statistically insignificant differences at the chosen level of significance ($\alpha = 0.05$) between the PPA and PBT specimens before the exposure at −30 °C (0.065 ± 1.435 kJ·m^{-2}) when the recyclate accounted for 30 wt.%, at −30 °C (-0.015 ± 1.492 kJ·m^{-2}) when the recyclate accounted for 50 wt.%, at −30 °C (-0.333 ± 1.403 kJ·m^{-2}) when the recyclate accounted for 60 wt.%, and at −30 °C (−0.287 ± 1.317 kJ·m^{-2}) when the recyclate accounted for 20 wt.%. On the other hand, the maximum difference in the value of the unnotched impact strength between PPA and PBT before the exposure was found when the recyclate accounted for 80 wt.% and the specimen temperature was +23 °C (−7.360 ± 2.191 kJ·m^{-2}), followed by the difference in specimens with the recyclate accounting for 20 wt.% at +23 °C (−5.217 ± 1.925 kJ·m^{-2}). In addition to the negative differences, there were situations in which the unnotched impact strength values of the PPA materials were higher than those of the PBT materials, both in specimens in which the recyclate accounted for 10 wt.% and, at the same time, with specimen temperatures of −30 °C (2.131 ± 0.867 kJ·m^{-2}) and +23 °C (1.553 ± 1.332 kJ·m^{-2}).

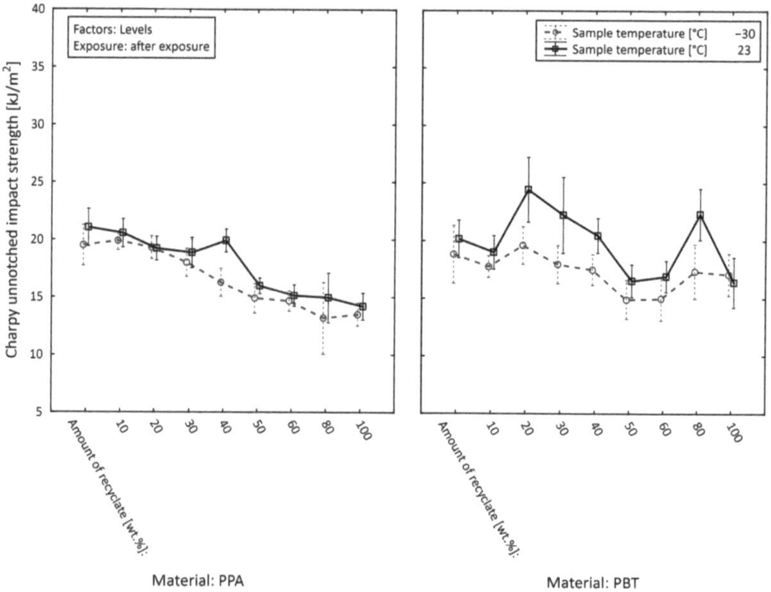

Figure 16. Effect of the materials on the impact test value of the unnotched impact strength depending on the amount of the recycled material and the specimen temperature after the exposure.

When, in the same way, we analyzed the value of the notch toughness, a_{cU} (kJ·m^{-2}), between the PPA and PBT materials in the state after the exposure, we arrived at the conclusion that the difference between the studied materials (PPA and PBT) reached an average value of −3.484 kJ·m^{-2}, which led us to a similar conclusion that the notch toughness value of PBT is on average higher than that of the PPA materials. However, in the state after the exposure, we found significantly higher differences than in the previous case. The maximum value of the difference in unnotched impact strength (−9.925 ± 2.604 kJ·m^{-2}) was recorded when the recyclate accounted for 100 wt.% and the specimen temperature was +23 °C. In terms of the sizes of the differences, the second most significant, when the recyclate accounted for 50 wt.% and the specimen temperature was +23 °C (−9.420 ± 2.203 kJ·m^{-2}), was followed by the difference when the recyclate accounted for 60 wt.% and the specimen temperature was +23 °C (−7.160 ± 2.030 kJ·m^{-2}), the difference when the recyclate accounted for 60 wt.% and the specimen temperature was −30 °C (−6.915 ± 1.721 kJ·m^{-2}), and the difference when the recyclate accounted for 100 wt.% and the specimen temperature was −30 °C (−6.588 ± 2.254 kJ·m^{-2}). It was also interesting that at lower levels of the recycled material, in most cases, the differences were statistically insignificant. When the recyclate accounted for 0 wt.%, the difference value with a specimen temperature of −30 °C was −0.465 ± 2.335 kJ·m^{-2}, and with a specimen temperature of +23 °C it was 0.445 ± 2.394 kJ·m^{-2}. When the recyclate accounted for 20 wt.% and the specimen temperature was −30 °C, it was −1.034 ± 2.159 kJ·m^{-2}, and with a specimen temperature of +23 °C, it was 0.199 ± 2.564 kJ·m^{-2}. When the recyclate accounted for 30 wt.% and the specimen temperature was −30 °C, it was −1.443 ± 1.522 kJ·m^{-2}, and when the specimen temperature was +23 °C, it was −2.928 ± 1.923 kJ·m^{-2}. The last value of the recycled material amount, with statistically insignificant differences in the values of unnotched impact strength at both specimen temperatures, was when the recyclate accounted for 40 wt.%. For the specimen temperature of −30 °C, it was specifically 1.120 ± 1.456 kJ·m^{-2}, and for the specimen temperature of +23 °C, it was 1.024 ± 1.769 kJ·m^{-2}. With further increases in the amount of recycled material, the differences were already significant at the selected level of significance (α = 0.05).

Author Contributions: J.D. conceived and designed the experiments and wrote the paper; M.G. and T.S. analyzed and evaluated data, processed data, and performed the experiments and measurements. All authors have read and agreed to the published version of the manuscript.

Funding: This research received no external funding.

Institutional Review Board Statement: Not applicable.

Informed Consent Statement: Not applicable.

Data Availability Statement: Not applicable.

Acknowledgments: This paper has been eleborated in the framework of the project KEGA 063TUKE-4/2021.

Conflicts of Interest: The authors declare no conflict of interest. The founding sponsors had no role in the design of the study; in the collection, analyses, or interpretation of the data; in the writing of the manuscript; or in the decision to publish the results.

References

1. Tomar, N.; Maiti, S.N. Mechanical properties of PBT/ABAS blends. *J. Appl. Polym. Sci.* **2007**, *104*, 1807–1817. [CrossRef]
2. Szostak, M. Mechanical and thermal properties of PET/PBT blends. *Mol. Cryst. Liq. Cryst.* **2004**, *416*, 209–215. [CrossRef]
3. Sharma, R.; Jain, P.; Sadhu, S.D.; Kaur, B. Mechanical and thermal properties of impact modified PBT blends and impact modified PBT nanocomposites. *J. Polym. Eng.* **2013**, *33*, 489–500. [CrossRef]
4. Ambrosio, J.D.; Larocca, N.M.; Pessan, L.A.; Hage, E. Influence of the Process Parameters of an Intermeshing Co-rotating Twin Screw Extruder on the Morphology and Notched Izod Impact Strength of PBT/ABS/MGE Blends. *Polym. Eng. Sci.* **2010**, *50*, 2382–2391. [CrossRef]
5. Ajorloo, M.; Ghodrat, M.; Kang, W.H. Incorporation of recycled polypropylene and fly ash in polypropylene-based composites for automotive applications. *J. Polym. Environ.* **2020**, *29*, 1298–1309. [CrossRef]
6. Guo, Y.; Zhu, S.; Chen, Y.; Li, D. Analysis and identification of the mechanism of damage and fracture of high-filled wood fiber/recycled high-density polyethylene composites. *Polymers* **2019**, *11*, 170. [CrossRef] [PubMed]
7. Ignatyev, I.A.; Thielemans, W.; Vander Beke, B. Recycling of Polymers: A Review. *ChemSusChem* **2014**, *7*, 1579–1593. [CrossRef]
8. Manas, D.; Manas, M.; Gajzlerova, L.; Ovsik, M.; Kratky, P.; Senkerik, V.; Skrobak, A.; Danek, M.; Manas, M. Effect of low doses beta irradiation on micromechanical properties of surface layer of injection molded polypropylene composite. *Radiat. Phys. Chem.* **2015**, *114*, 25–30. [CrossRef]
9. Jiun, Y.L.; Tze, C.T.; Moosa, U.; Tawawneh, M.A. Effects of Recycling Cycle on Used Thermoplastic Polymer and Thermoplastic Elastomer Polymer. *Polym. Polym. Compos.* **2016**, *24*, 735–739. [CrossRef]
10. Liu, F.; Li, T.D.; Xu, F.Y.; Li, J.Q.; Jiang, S.F. Microstructure, tensile property, and surface quality of glass fiber-reinforced polypropylene parts molded by rapid heat cycle molding. *Adv. Polym. Technol.* **2020**, *3*, 1–15. [CrossRef]
11. Hugo, A.M.; Scelsi, L.; Hodzic, A.; Jones, F.R.; Dwyer-Joyce, R. Development of recycled polymer composites for structural applications. *Plast. Rubber Compos.* **2011**, *40*, 317–323. [CrossRef]
12. Atagur, M.; Seki, Y.; Oncu, O.; Sever, K.; Seki, Y.; Sarikanat, M.; Altay, L. Evaluating of reinforcing effect of Ceratonia Siliqua for polypropylene: Tensile, flexural and other properties. *Polym. Test.* **2020**, *89*, 6607. [CrossRef]
13. Lin, X.; Wu, C.Q.; Ren, D.Y. Tensile properties and die swell behaviours of highly filled polypropylene nanocomposites. *Plast. Rubber Compos.* **2020**, *49*, 47–56. [CrossRef]
14. Gel'atko, M.; Hatala, M.; Botko, F.; Vandzura, R.; Hajnys, J. Eddy Current Testing of Artificial Defects in 316L Stainless Steel Samples Made by Additive Manufacturing Technology. *Materials* **2022**, *15*, 6783. [CrossRef] [PubMed]
15. Razak, Z.; Sulong, A.; Muhamad, N.; Haron, C.H.C.; Radzi, M.K.F.M.; Ismail, N.F.; Tholibon, D.; Tharazi, I. Effects of thermal cycling on physical and tensile properties of injection moulded kenaf/carbon nanotubes/polypropylene hybrid composites. *Compos. Part B Eng.* **2019**, *168*, 159–165. [CrossRef]

16. Mital'ova, Z.; Botko, F.; Vandzura, R.; Litecka, J.; Mital, D.; Simkulet, V. Machining of Wood Plastic Composite Using AWJ Technology with Controlled Output Quality. *Machines* **2022**, *10*, 566. [CrossRef]
17. Khademieslam, H.; Kalagar, M. Evaluation of the bending strength, impact strength, and morphological properties of wheat straw fiber/paper mill sludge/polypropylene composites. *Bioresources* **2016**, *11*, 3914–3922. [CrossRef]
18. Panda, A.; Dyadura, K.; Valicek, J.; Harnicarova, M.; Kusnerova, M.; Ivakhniuk, T.; Hrebenyk, L.; Sapronov, O.; Sotsenko, V.; Vorobiov, P.; et al. Ecotoxicity Study of New Composite Materials Based on Epoxy Matrix DER-331 Filled with Biocides Used for Industrial Applications. *Polymers* **2022**, *14*, 3275. [CrossRef]
19. Hamada, H.; Fujihara, K.; Harada, A. Influence of sizing conditions on bending properties of continuous glass fiber reinforced polypropylene composites. *Compos. Part A Appl. Sci. Manuf.* **2000**, *31*, 979–990. [CrossRef]
20. Rusinko, J.; Baron, P.; Panda, A.; Kocisko, M. Research of mold cooling process optimization for production of expanded polystyrene with the application of experimental measurements on the test form. *Adv. Mech. Eng.* **2021**, *13*, 41012. [CrossRef]
21. Kim, J.; Cho, D. Effects of waste expanded polypropylene as recycled matrix on the flexural, impact, and heat deflection temperature properties of kenaf fiber/polypropylene composites. *Polymers* **2020**, *12*, 2578. [CrossRef] [PubMed]
22. Behalek, L.; Novak, J.; Brdlik, P.; Boruvka, M.; Habr, J.; Lenfeld, P. Physical Properties and Non-Isothermal Crystallisation Kinetics of Primary Mechanically Recycled Poly(l-lactic acid) and Poly(3-hydroxybutyrate-co-3-hydroxyvalerate). *Polymers* **2021**, *13*, 3396. [CrossRef] [PubMed]
23. Wang, J.; Chen, D.J. Flexural properties and morphology of microcellular-insert injection molded all-polypropylene composite foams. *Compos. Struct.* **2018**, *187*, 403–410. [CrossRef]
24. Svetlik, J.; Malega, P.; Rudy, V.; Rusnak, J.; Kovac, J. Application of Innovative Methods of Predictive Control in Projects Involving Intelligent Steel Processing Production Systems. *Materials* **2021**, *14*, 1641. [CrossRef]
25. Khalili, S.M.R.; Farsani, R.E.; Soleimani, N.; Hedayatnasab, Z. Charpy impact behavior of clay/basalt fiber-reinforced polypropylene nanocomposites at various temperatures. *J. Thermoplast. Compos. Mater.* **2016**, *29*, 1416–1428. [CrossRef]
26. Milosevic, M.; Cep, R.; Cepova, L.; Lukic, D.; Antic, A.; Djurdjev, M. A Hybrid Grey Wolf Optimizer for Process Planning Optimization with Precedence Constraints. *Materials* **2021**, *14*, 7360. [CrossRef]
27. Hernandez-Diaz, D.; Villar-Ribera, R.; Espinach, F.X.; Julian, F.; Hernandez-Abad, V.; Delgado-Aguilar, M. Impact properties and water uptake behavior of old newspaper recycled fibers-reinforced polypropylene composites. *Materials* **2020**, *13*, 1079. [CrossRef]

materials

MDPI

Article

Fe–Al–Si-Type Iron Aluminides: On the Strengthening by Refractory Metals Borides

Věra Vodičková [1], Martin Švec [2,*], Pavel Hanus [1], Šárka Bukovská [2] and Petra Pazourková Prokopčáková [1]

[1] Department of Material Science, Faculty of Mechanical Engineering, Technical University of Liberec, 46117 Liberec, Czech Republic
[2] Department of Technology, Faculty of Mechanical Engineering, Technical University of Liberec, 46117 Liberec, Czech Republic
* Correspondence: martin.svec@tul.cz; Tel.: +420-485353308

Abstract: The effect of boron addition into Fe–28Al–5Si–X (X = -, 2Mo, or 2Ti) on the structure and high-temperature yield stress was investigated. Generally, the alloying of binary Fe_3Al-type iron aluminides by silicon significantly improves high-temperature mechanical properties by solid-solution strengthening. On the other hand, the workability and ductile properties at room or slightly elevated temperatures get worse with the increasing silicon content. Boron alloying together with titanium or molybdenum alloying is one of the ways to improve the workability of this type of alloy and, at the same time, ensure the formation of a sufficient amount of secondary phase particles required for effective strengthening. In this paper, the influence of 1 at. % of boron on high-temperature yield stress is evaluated in response to structural changes and compared with results obtained previously on the same type of alloy (Fe–28Al–5Si–2X, X= -, Mo, or Ti) but without boron alloying. It can be concluded that the network structure of borides of refractory metals formed due to boron alloying works more effectively for alloy hardening at higher temperatures than a mixture of silicides and carbides present in the boron-free alloy of the same composition.

Keywords: Fe_3Al-based cast iron aluminide; titanium; molybdenum and boron addition; grain size; heat treatment; high-temperature yield stress

Citation: Vodičková, V.; Švec, M.; Hanus, P.; Bukovská, Š.; Pazourková Prokopčáková, P. Fe–Al–Si-Type Iron Aluminides: On the Strengthening by Refractory Metals Borides. *Materials* **2022**, *15*, 7189. https://doi.org/10.3390/ma15207189

Academic Editor: Frank Czerwinski

Received: 5 September 2022
Accepted: 13 October 2022
Published: 15 October 2022

1. Introduction

Iron aluminides belong to the group of intermetallics for structural applications, but their potential has not been fully developed yet. Due to their excellent corrosion resistance, they are considered a material capable of replacing high-alloy steels for high-temperature applications (f.e. brake discs and furnace grids). Unfortunately, too low ductility at room temperature-causing problematic workability-hinders their wider use. The ordered iron aluminides possess low room temperature ductility even though their crystallographic characteristic (for example, more than five independent slip systems) should be the assumption for good. Both Fe_3Al and FeAl are reasonably ductile at compression, while their tensile ductility is limited at temperatures below brittle-to-ductile transition [1]. The ductility of B2 type Fe40Al alloy can increase with decreasing grain size (<10 μm) [2].

In general, the brittleness of iron aluminides can result from grain boundary properties, and concurrently both Fe_3Al and FeAl are prone to moisture or hydrogen embrittlement. The reaction with moisture, forming Al_2O_3 and atomic hydrogen, and causing brittle cleavage crack propagation, is why the room-temperature ductility is up to 4% maximum in the presence of moisture [3–5]. The binary Fe_3Al alloy shows a cleavage mode of fracture independent of ductility or environment. With aluminum content increasing to 36.5 at. %, the fracture mode is changing from transgranular to intergranular cleavage [1].

If the low ductility of iron aluminides is caused by the environmental effect, it can be improved using alloying. Among others, the effect of chromium on the ductility of iron aluminides was investigated [5]. It is known that small addition of boron can be used to increase the ductility of B2 FeAl [6–8]. It was also found that the segregation of boron atoms to the grain boundaries can improve grain boundary cohesion very effectively [9,10].

The mechanism of boron segregation has been intensively studied in FeAl-type intermetallic alloys. There was confirmed that boron segregation has two different origins-non-equilibrium segregation (resulting from an attractive interaction between boron atoms and excess vacancies) working during the quench from high temperature and equilibrium segregation, which is explained by strong repulsive interactions between segregated boron atoms in low-temperature annealed materials. It was also found that bulk properties of FeAl-type iron aluminides are strongly modified by added boron due to the nano-heterogenous structure formation [8]. So, the mechanism of grain boundary strengthening in FeAl-type intermetallic alloys had to be thoroughly investigated on the atomic scale [11–13]. A very detailed investigation using a three-dimensional atom probe showed that boron plays a significant role in the formation and stabilization of stacking faults present in the structure of intermetallics. The existence of B-enriched/Al-depleted stacking faults has been demonstrated. This phenomenon of boron enrichment was also observed in Cottrell atmospheres or antiphase boundaries [13].

When the amounts of added boron are higher, there was found that this beneficial effect of boron disappears when borides precipitate along grain boundaries [14]. Nevertheless, for Fe_3Al-based alloys doped with Mo/W–Ti–B was noticed that precipitation of borides at grain boundaries maintained ductility at comparably low temperatures [15,16].

In the case of Fe_3Al-based iron aluminides, boron addition can also have a similar effect on other mechanical properties (namely ultimate strength). In the past, carbon alloying or boron microalloying together with alloying of the elements having low solubilities in Fe_3Al-based aluminides had been applied as the most feasible way to achieve precipitation strengthening due to carbides or borides formation [1,17–21].

Currently, the concept of strengthening by incoherent precipitates—borides—has also been successfully developed and applied to improve the high-temperature mechanical properties of Fe_3Al-based iron aluminides [22,23]. In contrast with carbides, which tend to coarsen rapidly at high temperatures and, due to the shape of particles (plates, needles), can act as sources for crack nucleation, borides can prevent the coarsening of the Fe–Al matrix by pinning grain boundaries [15,24]. Compared to binary $D0_3$-ordered iron aluminides, Ti and B-doped aluminides show significantly higher flow stresses and creep resistances. Moreover, the order of the matrix can play an important role-it was shown that the presence of the DO_3 ($L2_1$)-order at higher temperatures increases the alloy strength [15].

If the strengthening by borides precipitated in the structure acts together with solid-solution strengthening, the effect of strengthening can be more significant, and the high-temperature characteristics such as creep rate or high-temperature yield stress can be improved.

The article is focused on evaluating the structure and comparison of HT mechanical properties of Fe–Al–Si (-X) intermetallic alloys doped with boron (X = refractory metal). The strengthening by Si soluted in the matrix together with the fine incoherent boron-based precipitates could be beneficial for HT properties of alloy intended for HT applications.

2. Materials and Methods

Vacuum induction melting and casting were used for the preparation of the alloys. The nominal chemical composition of the investigated alloys is summarized in Table 1.

Table 1. The nominal chemical composition of investigated alloys [at. %].

Alloy	Fe	Al	Si	Mo	Ti	B
Fe–28Al–5Si–1B	Bal.	28	5	-	-	1
Fe–28Al–5Si–2Mo–1B	Bal.	28	5	2	-	1
Fe–28Al–5Si–2Ti–1B	Bal.	28	5	-	2	1

Scanning electron microscope (SEM) Tescan Mira 3 was used to study the sample's microstructure. The microstructure of alloys was studied in the state after oxide polishing. The local chemical composition of phases was determined by Energy Dispersive X-ray Spectroscopy (EDX) using Oxford UltimMax 65 detector. The phase compositions of alloys were verified by X-ray diffraction (XRD) using an X'Pert^3 Powder diffractometer in Bragg-Brentano geometry (Co Kα radiation, λ = 1.78901 Å). The grain size of alloys and the presence of γ iron were determined using Electron Backscatter Diffraction (EBSD) performed by Oxford Instruments Symmetry detector. The process parameters for each image are described in the manuscript.

The structure of the three investigated samples was studied in an as-cast state, after heat treatment at 800 $\pm$ 5 °C for 100 hrs and after a compression test at 600 °C. The temperature and time of this stabilization annealing were chosen concerning the expected maximum work temperature of this type of material. The equilibrium was reached after annealing at a selected temperature and time. Annealing was carried out in the vacuum furnace.

The values of high-temperature compression yield stress $\sigma_{0.2}$ were measured by the TESTOMETRIC FS100CT device at temperatures of 20, 600, 700, and 800 °C. The accuracy of the temperature chamber of the device was $\pm$1 °C, and the used initial strain rate was 1.5×10^{-4} s^{-1}. Both states (as-cast and after heat treatment at 800 °C for 100 h) were tested at all temperatures. Samples for high-temperature tests with dimensions of 6 $\times$ 6 $\times$ 8 mm were prepared by spark machining.

3. Results

3.1. Structure of Investigated Alloys

All combinations of additives used (boron only, molybdenum–boron, and titanium–boron) caused significant grain refinement (see Figure 1) in comparison to binary Fe–28Al alloy as well as to ternary Fe–28Al5Si alloy, as the grain size of both is usually reported in the order of hundreds of micrometers [25] or of units of millimeters (see Figure 1). The grain size of boron-doped alloys was in the range of 130–140 μm for all investigated alloys in the as-cast state.

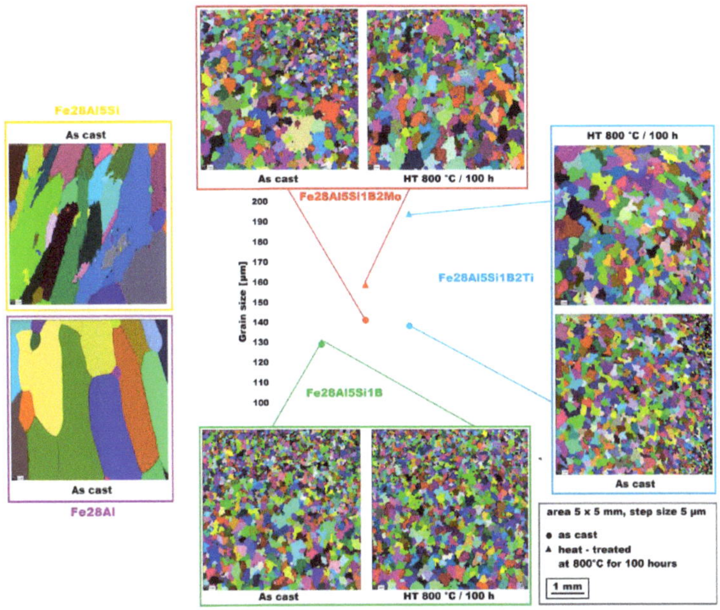

Figure 1. The grain size of investigated alloys in the as-cast state and also after heat treatment at 800 °C for 100 h (the Fe28Al and Fe28Al5Si alloy added for comparison).

The structure of the Fe–28Al–5Si–1B in the as-cast state shows a dendritic character (see Figure 2A). The secondary phase particles are distributed in clusters along dendrite boundaries and also inside dendrites (see Figure 2A,B). This phase was determined by EDX and by XRD as Fe_2B iron borides (see XRD diffractogram in Figure 3). The stabilization annealing at 800 °C for 100 h (Fe–28Al–5Si–1B HT 800/100 alloy) has no significant effect on the distribution or type of secondary phase (see Figure 4A,B). Additionally, the grain size remains unchanged after annealing (see Figure 1), namely 130 μm for both states.

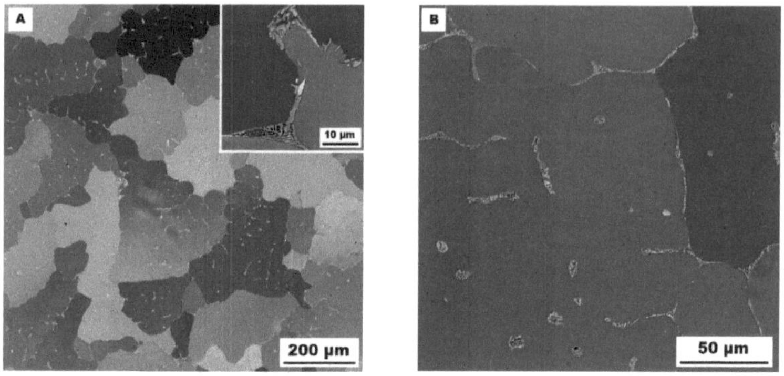

Figure 2. The structure of Fe–28Al–5Si–1B alloy in the as-cast state (SEM, 10 kV, BSE). (**A**) overview of the structure, (**B**) detail of the structure.

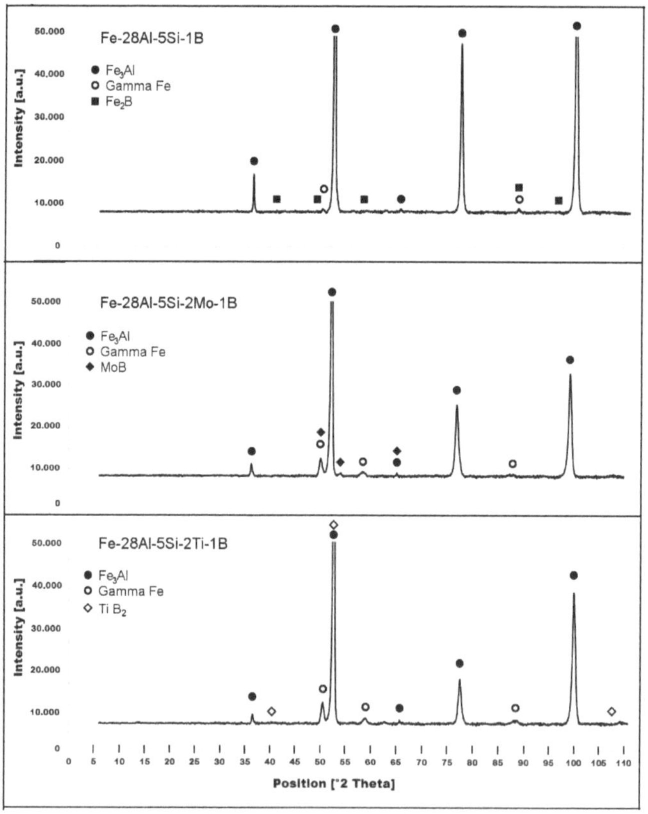

Figure 3. XRD diffractograms of investigated alloys in the as-cast state.

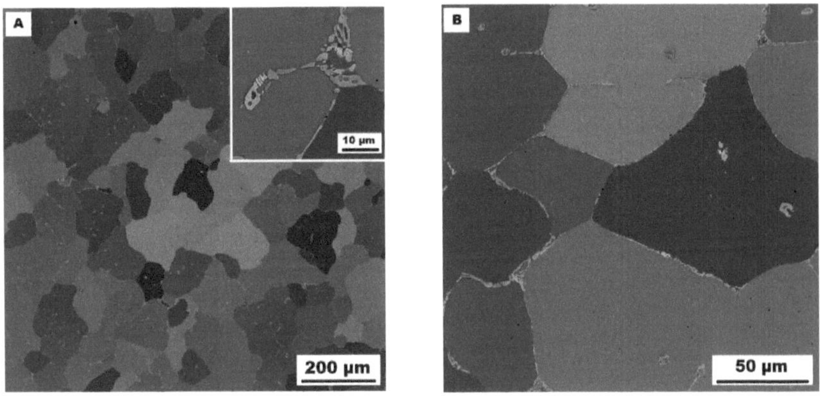

Figure 4. The structure of Fe–28Al–5Si–1B HT 800/100 alloy (SEM, 10 kV, BSE). (**A**) overview of the structure, (**B**) detail of the structure.

The character of the structure of Fe–28Al–5Si–2Mo–1B in the as-cast state is also dendritic (see Figure 5A,B). The secondary phase is distributed in the form of the "like-eutectics" (lamellar) areas in the interdendritic spaces and also in the form of the individual fine particles inside the dendrites. Based on EDX and XRD measurements, these particles were identified as MoB molybdenum borides (see Figure 3). It is necessary to notice that the verification of phase by XRD is not quite unambiguous due to the overlapping of phase peaks with Fe_3Al matrix peaks, but EDX results support this identification. Consistent with XRD analysis, very small areas of gamma-iron were found in the structure using EBSD. This fine-grained phase (grain size in the order of hundreds of nanometers) was distributed very rarely with particles of molybdenum borides, as can be seen in Figure 6. Heat treatment has no significant influence on the stability or distribution of the second phase in the Fe-28Al-5Si-2Mo-1B HT 800/100 alloy (see Figure 7A,B). Only the difference in the grain size was observed-it increases from 141 μm to 159 μm after annealing (see Figure 1).

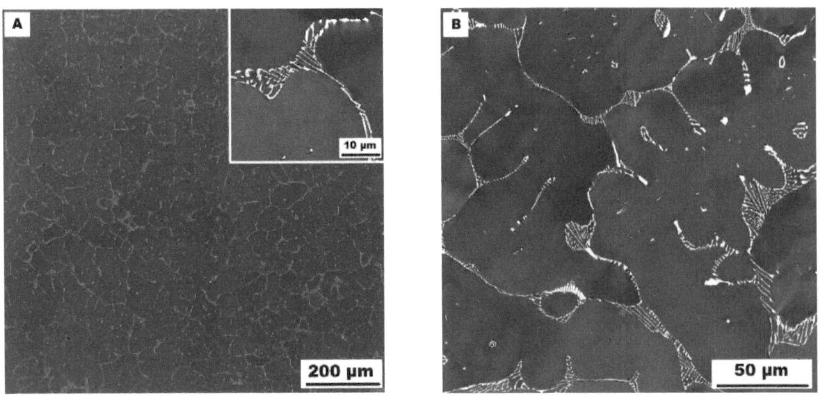

Figure 5. The structure of Fe–28Al–5Si–2Mo–1B alloy in the as-cast state (SEM, 10 kV, BSE). (**A**) overview of the structure, (**B**) detail of the structure.

The dendritic areas are present in the structure of Fe–28Al–5Si–2Ti–1B alloy in an as-cast state (see Figure 8A,B). The secondary phase particles are distributed in the form of needles or oval-shaped "chains" (size range 0.5–20 μm) along dendrite boundaries and also inside dendrites. The precipitates were determined by EDX and XRD as TiB_2 titanium borides (Figure 3).

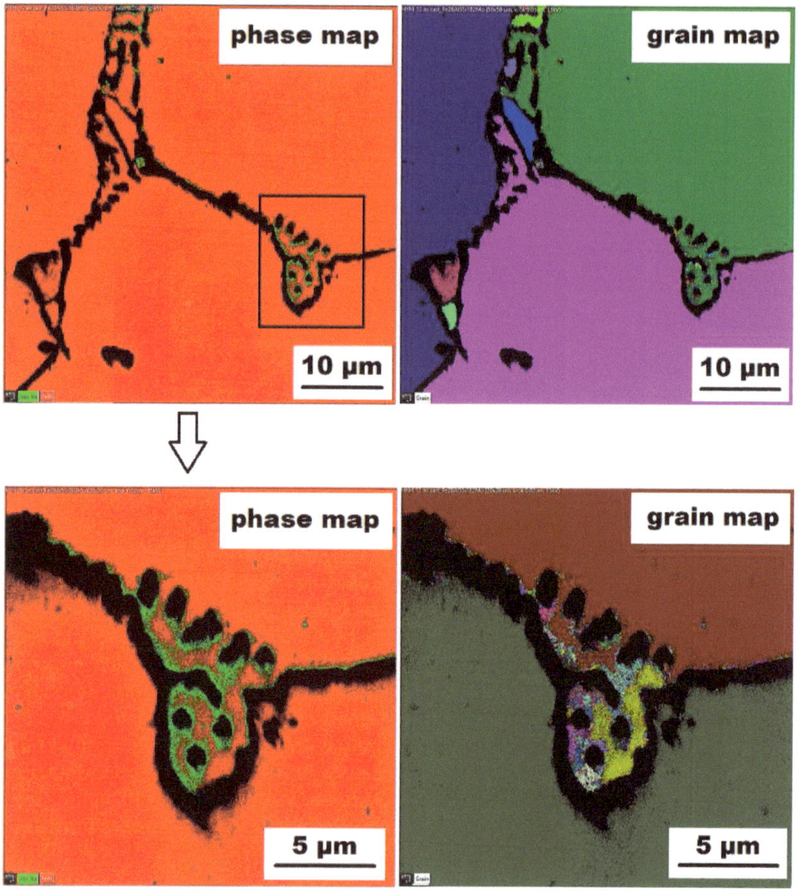

Figure 6. The EBSD phase and grain maps showing the gamma-iron presence in Fe–28Al–5Si–2Mo–1B as-cast alloy (green areas in phase maps), black areas—borides.

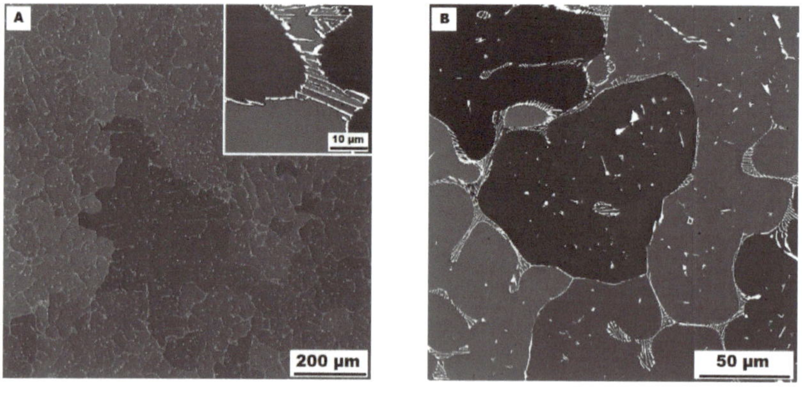

Figure 7. The structure of Fe–28Al–5Si–2Mo–1B HT 800/100 alloy (SEM, 10 kV, BSE). (**A**) overview of the structure, (**B**) detail of the structure.

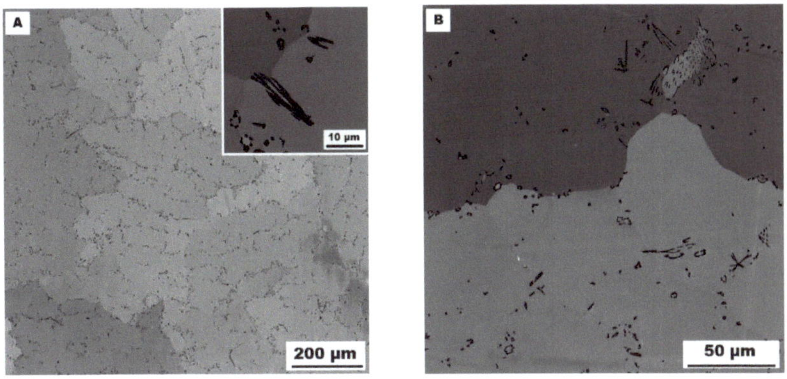

Figure 8. The structure of Fe–28Al–5Si–2Ti–1B alloy in the as-cast state (SEM, 10 kV, BSE). (**A**) overview of the structure, (**B**) detail of the structure.

Similar to the as-cast alloy Fe–28Al–5Si–2Mo–1B, the presence of "nanoscale" areas of gamma-iron (previously identified by X-ray diffraction, Figure 3) was also confirmed by EBSD in the as-cast Fe–28Al–5Si–2Ti–1B alloy, Figure 9.

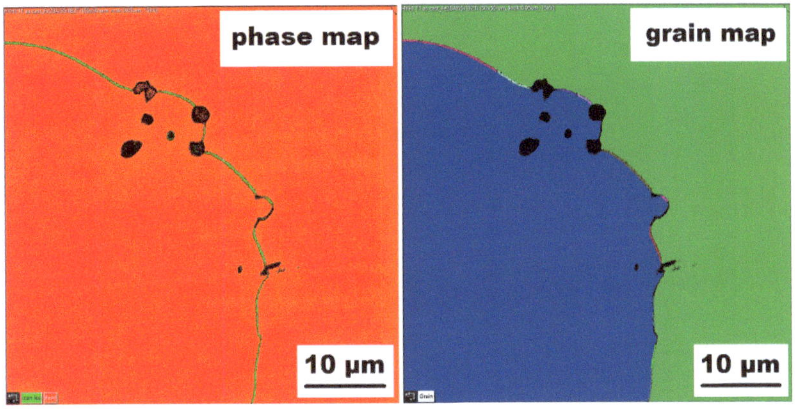

Figure 9. The EBSD phase and grain maps showing the gamma-iron presence in Fe–28Al–5Si–2Ti–1B as-cast alloy (green areas in phase maps).

In contrast to Mo-B doped alloy, these areas were located along the grain boundaries in Ti-B doped alloy, see Figure 9. After heat treatment (alloy **Fe–28Al–5Si–2Ti–1B HT 800/100**), the distribution and type of secondary phase remain without changes (see Figure 10A,B). A significant change was observed in the grain size, which increased from 138 µm to 194 µm after annealing (see Figure 1).

3.2. High-Temperature Yield Stress

The values of yield stress in compression at room and elevated temperatures are summarized in Figure 11. They were measured for both the as-cast state and heat-treated state (HT 800/100) of all investigated alloys. In the as-cast state, all three alloys maintain high values of yield stress up to 600 °C (comparable to room temperature values). At 700 °C, the alloy doped with titanium and boron in an as-cast state (full blue line in Figure 11) shows a relatively high value of yield stress in compression (460 MPa).

It is evident from the chart in Figure 11 that the yield stress values decreased after annealing (HT 800/100) at all tested temperatures for all three investigated alloys.

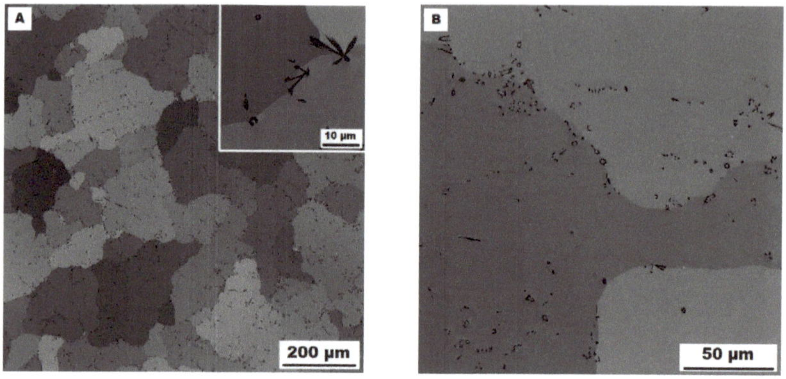

Figure 10. The structure of Fe–28Al–5Si–2Ti–1B HT 800/100 alloy (SEM, 10 kV, BSE). (**A**) overview of the structure, (**B**) detail of the structure.

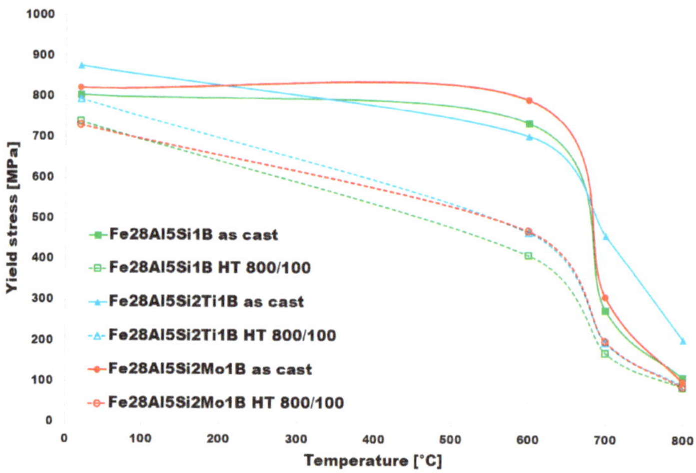

Figure 11. The values of the high–temperature yield stress of investigated alloys (as-cast states and also annealed states).

The difference between yield stress values for as-cast and heat-treated states appears to be most pronounced at 600 °C for Fe–28Al–5Si–1B and Fe–28Al–5Si–2Mo–1B alloys-in the case of Fe–28Al–5Si–2Ti–1B alloy, the maximum decreases in yield stress values are comparable at 600 °C and 700 °C.

In comparison to the same type of alloys doped with titanium or molybdenum only so, without boron [25], the values of yield stress up to 600 °C appear to be of a similar level to these. In the higher temperature range, the yield stress values of Fe–28Al–5Si–2Ti–1B alloy are higher (almost doubled at 800 °C) compared to boron-free alloy in [25].

4. Discussion

The relevance of phase equilibria for the strengthening of respective alloys was discussed in detail in [17]. For Fe–Al-type alloys, mechanisms based on strengthening mechanisms derived from the phase diagrams themselves play a significant role, i.e., solid-solution hardening, strengthening by incoherent or coherent particles, and ordering. Nevertheless, grain refinement strengthening through alloying-as shown here-is also available for Fe–Al-based alloys.

Solid-solution strengthening (strengthening by additive atoms) occurs when the presence of an impurity atom in the base metal lattice causes strain in the crystal lattice. An example can be the Fe–Al–Cr system, where within a wide range of concentrations, only solid-solution hardening is possible.

The strengthening of coherent or non-coherent precipitates is a concept frequently used for increasing mechanical properties, not only in the aluminides field. The precipitates affect the strengthening significantly because they work as dislocation motion obstacles. Principally, the dislocation can overcome the coherent precipitate. The non-coherent precipitate can stop the dislocation motion, but when sufficiently large stress is applied, the dislocation can also overcome this obstacle, i.e., by the Orowan or Friedel mechanism.

The solubility in solid solution for the third element within the Fe–Al phases is limited in many Fe–Al–X systems, and the possibility of strengthening Fe–Al-based alloys by precipitation of coherent (Fe–Al–Ni system) or non-coherent phase (Fe–Al–Nb, Fe–Al–Zr system) can be used [17].

Additionally, immobile dislocations located in the matrix (dislocation structure includes grain boundaries, subgrains boundaries, and arrangement of dislocations within grains) can be considered effective obstacles. Then the contribution of dislocation strengthening depends on the energy necessary to push dislocation through obstacles of dislocation origin.

For strengthening Fe–Al-based alloys, there is an additional possibility of strengthening by the ordering of crystal structure, e.g., by stabilization of the $D0_3$ structure to higher temperatures.

In the case of alloys investigated in this article, the overall strengthening should include the contribution of the solid-solution strengthening by silicon atoms due to the low solubility of silicon in the Fe_3Al matrix and the contribution of non-coherent precipitates-borides of iron, titanium, or molybdenum. The role of boron could be questionable-although it participates in precipitate formation, its segregation on grain boundaries or on defects (as in the case of Fe–Al-type alloys) is not completely excluded. The strengthening by dislocation arrangement or by grain refinement could be co-operating too.

According to the simplest/linear model, the additive action of mechanisms could be assumed. However, for Fe–Al-based alloys, the contribution of individual mechanisms to the overall strengthening is not yet completely clear, as there are not as much data available as for other materials. The aspect of strengthening Fe–Al-based alloys at high temperatures is still open and an important issue.

It is evident from Figure 11 that materials with borides of refractory metals (blue and red lines) achieve higher values of yield stress in both states in comparison to alloy Fe–28Al–5Si–1B, except for the value at 600 °C for the as-cast alloys (here the yield strength of Fe–28Al–5Si–1B is higher than that of Fe–28Al–5Si-2Ti–1B, but these values appear to be comparable in the range of measurement errors). This fact could be explained by a noticeably higher density of fine precipitates in Ti/Mo doped alloy structures—compare Figures 2A, 5A and 8A or Figures 4A, 7A and 10A (it has not been quantified). In this case, the contribution of incoherent precipitates—borides—is crucial for high-temperature strengthening, while the effect of solid-solution hardening under the influence of Si atoms dissolved in the matrix should be the same for all investigated alloys. The contribution of solid-solution hardening to the overall strengthening would increase with a higher amount of silicon, as had been shown for silicon/molybdenum doped alloy [26]. The increasing of silicon content led to the enhancement of values of high-temperature yield stress in compression, mainly at temperatures up to 600 °C (the increase of about 160 MPa in the as-cast state)-see [26]. Nevertheless, the strengthening by borides proves to be more advantageous in terms of workability, which deteriorates as the silicon content increases.

To assess the strengthening mechanisms interactions during deformation, the samples deformed at 600 °C were subjected to structure- and EBSD analysis. The structures of samples after compression tests at 600 °C are shown in Figures 12–17. There can be stated from the comparison of EBSD Figures for as-cast and annealed states (compare Figure 12C to Figure 13C; Figure 14C to Figures 15C and 16C to Figure 17C) that the annealed samples (800 °C/100 h) have recrystallized during compression tests at 600 °C. On the other hand, the samples in the as-cast state show a deformation texture (see EBSD Figures). In both cases (in as-cast and annealed states), very small grains started to grow in narrow areas near the secondary particles. This phenomenon is more pronounced in as-cast states of samples—compare Figure 12A,B to Figure 13A,B; Figure 14A,B to Figure 15A,B, and Figure 16A,B to Figure 17A,B.

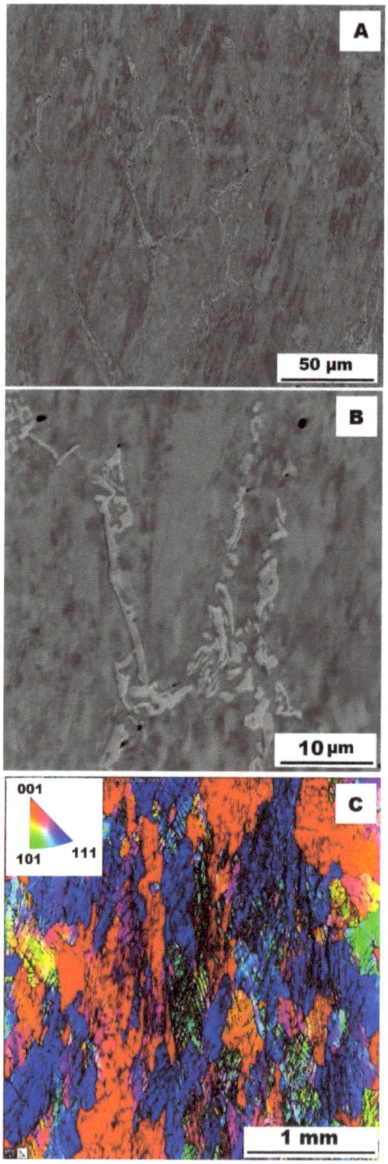

Figure 12. The structure of Fe–28Al–5Si–1B alloy in the as-cast state after compression test at 600 °C. (**A**) Overview of the structure-10 kV, BSE; (**B**) Detail of the structure-10 kV, BSE; (**C**) EBSD map of grains-area 3 × 3 mm, step size 2 µm, 20 kV.

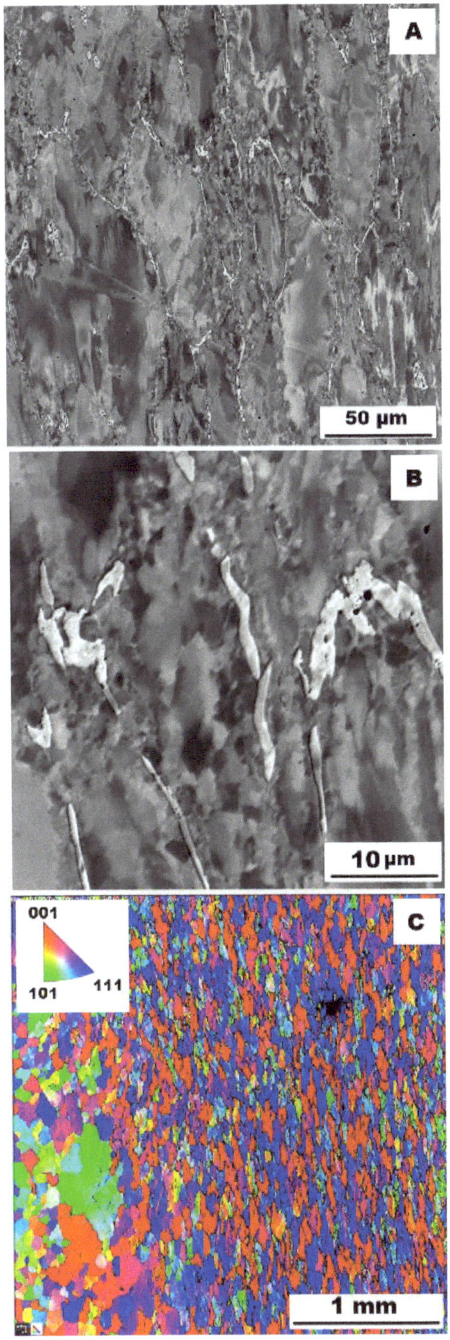

Figure 13. The structure of Fe–28Al–5Si–1B HT 800/100 alloy after compression test at 600°C. (**A**) Overview of the structure-10 kV, BSE. (**B**) Detail of the structure-10 kV, BSE. (**C**) EBSD map of grains-area 3 × 3 mm, step size 2 µm, 20 kV.

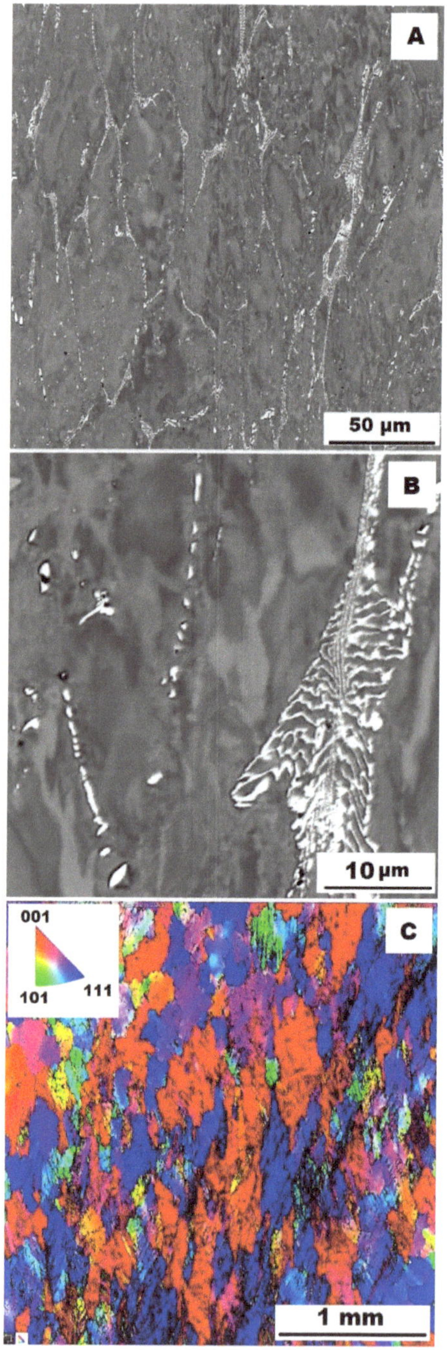

Figure 14. The structure of Fe–28Al–5Si–2Mo–1B alloy in the as-cast state after compression test at 600 °C. (**A**) Overview of the structure-10 kV, BSE. (**B**) Detail of the structure-10 kV, BSE. (**C**) EBSD map of grains-area 3 × 3 mm, step size 2 µm, 20 kV.

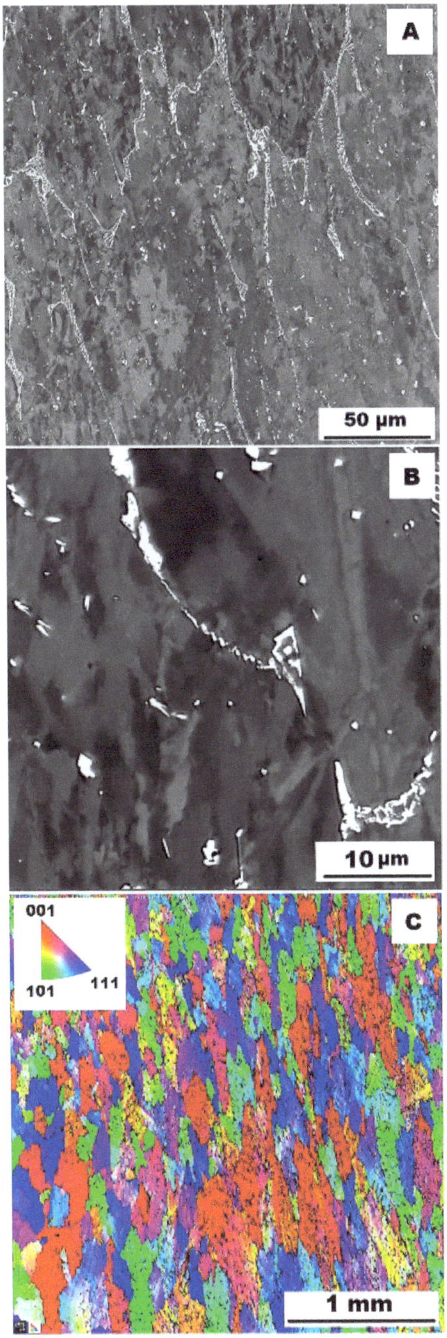

Figure 15. The structure of Fe–28Al–5Si–2Mo–1B HT 800/100 alloy after compression test at 600 °C. (**A**) Overview of the structure-10 kV, BSE. (**B**) Detail of the structure-10 kV, BSE. (**C**) EBSD map of grains-area 3 × 3 mm, step size 2 μm, 20 kV.

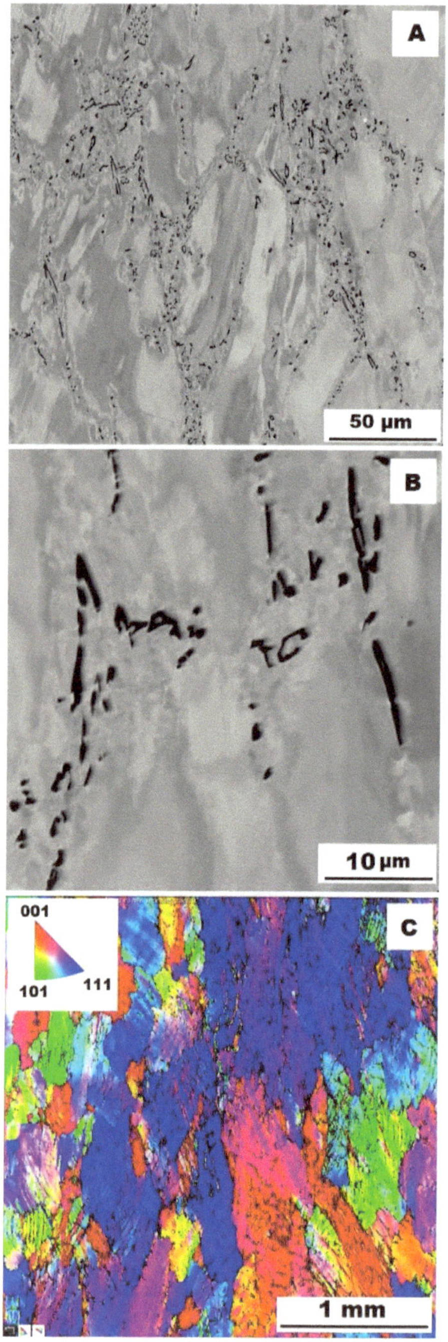

Figure 16. The structure of Fe–28Al–5Si–2Ti–1B alloy in the as-cast state after compression test at 600 °C. (**A**) Overview of the structure-10 kV, BSE. (**B**) Detail of the structure-10 kV, BSE. (**C**) EBSD map of grains-area 3 × 3 mm, step size 2 µm, 20 kV.

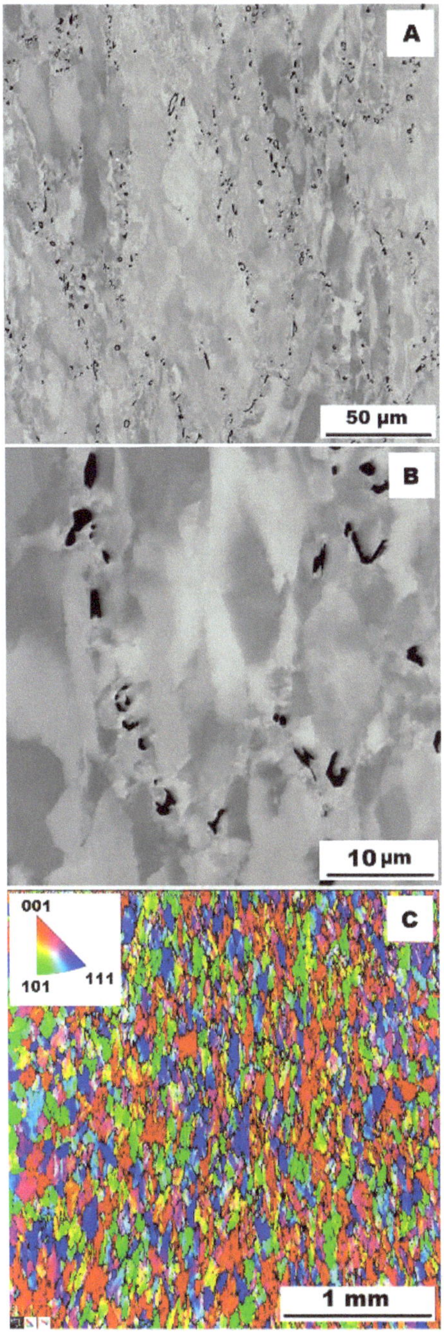

Figure 17. The structure of Fe–28Al–5Si–2Ti–1B HT 800/100 alloy after compression test at 600 °C. (**A**) Overview of the structure-10 kV, BSE. (**B**) Detail of the structure-10 kV, BSE. (**C**) EBSD map of grains-area 3 × 3 mm, step size 2 µm, 20 kV.

The decrease in yield stress values of alloys after annealing corresponds to Figures 12–17 very well. The deformed structure of as-cast samples shows higher values of yield stress due to a higher density of dislocations and, therefore, a higher stress level in the crystal lattice-the deformation strengthening is working. In annealed samples, recrystallization has taken place, so dislocation recovery has caused a loss of strengthening.

In both states, the formation of very small grains along phase boundaries could contribute to strengthening. In the as-cast state, the volume fraction of these areas is higher.

In sum, the structure recrystallized during compression tests, together with a lower amount of very small recrystallized grains near the secondary particles, can cause lower values of yield stress in the case of samples annealed at 800 °C for 100 h.

As far as the presence of gamma iron in the structure of investigated alloys, a similar phenomenon has only been described exceptionally for Fe–Al-type intermetallic alloys. In Fe–Al-type intermetallics manufactured by the SHS process, only the temporary presence of gamma iron during reactive sintering was confirmed using the in-situ XRD study of the formation of intermetallic phases [27].

The above-mentioned presence of gamma iron should be due to the limited diffusion of aluminum. There can be several reasons for this phenomenon. In Figure 6 (grain map-detail), the formation of superfine grains can be observed in the areas surrounding the "chain" of borides (or between them). The higher density of grain boundaries (and therefore the higher density of crystal lattice defects) in the areas close to boride chains should help the diffusion of Al atoms. Nevertheless, the limitations of aluminum atoms diffusion could be connected with boron alloying-as stated above (in the Introduction), it was confirmed in Fe–Al-type iron aluminides that boron atoms can segregate on grain boundaries, phase boundaries, or stacking faults. It is not excluded that a similar phenomenon may also occur in Fe_3Al-type iron aluminides. Provided that the major part of added boron is drained for borides formation, and the remaining small amount of boron tends to nanosegregation along with the interface, then boron atoms maybe preferentially occupy these grain boundaries (or defects) and deteriorate their "permeability" for diffusing atoms, thereby supporting the formation of aluminum-depleted submicron regions.

5. Conclusions

For Fe–Al–Si–X alloys, it has been verified that the addition of boron effectively prevents grain coarsening that occurs as a result of annealing.

The Ti-doped as-cast alloy reached significantly higher yield stress values at 700 and 800 °C compared to the Fe-28Al-5Si-1B alloy due to the formation of an almost continuous network structure of titanium borides. After annealing (HT 800/100), slightly higher yield stress values are achieved for both alloys doped with Ti or Mo at all tested temperatures.

Compared to the same alloys without boron [25] can be stated that the presence of borides in the structure of Ti and B doped alloy enhances high-temperature yield stress values in compression in higher temperatures, especially in the temperature range of 700°−800°C. So, in the case of Fe–28Al–5Si–2Ti–1B alloy, a continual network structure of titanium borides is more effective for alloy strengthening than a mixture of silicides and carbides in the form of residual eutectics-like areas for boron-free alloy [25].

During compression tests, the recrystallization occurs for all tested alloys annealed at 800°C for 100 h. Lower yield stress values in the case of annealed alloys are related to the recovery of dislocations due to the recrystallization during deformation. In both states—as-cast and annealed—the formation of regions of very small grains along the phase interfaces was noted during deformation, which may also contribute to strengthening.

Author Contributions: Conceptualization, V.V. and M.Š.; methodology, V.V. and M.Š.; formal analysis, M.Š., V.V., and P.H.; investigation, M.Š. and P.P.P.; data curation, P.P.P. and Š.B., writing—original draft preparation, V.V. and M.Š.; funding acquisition, P.H. All authors have read and agreed to the published version of the manuscript.

Funding: This research was funded by the project of by the Ministry of Education, Youth and Sports of the Czech Republic and the European Union-European Structural and Investment Funds in the frames of Operational Programme Research, Development, and Education-project Hybrid Materials for Hierarchical Structures, grant HyHi, Reg. No. CZ.02.1.01/0.0/0.0/16_019/0000843.

Institutional Review Board Statement: Not applicable.

Informed Consent Statement: Not applicable.

Data Availability Statement: Not applicable.

Acknowledgments: The authors wish to thank Pavel Novák, who kindly carried out the XRD phase analysis.

Conflicts of Interest: The authors declare no conflict of interest.

References

1. Mc Kamey, C.G. Iron Aluminides. In *Physical Metallurgy and Processing of Intermetallic Compounds*, 2nd ed.; Stoloff, N.S., Sikka, V.K., Eds.; Chapman & Hall: New York, NY, USA, 1996; pp. 351–391.
2. Morris, D.G.; Morris-Muñoz, M.A. The influence of microstructure on the ductility of iron aluminides. *Intermetallics* **1999**, *7*, 1121–1129. [CrossRef]
3. Liu, C.T.; Mc Kamey, C.G.; Lee, E.H. Environmental effects on room temperature ductility and fracture in Fe_3Al. *Scr. Metall. Mater.* **1990**, *24*, 385–390. [CrossRef]
4. Stoloff, N.S.; Liu, C.T. Environmental embrittlement of iron aluminides. *Intermetallics* **1994**, *2*, 75–87. [CrossRef]
5. Zamanzade, M.; Barnoush, A.; Motz, C. A review on the properties of iron aluminide intermetallics. *Crystals* **2016**, *6*, 10. [CrossRef]
6. Liu, C.T.; George, E.P.; Maziasz, P.J.; Schneibel, J.H. Recent advances in B2 iron aluminide alloys: Deformation, fracture and alloy design. *Mater. Sci. Eng. A* **1998**, *258*, 84–98. [CrossRef]
7. Bahadur, A. Enhancement of high-temperature strength and room temperature ductility of iron aluminides by alloying. *Mater. Sci. Technol.* **2003**, *19*, 1627–1634. [CrossRef]
8. Fraczkiewicz, A.; Gay, A.S.; Biscondi, M. On the boron effect in FeAl (B2) intermetallic alloys. *Mater. Sci. Eng. A* **1998**, *258*, 108–114. [CrossRef]
9. Liu, C.T.; George, E.P. Effect of aluminum concentration and boron dopant on environmental embrittlement in FeAl aluminides. *Mater. Res. Soc. Symp. Proc.* **1991**, *213*, 527–532. [CrossRef]
10. Fraczkiewicz, A. Influence of boron on the mechanical properties of B2-ordered FeAl alloys. *Mater. Trans. JIM* **2000**, *41*, 166–169. [CrossRef]
11. Cadel, E.; Lemarchand, D.; Gay, A.S.; Fraczkiewicz, A.; Blavette, D. Atomic Scale Investigation of Boron Nanosegregation in Fe Al Intermetallics. *Scripta Mater.* **1999**, *41*, 421. [CrossRef]
12. Lejček, P.; Fraczkiewicz, A. Boron segregation in intermetallics: On the possible origins of a low-level intergranular segregation. *Intermetallics* **2003**, *11*, 1053–1063. [CrossRef]
13. Cadel, E.; Fraczkiewicz, A.; Blavette, D. Suzuki effect on 001 stacking faults in boron-doped FeAl intermetallics. *Scripta Mater.* **2004**, *51*, 437–441. [CrossRef]
14. Briant, C.L. Intergranular and cleavage fracture. In *Intermetallic Compounds: Principles and Practice*, 2nd ed.; Westbrook, J.H., Fleischer, R.L., Eds.; John Wiley & Sons: UK, Chichester, 1995; Volume 1, pp. 895–910.
15. Krein, R.; Schneider, A.; Sauthoff, G.; Frommeyer, G. Microstructure and mechanical properties of Fe_3Al-based alloys with strengthening boride precipitates. *Intermetallics* **2007**, *15*, 1172–1182. [CrossRef]
16. Li, X.; Prokopcakova, P.; Palm, M. Microstructure and mechanical properties of Fe–Al–Ti–B alloys with additions of Mo and W. *Mater. Sci. Eng. A* **2014**, *611*, 234–241. [CrossRef]
17. Palm, M. Concepts derived from phase diagram studies for the strengthening of Fe–Al-based alloys. *Intermetallics* **2005**, *13*, 1286–1295. [CrossRef]
18. Schneider, A.; Falat, L.; Sauthoff, G.; Frommeyer, G. Mechanical properties of Fe–Al–M–C (MZTi, V, Nb, Ta) alloys with strengthening carbides and Laves phase. *Intermetallics* **2005**, *13*, 1256–1262.
19. Cieslar, M.; Karlík, M. Carbide formation in Zr-containing Fe3Al-based alloys. *Mater. Sci. Eng. A* **2007**, *462*, 289–293. [CrossRef]
20. Kratochvíl, P.; Vodičková, V.; Hanus, P.; Hakl, J.; Vlasák, T.; Pešička, J. High-temperature mechanical properties of Fe_3Al type alloy with combined additives TiB_2 and Zr. *Intermetallics* **2010**, *18*, 1365–1368. [CrossRef]
21. Kratochvíl, P.; Švec, M.; Vodičková, V. The Effect of Low Concentrations Nb and C on the Structure and High-Temperature Strength of Fe_3Al Aluminide. *Metall. Mater. Trans. A* **2017**, *48*, 4093–4096. [CrossRef]
22. Palm, M.; Stein, F.; Dehm, G. Iron Aluminides. *Annu. Rev. Mater. Res.* **2019**, *49*, 297–326. [CrossRef]
23. Michalcová, A.; Senčekova, L.; Rolink, G.; Weisheit, A.; Pešička, J.; Stobik, M.; Palm, M. Laser additive manufacturing of iron aluminides strengthened by ordering, borides or coherent Heusler phase. *Mater. Des.* **2017**, *116*, 481–494. [CrossRef]
24. Doucakis, T.; Kumar, K.S. Formation and stability of refractory metal diborides in an Fe_3Al matrix. *Intermetallics* **1999**, *7*, 765–777. [CrossRef]

25. Vodičková, V.; Švec, M.; Hanus, P.; Novák, P.; Záděra, A.; Keller, V.; Pazourková Prokopčáková, P. The Effect of Simultaneous Si and Ti/Mo Alloying on High-Temperature Strength of Fe_3Al-Based Iron Aluminides. *Molecules* **2020**, *25*, 4268. [CrossRef] [PubMed]
26. Švec, M.; Vodičková, V.; Hanus, P.; Pazourková Prokopčáková, P.; Čamek, L.; Moravec, J. Effect of Higher Silicon Content and Heat Treatment on Structure Evolution and High-Temperature Behaviour of Fe-28Al-15Si-2Mo Alloy. *Materials* **2021**, *14*, 3031. [CrossRef]
27. Novák, P.; Michalcová, A.; Marek, I.; Mudrová, M.; Saksl, K.; Bednarcík, J.; Zikmund, P.; Vojtěch, D. On the formation of intermetallics in Fe-Al system—An in situ XRD study. *Intermetallics* **2013**, *32*, 127–136. [CrossRef]

materials

Article

Comparison of the Influence of Two Types of Plasma Treatment of Short Carbon Fibers on Mechanical Properties of Epoxy Composites Filled with These Treated Fibers

Jana Novotná [1,*], Martin Kormunda [2], Jakub Perner [2] and Blanka Tomková [1]

[1] Department of Material Engineering, Faculty of Textile Engineering, Technical University of Liberec, 461 17 Liberec, Czech Republic
[2] Department of Physics, Faculty of Science, University of Jan Evangelista Purkyně in Ústí nad Labem (UJEP), České Mládeže 8, 400 96 Ústí nad Labem, Czech Republic
* Correspondence: jana.novotna3@tul.cz; Tel.: +420-739-546-853

Abstract: The interfacial interface between fibers and matrix plays a key role for epoxy matrix composites and short recycled randomly arranged fibers. This study used short recycled carbon fiber (RCF) as a filler. Plasma treatment was used for carbon fiber surface treatment. This treatment was performed using radio (RF) and microwave (MW) frequencies at the same pressure and atmosphere. Appropriate chemical modification of the fiber surfaces helps to improve the wettability of the carbon fibers and, at the same time, allows the necessary covalent bonds to form between fibers and the epoxy matrix. The effect of the plasma treatment was analyzed and confirmed by X-ray photoelectron spectroscopy, Raman microscopy, scanning electron microscopy, transmission electron microscopy and wettability measurements. Composite samples filled with recycled carbon fibers with low concentrations (1 wt%, 2.5 wt% and 5 wt%) and high concentrations (20 wt% and 30 wt%) were made from selected treated fibers. The mechanical properties (impact toughness, 3PB) were analyzed on these samples. It was found that the modulus of elasticity and bending stress increase with the increasing content of recycled carbon fibers. A more significant change in impact strength occurred in samples with low concentration.

Keywords: recycled carbon fiber (RCF); fiber-reinforced epoxy composites (FRE); plasma treatment; mechanical properties

check for updates

Citation: Novotná, J.; Kormunda, M.; Perner, J.; Tomková, B. Comparison of the Influence of Two Types of Plasma Treatment of Short Carbon Fibers on Mechanical Properties of Epoxy Composites Filled with These Treated Fibers. *Materials* **2022**, *15*, 6290. https://doi.org/10.3390/ma15186290

Academic Editor: Jozef Dobránsky

Received: 29 July 2022
Accepted: 7 September 2022
Published: 9 September 2022

1. Introduction

In recent years, the production of carbon fiber-filled composites has increased significantly, and thus the topic of recycling these materials has become increasingly important [1]. When developing new materials, it is necessary to consider the life cycle of the resulting products and to pay attention to the possibilities of reusing recycled raw materials. This work focuses on the use of short recycled carbon fibers as filler in epoxy composites. The shorter the fibers used as reinforcement components in composites, the more significant the role of their cross-sectional area entering the interfacial interaction. In the composites industry, the interface between fibers and matrix is of fundamental importance for the mechanical properties of fiber-reinforced composite materials. The optimal interfacial interface between fibers and matrix depends on the specific requirements of each application, and the possibility of modifying it is significant [2]. To date, there has been no published research report addressing the effect of plasma action, particularly concerning fiber ends and their cross-sections, as in this study.

Furthermore, this study analyzes the mechanical properties of epoxy composites reinforced with recycled short carbon fibers (RCF) with and without plasma treatment. Plasma treatment usually increases the adhesion between fibers and matrix, and the incorporation of the fibers into the matrix provides an ideal reinforcement effect. The mechanical properties of the RCF-filled composites were analyzed using impact toughness and flexural strength tests. These tests were chosen because, based on previous knowledge of short fiber reinforced composites, it can be assumed that the tested specimens have low ductility. In tension, the applied tensile force would act primarily on the matrix of the material, and short fiber reinforcement would not significantly help the tensile strength.

In general, the structure of carbon fibers varies depending on their precursor, with graphitic basal planes usually present on their surface, which exhibits poor wettability and low interaction with polymers [2]. Therefore, surface treatments are applied to the surface of CFs to improve their interaction with the matrix. Plasma deposition of carbon fibers is a dry reaction process; depending on the process conditions, it can have several effects that may overlap. The reason for these treatments is to increase the roughness of the fiber surface, remove impurities from the fiber surface, remove weakly bonded carbon planes or attach chemically active elements, which allows for better interfacial adhesion in composites [2–4]. The action of inert gases results in crosslinking, where two or more parallel polymer chains are linked by O_2 or He or Ar, and the by-products allow the formation of bonds to adjacent chains [5,6]. The surface properties of carbon fibers can be affected by foreign elements present on the surface of these fibers; therefore, in this study, much attention was paid to the analysis of the fiber surface using X-ray photoelectron spectroscopy (XPS).

Plasma treatment of additives can solve two basic requirements for the quality of composites: homogeneous dispersity; and a strong bond between the additive and the matrix. However, the operating conditions of plasma treatment need to be appropriately selected. For virgin CF, Morgan [7] reports the following operating conditions for a plasma reactor: 100 W, RF generator at a pressure of 2–50 Pa and an operating time of 20 s to 20 min [3,7,8]. Donnet [9] points out that different results can be achieved depending on whether the plasma is guided by radio frequency (RF) or microwave (MW).

In recycled CFs, surface changes also occur due to plasma action, as reported by [10,11]. These studies confirmed that plasma exposure causes a slight deepening of the surface roughness of recycled carbon fibers. Both studies involve a polypropylene matrix, whereas the more common epoxy matrix used in this work is an epoxy matrix. Study [9] notes a reduction in flexural strength of plasma treated RCF composite samples at higher plasma powers (200 W and 300 W) and recommends the use of 100 W for 70 s. Study [10] reports a concise plasma exposure time of 0.5 s or less at 0.8 kW at RF frequency. The experimental conditions chosen in this study are based on these data. The study by Altay [10] compares the mechanical properties of composites at 20 wt% concentration and the study by Hooseok [11] at 15% *v*/*v* concentration.

In a previous study [12], it was experimentally confirmed that along with the incorporation of new surface functional groups, changes in fiber topography could also occur, which will be further analyzed by scanning electron microscopy (SEM) and X-ray photoelectron spectroscopy (XPS).

2. Experiment

2.1. Materials

An epoxy matrix system made from a low viscosity bisphenol A-based epoxy resin L 285, together with a cyclic-aliphatic polyamine curing agent H 508, with a mixing ratio of 100:40 by weight, was used. The density of the resin was 1200 kg/m^3 and the density of the hardener was 1030 kg/m^3. Commercially available PAN-based Carbiso milled carbon fiber was used, with a bulk density of 400 g/L. The fibers were 7.0 $\pm$ 0.3 μm in diameter and 100 $\pm$ 9 μm in length. These fibers were manufactured by ELG Carbon Fiber (Coseley, UK). ELG uses dry carbon fiber waste from production and uncured prepregs and cured laminates containing CF at the end of life to produce these fibers. RCF Carbiso has been produced using a modified patented pyrolysis process. Because the fibers were recycled, they contained no sizing but may have contained trace metals (<0.5 g/1 kg).

2.1.1. Plasma Treatment of Recycled Carbon Fibers

Based on the available data presented in [7–10,13–15], the following conditions were selected for the plasma application: source power: 30 W and 100 W, time: 0.5–50 min. This work analyses two types of plasma treatments of recycled carbon fibers. The same pressure (100 Pa) is used for both treatments, but they differ in the treatment time and the frequency used.

Microwave Plasma Treatment

The first type of plasma treatment was carried out at microwave frequency (MW) under the influence of oxygen 200 sccm and hydrogen 50 sccm at a power of 100 W. A particular fluidization device adapted for very light bulk materials was used to treat the recycled carbon fibers. The device contained a unique hopper that ensures uniform dosing of the bulk filler to the plasma apparatus; the plasma action lasted 1–50 min. Small cracks were observed in the SEM images of the recycled plasma-treated carbon fibers (exposure time $\geq$10 min), as shown in Figure 1c. Based on a pilot series of specimens containing these plasma-treated recycled carbon fibers (results published in [12]), where the majority of the plasma-treated fiber-filled composite specimens showed a deterioration in mechanical property values compared to the untreated fiber-filled specimens, a series of plasma-treated fibers were fabricated where the treatment time was reduced to 1 min. After this change, no more destruction was found in the SEM images of the fibers, see Figure 1b.

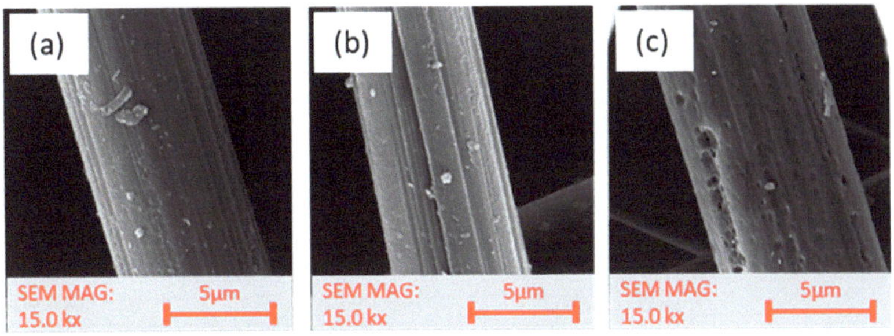

Figure 1. SEM photos of RCF: (**a**) without treatment; (**b**) MW plasma-treated for 1 min; and (**c**) MW plasma-treated for 10 min.

Radiofrequency Plasma Treatment

Based on the recommendation already published by Donnet [9], an apparatus for the treatment of powder fillers by radiofrequency (RF) plasma treatment was developed. The duration of this treatment was 30 s. No defects were observed in the SEM images of the recycled fibers treated by radiofrequency plasma treatment, as shown in Figure 2. The recycled carbon fibers were plasma treated in a fluidized bed reactor. The working gas (air) flowed into the reactor at its bottom, fluidizing the powder. Two outer ring electrodes were attached to the bottom of the reactor 1 cm apart to ignite the plasma discharge. The electrodes were connected to a Dressler Cesar 133 RF generator via a matching network. The airflow rate was 150 sccm.

The plasma treatment times (1xcycle = 30 s, 3xcycles = 90 s) and the applied power of 30 W and 100 W were compared. Based on the XPS analysis showing the elemental composition of the plasma-treated recycled carbon fibers shown in Figure 5 samples prepared using 100 W for 30 s were selected for application to epoxy composites.

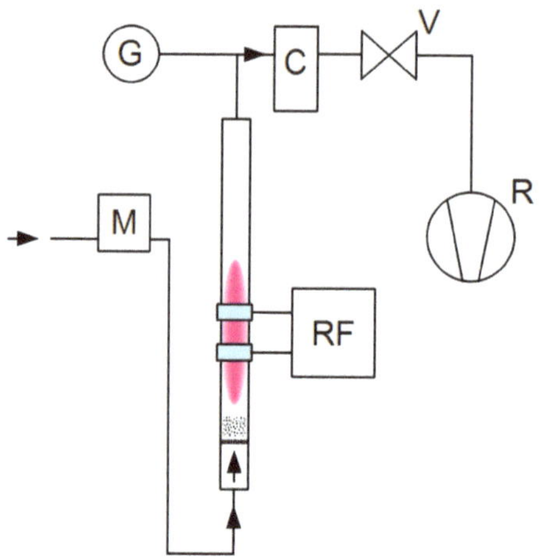

Figure 2. Diagram of RF reactor: G—vacuum gauge, C—cyclone, V—valve, M—mass flow controller, RF—generator with matching unit, R–rotary vane pump.

2.1.2. Characterization of Recycled Carbon Fibers

SEM. Characterization of the effect of plasma treatments on recycled carbon fibers was performed by scanning electron microscopy (SEM; VEGA 3 TESCAN (TESCAN ORSAY HOLDING, a.s., Brno, Czech Republic); ZEISS NEOPHOT 32 (Carl Zeiss Microscopy GmbH), Jena, Germany)

TEM. High-resolution transmission electron microscopy (HR-TEM) images were obtained using an FEI Titan electron microscope operating at 80 kV (Titan, Thermo Fischer Scientific, London, UK).

Wetting measurements were performed by determining the water contact angle (WCA). It was measured using a DSA30 droplet shape analyzer (Krüss, Berlin, Germany) with a water volume of 3 μL; RCF was spread on adhesive tape to measure the contact angle. Droplets were applied to each prepared sample and then the approximate WCA value was calculated.

XPS analysis. X-ray photoelectron spectra (XPS) were recorded using a Phoibos 100 hemispherical analyzer (from Specs) operated in FAT mode. An AlKα spectral line with an energy of 1486.6 eV was used and the spectra were referenced to aliphatic carbon bonds at 285 eV. Resolution spectra for a transition energy of 10 eV were used and the CasaXPS software (including RSF factor) [16] was used for calculations.

2.2. Preparation of FRE Composite Samples

For low-fiber concentrations (1 wt%, 2.5 wt% and 5 wt%), the epoxy, hardener and fibers were mixed for 10 min using a magnetic stirrer. At higher concentrations (20 wt% and 30 wt%), manual mixing was used. The plasma-treated RCF could be more easily mixed into the epoxy-hardener mixture, and magnetic stirring was also used for the 20 wt% concentration. The mixture was then poured into molds and left at room temperature for 24 h. Subsequently, the mixture was left in the oven for 15 h at 60 °C. The overall sample preparation process is presented in [17].

2.3. Characterization of FRE Composite Samples

SEM analysis. Characterization of the fracture surfaces of the FRE composites was performed by scanning electron microscopy (SEM; VEGA 3 TESCAN); the images are shown in Figure 6.

The impact toughness was tested according to ASTM D 256. A schematic of the test equipment is shown in Figure 3. A 2.7 J pendulum was used at an impact velocity of 3.46 m^2/s. The dimension of each specimen was 100 × 10 × 3 mm. The specimen thus prepared was placed on two supports with the larger specimen facedown and then broken with a pendulum hammer on the narrower side of the "edgewise" body. Samples with the same fiber concentration were measured 10 times, and the arithmetic mean and standard deviation was calculated from the data obtained. The impact toughness measurements A [J/m^2] are shown in Figure 8. The following formula was used for the calculation:

$$A = \frac{E_c}{h \cdot b}, \tag{1}$$

where E_c [J] is the impact energy required to break the sample, h [m] is thickness and b [m] is width of the composite samples.

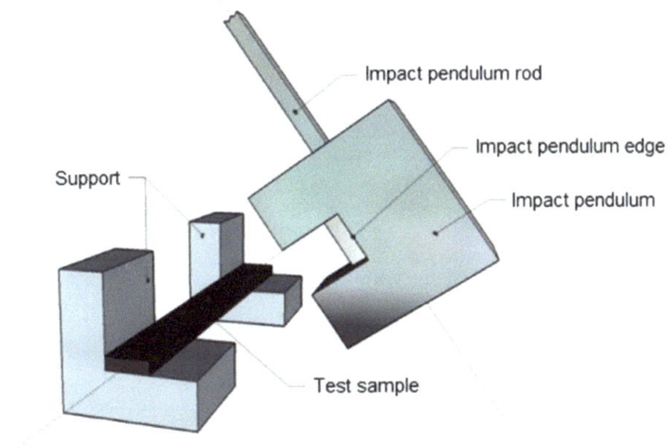

Figure 3. Schematic of the Charpy impact test.

The flexural strength was measured according to EN ISO 14125. A schematic of the test equipment is shown in Figure 4. The three-point bending test was carried out on a Tiratest 2300 testing machine. The dimension of each specimen was 100 × 10 × 3 mm. The movement of the standing transom, which applied the deformation to the test specimen, was set to a maximum displacement of 1.5 mm. Samples with the same fiber concentration were measured 10 times, and the arithmetic mean and standard deviation were calculated from the data obtained. The resulting bending stress σ [MPa] and elastic modulus E [GPa] are shown in Figure 8, and are valid:

$$\sigma = \frac{3 \cdot F \cdot l}{2\, b \cdot h^2} \tag{2}$$

where F [N] is the loading force and l [mm] is the distance between the supports. To calculate the elastic modulus:

$$E = \frac{F \cdot l^3}{4 \cdot s \cdot h \cdot b^3} \tag{3}$$

where s is the deflection [mm].

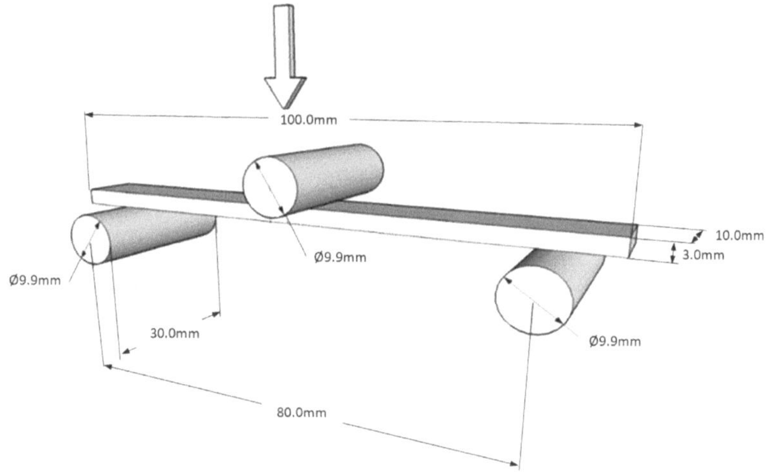

Figure 4. Schematic of the three-point bending test.

3. Results and Discussion

3.1. Analysis and Morphology of Fibers

SEM images. Together with XPS analysis, these images helped select a suitable filler type for the fabrication of the composite samples. Comparing the SEM images of RCF in Figure 1, we can see the fiber surface destruction in the plasma-treated fibers that occurred during MW plasma treatment after 10 min exposure time, similar to [12,17,18]. After MW plasma treatment after an exposure time of 1 min, deepening and highlighting the striated fiber surface occurred. Similar highlighting of the striated surface was also reported in [19–21].

Comparing the SEM images of RCF, Figure 5a with the MW plasma-treated images in Figure 5b and the RF plasma-treated images in Figure 5c, we can see that after MW plasma treatment, the striated surface of the fibers was highlighted. After both two types of plasma treatments, there was a loss of minor impurities formed on the recycled fibers during the recycling process. The SEM images (Figure 5, at the bottom) show the ends of the filaments, and when compared with each other no significant changes are apparent. The original assumption of the effect of plasma treatments on cross sections and fiber ends was not observed. However, the removal of impurities is noticeable.

TEM images. Due to the nature of TEM imaging, we focused on the ends of the fibers, where details of the longitudinal arrangement of the carbon fiber surface can be seen. No significant changes were observed between fibers with and without plasma treatment at the fiber ends. In Figure 6, the turbostratic structure of the transverse fracture arrangement of the fibers can be seen. Therefore, the original assumption of the influence of plasma treatments on the cross sections and fiber ends was not confirmed.

Measurement of wettability. The contact angle of WCA recycled carbon fiber is affected by plasma action. Figure 7 shows that it can be observed that the plasma treatment has a significant effect on the changes in the hydrophobic character of the recycled carbon fibers; both the type of plasma treatment and the duration of this treatment influence results. In the case of MW plasma, the WCA value decreased with a longer plasma treatment time. Untreated fibers exhibit hydrophobic behavior, while plasma treated fibers exhibit hydrophilic behavior.

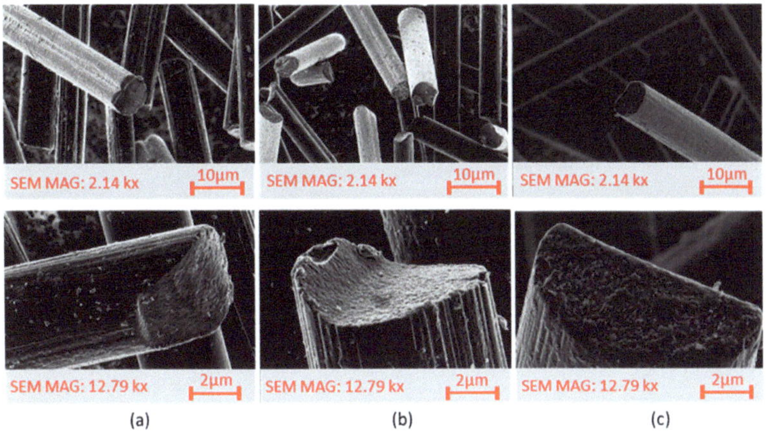

Figure 5. SEM photos of RCF (longitudinal views in the upper part of the figure, cross-sectional details in the lower part of the figure): (**a**) RCF; (**b**) MW plasma-treated for 1 min; and (**c**) RF plasma-treated for 30 s.

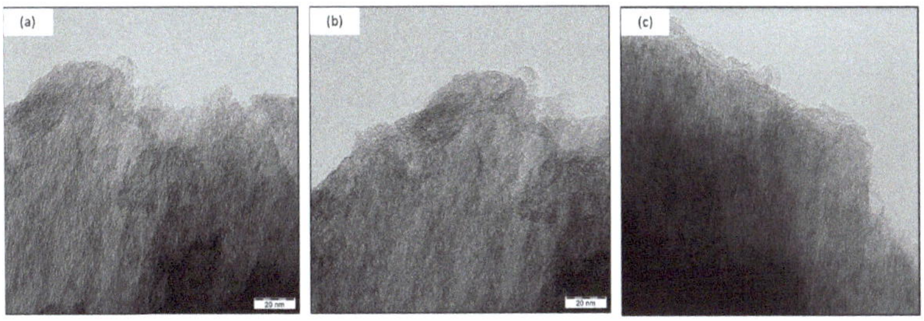

Figure 6. TEM images of the ends of the used RCF: (**a**) RCF; (**b**) MW plasma-treated for 1 min; and (**c**) RF plasma-treated for 30 s.

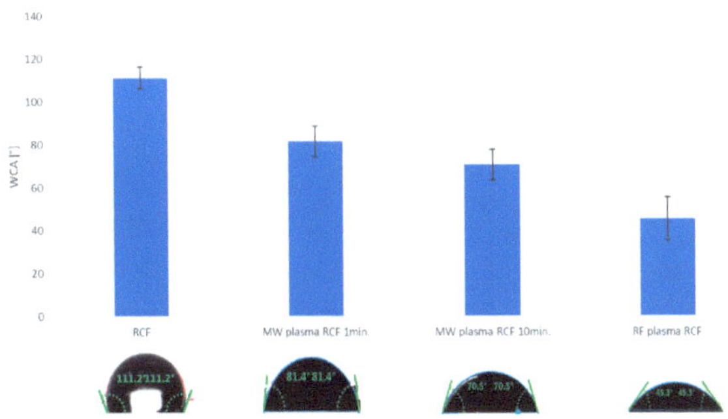

Figure 7. Measurement of wettability of RCF.

Due to the roughness of the surface, the measured contact angle is only indicative of the plasma treatment trend. The measured values show that plasma treatment of CF has a significant effect on the change of hydrophilic properties. Untreated fibers show hydrophobic behavior, while plasma treated fibers show hydrophilic behavior. The increased hydrophilicity of the fibers is probably due to the binding of oxygen functional groups to the fiber surface, together with the higher surface roughness caused by the plasma treatment.

However, in general, roughness is considered a relatively unimportant factor causing contact angle hysteresis [2]. The hysteresis is dominated by the heterogeneity of the surface chemistry; thus the reduced hysteresis may be due to the more uniform chemical composition of the plasma-treated CF surface.

Comparing the WCA results shown in Figure 7, we see that microwave plasma's longer plasma exposure time reduces the WCA angle value, i.e., the fibers are more wettable after more extended MW treatment. Nevertheless, considering the destructions demonstrated by SEM analysis, fibers with an MW treatment time of 1 min were chosen for application to composites.

Raman spectroscopy was used to assess the effect of plasma treatment on the structural changes of carbon fibers. The Raman spectra (RS) for different types of carbon structures show two peaks at approximately 1355 and 1580 cm^{-1}, corresponding to the defective carbon (D band) and graphite mode (G band) [22]. Figure 8 shows the RS of RCF and the plasma-treated RCF. The intensities of the D and G oscillations are related to the graphite crystal size and the proportion of the amorphous carbon phase. These peaks correspond to the usual D and G peaks in PAN-based carbon fibers reported by Fitzer and Rozploch [23]. Recycling of carbon fibers does not change their RS spectra and thus probably does not significantly change their surface structure. Comparing the RS results of RCF and those of plasma-treated RCF, it can be concluded that plasma treatment did not fundamentally change the crystalline structure of RCF because plasma only produces a surface effect in the nanometer range.

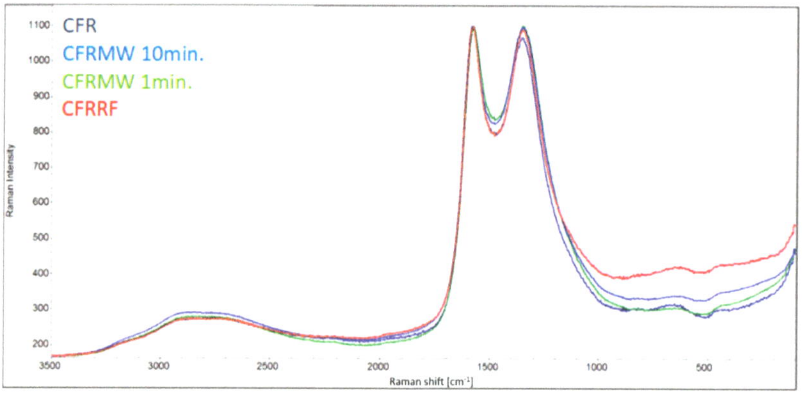

Figure 8. RS used RCF.

XPS analysis. The plasma treatments modify surface of carbon fibers significantly. The main observed effect is functionalization of fibers by introducing oxygen functionalities on the surface. The plasma treatment increased oxygen content twice from 10% on untreated fibers up to approximately 20% on plasma treated in both types of discharges. The nitrogen content is not significantly changed by the plasma treatment.

The detailed study of chemical bonding on the surface was made on high resolution photoelectron spectra, see Figure 9. Carbon related photoelectron peak C1s consists of multiple components because carbon bonds to carbon, nitrogen and oxygen are expected. There are many possible types of bonds, thus the components shall be interpreted as not comprehensive list. The main components were identified as follow: C–C at 284.5 eV, C–O–C ether-like at 286.2 eV, C=O carbonyl-like at 287.4 eV, C=O–O carboxyl/ester-like at 288.1 eV and p-p shake-up at 291.8 eV in agreement with literature [10,24]. The C to N bonds should also be present in C1s peak but due to the wide range of possible binding energies of this components, and lower content of nitrogen compare to oxygen, we have neglected C–N bonds in C1s spectra; moreover, the C–N components are overlapped by C–O components.

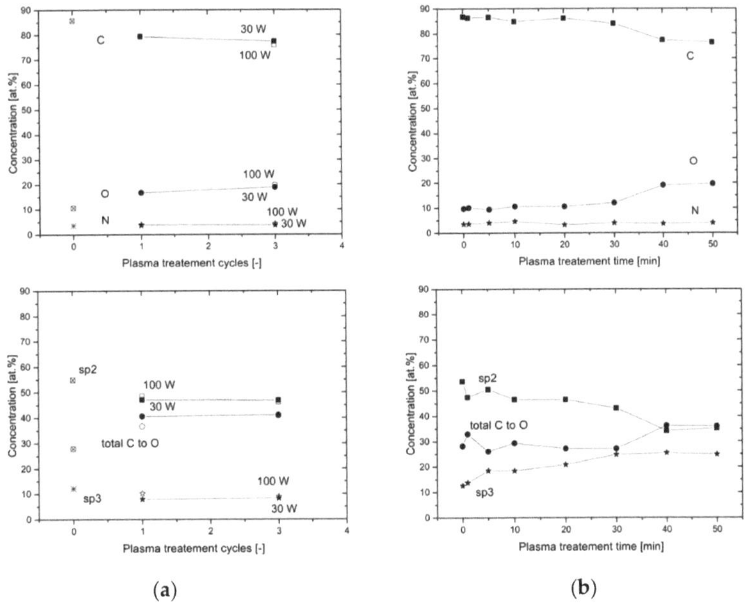

Figure 9. XPS analysis of plasma-treated RCF: (**a**) RF plasma-treated (dotted line = fibers treated at 30 W, solid line = fibers treated at 100 W); and (**b**) MW plasma-treated.

Literature reported Raman analyses showed both G and D bands typical for sp2 and defects/sp3 hybridization [25]. Therefore, we have introduced a sp3 related component to the C1s component model at binding energy about 0.8 eV shifted from the main C–C peak, then the original C–C peak acts as a sp2 related component. When the model without the sp3 component was used, the C–C component was approximately in the amount of sp2 + sp3 components. The Raman analyses in literature [25] shows an increase in ratio D/G for the plasma treated samples and a decrease for thermal treated samples. Although the C1s component for sp3 and sp2 hybridization are not equal to the D and G bands, we observe similar behavior for the MW plasma treatment ratio sp3/sp2 from XPS spectra is increasing from 0.23 up to 0.7 when RF plasma treatment leads to a decrease in ratio sp3/sp2 to 0.17 for some conditions. The RF plasma treatment is similar to thermal treatment for some materials [26]. This is probably because higher energy particles are created in the discharge and they can modify the fibers more inside with more effective heating effect than low-energy particles in the MW plasma, with dominant surface effect only.

The nitrogen related N 1 s peek has two components [24] located at around 398.5 eV and 400.5 eV binding energy. Nitrogen present in the carbon fibers is from the polyacrylonitrile (PAN) used as precursor. Even after carbonization at temperatures above 1000 °C a small amount of nitrogen remains in the fibers, and it is mostly incorporated in the graphite-like structure in two ways: either at the center of three aromatic rings, bonded to three carbon atoms (graphitic nitrogen); or bonded to two carbon atoms at the edges of the graphite-like sheets or at defects (pyridine-like nitrogen). XPS study shows significant differences between plasma treated fibers [24]. The lower binding energy component represents the pyridine-like nitrogen and this type of nitrogen is reduced by both of the studied plasma treatments. The pyridine-like nitrogen is reduced from an initial 23% to 19% in 100 W and longer 30 W RCF plasma-treated RF, and to 12% in RCF plasma-treated

MW for treatment times longer than 30 min. Therefore, we can control the N1s content on the surface in this range by plasma conditions.

The oxygen photoelectron peek O1s can be successfully fitted by three components. Where the component at lowest energy 531 eV is O=C, component at 532.9 eV is O–C similarly as in polymers [24], the final weak component about 536.4 eV is adsorbed water. The plasma treatments are significantly reducing the amount of lowest energy component O=C from an initial 40% to approximately 17% for MW or 20% for RF plasma. The simultaneously O–C component is increased in relative and also in absolute values because of the increase in the total amount of oxygen.

3.2. *Characteristics of FRE Composites*

Characteristics of fracture surfaces. SEM analyzed the surface of the fracture surfaces after impact toughness measurements. Figure 10a is a snapshot of the pure resin without filler. Figure 10b–d are microphotographs of the composites filled with recycled carbon fibers at 1 wt%.

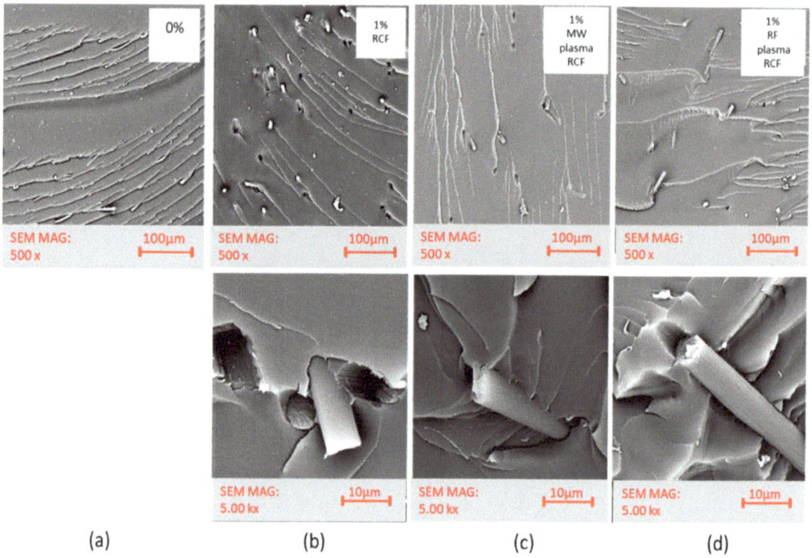

Figure 10. SEM photos of FRE after impact toughness measurements (view of the total fracture surface in the upper part of the figure, details of the fiber-resin interfacial interface in the lower part of the figure): (**a**) neat resin; (**b**) RCF–1 wt%; (**c**) MW plasma-treated RCF–1 wt%; and (**d**) RF plasma-treated RCF–1 wt%.

After impact toughness measurements on FRE composite specimens, the fracture surface images obtained by SEM were analyzed. The dispersion state of the recycled carbon fibers in the epoxy resin was evaluated. Figure 10a, pure resin, shows a brittle homogeneous fracture surface. The images of the composites containing filler in the form of RCF, in Figure 10b–d, show fibers that are dispersed continuously in the matrix. According to morphology of the carbon fibers, these short fibers can transmit stress, which leads to the absorption of external energy during mechanical testing. When cracks form in the composite, the dispersed fibers of the modified RCFs can prevent the propagation of these cracks, but only to a certain extent and at lower fiber concentrations, as also reported by Altay [10]. The improved mechanical properties of epoxy composites with plasma modified RCFs are due to the inhibition of crack growth, which was also confirmed in [27,28].

Impact toughness of FRE composites. The impact toughness results are shown in Figure 11. One of the possible disadvantages of using micro fillers in epoxy matrices is the potential formation of microscopic cracks. Microcracks in the materials can propagate under repeated loading or a combination of different loads and lead to the destruction of the composite, as pointed out by Teh [29]. If the measured values of the impact toughness of composites filled with recycled carbon fiber are compared with the impact toughness of pure resin (in Figure 11), the RCF plasma treatment improves the impact toughness of the samples with 1 wt% RCF. In contrast, plasma treatment of fibers has no significant effect on samples with high filler concentrations. Low-fiber concentrations (1 wt%–2.5 wt%) increased the tested composites' impact toughness values compared to the unfilled resin's impact toughness values. Compared to other studies [27–29], there was no stable increase in impact toughness values for the composites filled with recycled carbon fibers; the composites with the lowest filler concentration achieved the best results, while higher concentrations resulted in a deterioration of impact toughness compared to the impact toughness value of unfilled resin. This is probably due to the shape of the filler in the form of cylinders, and it may also be due to the formation of clusters after reaching a suitable fiber concentration. In contrast, at higher fiber concentrations, the influence of micro-layers reduces the impact toughness value, as also pointed out by [28].

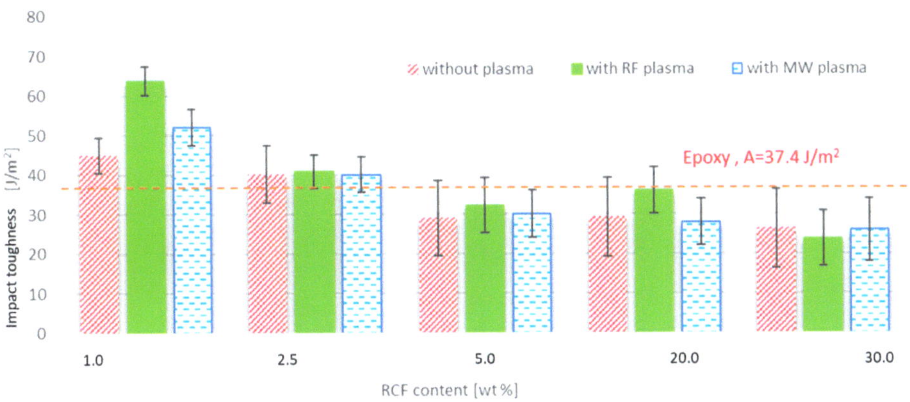

Figure 11. The impact toughness of FRE composites.

Three-point bending. The three-point bending results are shown in Figure 12. Flexural strength is the ability of material to resist a bending force applied perpendicular to the longitudinal axis of the composite specimen. The stress induced by the bending load is then a combination of compressive and tensile stresses [29]. Figure 12 compares the tested composite specimen elastic modulus and bending stress values. It can be seen that the addition of fibers to the resin improves the elastic modulus and bending stress for all fiber concentrations and types. The effect of plasma treatments is negligible except at the lowest concentrations. At low concentrations of 1 wt%, an increase in elastic modulus is observed, namely by 10% for samples with RCF fibers without plasma treatment, by 29% for samples with RCF with RF plasma treatment, and 21% for samples with RCF with MW plasma treatment. The improvement in bending stress values at the same concentration was 15% for samples with RCF without plasma treatment, 31% for samples with RCF with RF plasma treatment, and 26% for samples with RCF with MW plasma treatment. The scatter of the elastic modulus and bending stress values is lower for samples with fibers treated with both types of plasma treatments than for samples with untreated fibers. In this study, the samples are prepared using magnetic and manual mixing, which can be a problem at higher-fiber concentrations and can lead to increased fiber agglomeration and, indirectly, higher values of flexural modulus.

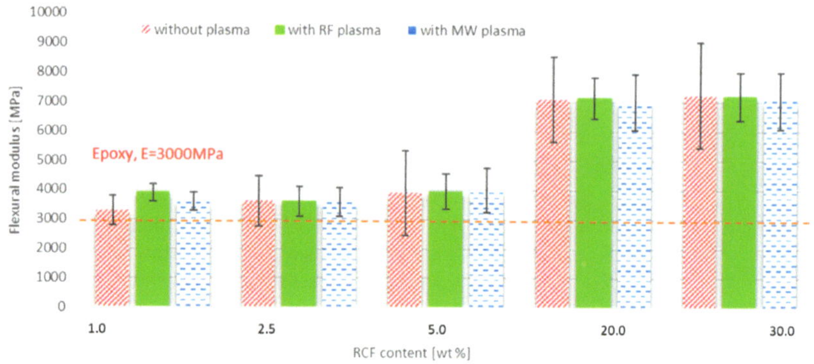

(a)

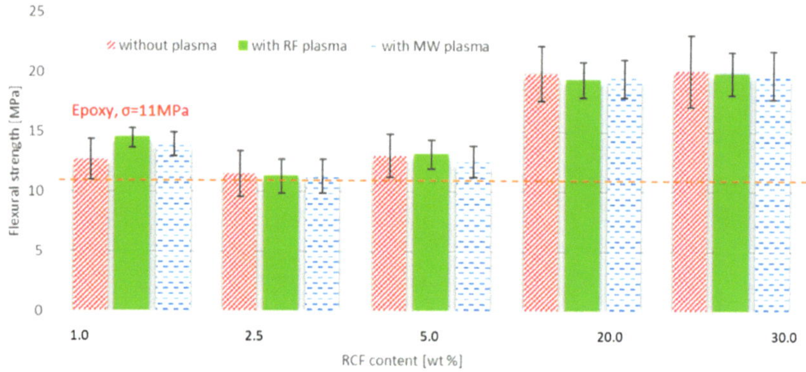

(b)

Figure 12. The flexural strength of FRE composites: (**a**) flexural modulus; and (**b**) flexural strength.

4. Conclusions

Fiber-reinforced polymers, particularly carbon fiber reinforced epoxies, are expected to account for more than 50% of the structural weight of motor vehicles in the future. As the number of applications increase, the amount of ancient carbon fiber structures will grow, and recycling of these materials will become more important [2]. Therefore, in this work, attention was paid to plasma modifications of short RCFs, to determine the suitability of applying these modifications to improve the mechanical properties of epoxy composites filled with these modified fibers.

This study analyzed the mechanical properties (represented by impact toughness and three-point bending) of epoxy composites filled with short RCFs. Two types of plasma treatment were used to modify the RCF surface. These treatments were performed using radio frequency (RF) and microwave (MW) radiation at the same pressure and atmosphere. WCA measurements and XPS analysis confirmed that plasma treatment changes the surface layers of carbon fibers, as reported in studies [10,11]. The SEM and TEM images reported in this study focused on the region of the fiber ends, and their cross-sections, since these areas form a significant part of the interfacial interface in short fiber-filled composites. As expected, based on the pilot study [12], the use of both types of plasma treatment increases the wettability of the recycled carbon fibers. Both types of plasma treatments also contribute to the ease of fiber introduction into the epoxy matrix.

The suitability of plasma treatment was verified in mechanical tests performed on composite samples prepared from these fibers. It was found that the modulus and bending stress increased with increasing recycled carbon fiber content, as also reported in the study [10,11]. However, the difference in the effect of plasma treatment on the increase in modulus and bending stress for samples with higher concentrations are negligible, compared to the data reported in [10,11]. This is probably due to the effect of the technology used to prepare the composite samples, or in this study, the samples are prepared using magnetic and manual mixing, which can be a problem for higher fiber concentrations and can lead to increased fiber agglomeration; whereas in [10,11] injection molding technology is used, but this is also due to the type of polymer matrix (PP) used.

Furthermore, in this work, the spectrum of different concentrations of epoxy-filled RCF composites has been analyzed. The increase in the adhesion between fibers and matrix is due to the increase in the surface roughness of the fibers and the oxidation of their surface. For impact toughness, it was found that there is an optimum limit. A suitable balance between elasticity and toughness was achieved for composite samples with 1 wt% RCF. and both types of plasma treatment further improved the tested mechanical properties by this concentration of RCF. However, RF plasma treatment of the fibers had a more significant effect on the improvement of impact toughness and modulus.

The selected RF plasma treatment procedure is readily applicable for plasma treatment of short recycled fibers and represents a significant advantage (for the lowest concentration samples) over the use of untreated RCF. FRE composites reinforced with plasma-modified RCF fibers can be obtained for various applications in the composites industry.

Author Contributions: Conceptualization, J.N. and M.K.; methodology, B.T.; software, J.N.; validation, J.N., M.K. and B.T.; formal analysis, B.T.; investigation, B.T., M.K. and J.P.; resources, J.N.; data curation, J.N., J.P.; writing—original draft preparation, J.N.; writing—review and editing, B.T.; visualization, J.N., J.P.; supervision, M.K.; project administration, J.N.; funding acquisition, B.T. All authors have read and agreed to the published version of the manuscript.

Funding: The authors are grateful for financial support from the Student Research Project 2020 No. 6050, supported by Czech Ministry of Education, and by Faculty of Textile, Technical University of Liberec. The authors acknowledge the assistance provided by the Research Infrastructure NanoEnviCz, supported by the Ministry of Education, Youth and Sports of the Czech Republic under Project No. LM2018124.

Institutional Review Board Statement: Not applicable.

Informed Consent Statement: Not applicable.

Data Availability Statement: Not applicable.

Acknowledgments: The authors are grateful for financial support from the Student Research Project 2020 No. 6050, supported by Czech Ministry of Education, and by Faculty of Textile, Technical University of Liberec. The authors acknowledge the assistance provided by the Research Infrastructure NanoEnviCz, supported by the Ministry of Education, Youth and Sports of the Czech Republic under Project No. LM2018124.

Conflicts of Interest: The authors declare no conflict of interest.

References

1. Borjan, D.; Knez, Ž.; Knez, M. Recycling of Carbon Fiber-Reinforced Composites—Difficulties and Future Perspectives. *Mater* **2021**, *14*, 4191. [CrossRef] [PubMed]
2. Bismarck, A.; Song, B.; Springer, J. Contact Angle Measurements on Fibers and Fiber Assemblies, Bundles, Fabrics, and Textiles. In *Surface and Interfacial Tension*; CRC Press: Boca Raton, FL, USA, 2004.
3. Corujeira-Gallo, S.; Dong, H. Effect of microstructure on the plasma surface treatment of carbon fibers. *J. Compos. Mater.* **2017**, *51*, 3239–3256. [CrossRef]
4. Morra, M.; Occhiello, E.; Garbassi, F.; Nicolais, L. Surface studies on untreated and plasma-treated carbon fibers. *Compos. Sci. Technol.* **1991**, *42*, 361–372. [CrossRef]
5. Gravis, D.; Moisan, S.; Poncin-Epaillard, F. Surface characterization of plasma-modified carbon fiber: Correlation between surface chemistry and morphology of the single strand. *Surf. Interfaces* **2020**, *21*, 100731. [CrossRef]

6. Scheffler, C.; Wölfel, E.; Förster, T.; Poitzsch, C.; Kotte, L.; Mäder, G. Influence of microwave plasma treatment on the surface properties of carbon fibers and their adhesion in a polypropylene matrix. In IOP Conference Series: Materials Science and Engineering, Proceedings of the 37th Risø International Symposium on Materials Science, Risø, Denmark, 5–8 September 2016; IOP Publishing Ltd.: Bristol, UK, 2016; Volume 139, p. 139. [CrossRef]
7. Morgan, P. *Carbon Fibers and Their Composites*; Taylor & Francis Group: Boca Raton, FL, USA, 2005; ISBN 0824709837/9780824709839.
8. Chang, T. Plasma surface treatment in composites manufacturing. *J. Ind. Technol.* **1998**, *15*, 1–7.
9. Donnet, J.; Dhami, T.L.; Dong, S.; Brendlé, M. Microwave plasma treatment effect on the surface energy of carbon fibers. *In: J. Phys. D Appl. Phys.* **1987**, *20*, 269. [CrossRef]
10. Altay, L.; Bozaci, E.; Atagur, M.; Sever, K.; Tantug, G.S.; Sarikanat, M.; Seki, Y. The effect of atmospheric plasma treatment of recycled carbon fiber at different plasma powers on recycled carbon fiber and its polypropylene composites. *J. Appl. Polym. Sci.* **2019**, *136*, 47131. [CrossRef]
11. Lee, H.; Ohsawa, I.; Takahashi, J. Effect of plasma surface treatment of recycled carbon fiber on carbon fiber-reinforced plastics (CFRP) interfacial properties. *Appl. Surf. Sci.* **2015**, *328*, 241–246. [CrossRef]
12. Novotná, J.; Tomková, B.; Šourková, H. Influence of the use of oxygen plasma for the compatibility of carbon fibers with the epoxy matrix. *Adhes. Adhesives* **2022**, *30*, 1–11.
13. Yuan, L.Y.; Chen, C.S.; Shyu, S.S.; Lai, J.Y. Plasma surface treatment on carbon fibers. Part 1: Morphology and surface analysis of plasma etched fibers. *Compos. Sci. Technol.* **1992**, *45*, 1–7. [CrossRef]
14. Jin, P.S.; Chang, Y.; Moon, C.; Seo, D.; Im-soon, I.S.; Kim, Y. A study of atmospheric plasma treatment on surface energetics of carbon fibers. *Bull. Korean Chem. Soc.* **2010**, *31*, 335–338. [CrossRef]
15. Weisweiler, W.; Schlitter, K. Surface Modification of Carbon Fibers by Plasma Polymerization. In *Carbon Fibers Filaments and Composites*; Springer: Berlin/Heidelberg, Germany, 1990; pp. 263–274. [CrossRef]
16. Fairley, N.; Fernandez, V.; Richard-Plouet, M.; Guillot-Deudon, C.; Walton, J.; Smith, E.; Flahaut, D.; Greiner, M.; Biesinger, M.; Tougaard, S. Systematic and collaborative approach to problem solving using X-ray photoelectron spectroscopy. *Appl. Surf. Sci. Adv.* **2021**, *5*, 100112. [CrossRef]
17. Novotná, J.; Müllerová, J.; Tomková, B. Dielectric analysis of composite materials with recycled carbon fibers. Influence of Filler Size and Structure on the electrical Properties of Carbon fiber reinforced composites. In Proceedings of the 12th International Conference on Nanomaterials - Research & Applicatio, Brno, Czech Republic, 21–23 October 2020. [CrossRef]
18. Tang, L.; Kardos, J.L. A review of methods for improving the interfacial adhesion between carbon fiber and polymer matrix. *Polym. Compos.* **1997**, *18*, 100–113. [CrossRef]
19. Tiwari, S.; Bijwe, J. Surface Treatment of Carbon Fibers—A Review. *Procedia Technol.* **2014**, *14*, 505–512. [CrossRef]
20. Donnet, J.B.; Brendle, M.; Dhami, T.L.; Bahl, O.P. Plasma treatment effect on the surface energy of carbon and carbon fibers. *Carbon* **1986**, *24*, 757–770. [CrossRef]
21. Dilsiz, N.; Erinç, N.K.; Bayramli, E.; Akovali, G. Surface energy and mechanical properties of plasma-modified carbon fibers. *Carbon* **1995**, *33*, 853–858. [CrossRef]
22. Montes-Morán, M.A.; Young, R.J. Raman spectroscopy study of HM carbon fibers: Effect of plasma treatment on the interfacial properties of single fiber/epoxy composites. *Carbon* **2002**, *40*, 845–855. [CrossRef]
23. Fitzer, E.; Rozploch, F. Laser Raman spectroscopy for determination of the C–C bonding length in carbon. *Carbon* **1988**, *26*, 594–595. [CrossRef]
24. Hinterreiter, A.P.; Duchoslav, J.; Kehrer, M.; Truglas, T.; Lumetzberger, A.; Unterweger, C.; Fürst, C.; Stifter, D. Determination of the surface chemistry of ozone-treated carbon fibers by highly consistent evaluation of X-ray photoelectron spectra. *Carbon* **2019**, *146*, 97–105. [CrossRef]
25. Nohara, L.B.; Petraconi Filho, G.; Nohara, E.L.K.; Urban, M.; Rezende, M.C. Evaluation of carbon fiber surface treated by chemical and cold plasma processes. *Mater. Res.* **2005**, *8*, 281–286. [CrossRef]
26. Čapková, P.; Matoušek, J.; Rejnek, J.; Bendlová, N.; Pavlík, J.; Kormunda, M.; Šplíchalová, L.; Pilařová, V. Effect of plasma treatment on structure and surface properties of montmorillonite. *Appl. Clay Sci.* **2016**, *129*, 15–19. [CrossRef]
27. Lee, S.; Jeong, E.; Lee, M.Y.; Lee, M.; Lee, Y. Improvement of the mechanical and thermal properties of polyethersulfone-modified epoxy composites. *J. Ind. Eng. Chem.* **2016**, *33*, 73–79. [CrossRef]
28. Lam, C.; Cheung, H.; Lau, K.; Zhou, L.; Ho, M.; Hui, D. Cluster size effect in hardness of nanoclay/epoxy composites. *Compos. Part B 3* **2005**, *36*, 263–269. [CrossRef]
29. Teh, P.L.; Jaafar, M.; Akil, H.M.; Seetharamu, K.N.; Wagiman, A.N.R.; Beh, K.S. Thermal and mechanical properties of particulate fillers filled epoxy composites for electronic packaging application. *Polym. Adv. Technol.* **2008**, *19*, 308–315. [CrossRef]

materials

Article

New Materials Based on Polyvinylpyrrolidone-Containing Copolymers with Ferromagnetic Fillers

Oleksandr Grytsenko [1], Ludmila Dulebova [2,*], Emil Spišák [2] and Bohdan Berezhnyy [1]

1 Department of Chemical Technology of Plastics Processing, Lviv Polytechnic National University, 12, St. Bandera Str., 79013 Lviv, Ukraine; oleksandr.m.grytsenko@lpnu.ua (O.G.); bohdan-volodymyr.berezhnyi.mnkhtm.2021@lpnu.ua (B.B.)

2 Department of Technologies, Materials and Computer Aided Production, Technical University of Kosice, 74 Mäsiarska, 04001 Kosice, Slovakia; emil.spisak@tuke.sk

* Correspondence: ludmila.dulebova@tuke.sk; Tel.: +421-55-602-3544

Abstract: The article investigates the peculiarities of the effect of ferromagnetic fillers (FMFs) of various natures (Ni, Co, Fe, FeCo, $SmCo_5$) on the formation of the structure and properties of 2-hydroxyethylmethacrylate (HEMA) with polyvinylpyrrolidone (PVP) copolymers. The composites were characterized using FTIR-spectroscopy, SEM, DMTA, magnetometry of vibrating samples, specific electrical resistivity studies, and mechanical and thermophysical studies. The formation of a grafted spatially crosslinked copolymer (pHEMA-gr-PVP) was confirmed and it was established that the FMF introduction of only 10 wt.% into the copolymer formulation increased the degree of crosslinking of the polymer network by three times. The surface hardness of composites increased by 20–25%. However, the water content decreased by 16–18% and lay within 42–43 wt.%, which is a relatively high number. The heat resistance of dry composites was characterized by Vicat softening temperature, which was 39–42 °C higher compared to the unfilled material. It was established that the obtained composites were characterized by a coercive force of 200 kA $\times$ m^{-1} and induction of a magnetic field at the poles of 4–5 mT and 10–15 mT, respectively. The introduction of FMF particles into pHEMA-gr-PVP copolymers, which, in the dry state, are dielectrics, provides them with electrical conductivity, which was evaluated by the specific volume resistance. Depending on the FMF nature and content, as well as their orientation in the magnetic field, the resistance of filled materials could be regulated within 10^2–10^6 Ohm·m. Therefore, the modification of HEMA with PVP copolymers by ferromagnetic fillers of various natures provides the possibility of obtaining materials with unique predicted properties and expands the fields of their use, for instance as magnetic sorbents for various applications, as well as the possibilities associated with their being electrically conductive materials that can respond by changing of electrical conductivity, depending on various factors.

Keywords: hydrogels; composite hydrogels; ferromagnetic fillers; magnetic hydrogels; copolymers; polyvinylpyrrolidone; 2-hydroxyethylmethacrylate

Citation: Grytsenko, O.; Dulebova, L.; Spišák, E.; Berezhnyy, B. New Materials Based on Polyvinylpyrrolidone-Containing Copolymers with Ferromagnetic Fillers. *Materials* **2022**, *15*, 5183. https://doi.org/10.3390/ma15155183

Academic Editor: Franz E. Weber

Received: 22 June 2022
Accepted: 22 July 2022
Published: 26 July 2022

1. Introduction

The processes of modification of polymeric materials with metal fillers of different natures are the subject of numerous studies, the result of which is the creation of new materials [1] with unique properties for different fields of science [2] and practice [3]. Nowadays, metal-filled hydrophilic polymers and hydrogels based on them are a new advanced class of polymer composite materials [4,5]. Polymer hydrogels are characterized by a unique porous structure [6] with the presence of hydrophilic functional groups, which provides swelling of the polymer matrix in water and other polar solvents, as well as high permeability to dissolved low molecular weight substances [7]. Composite metal-filled polymer hydrogels consist of at least two components, a hydrogel matrix and a fine metal filler. Basically, the hydrogel matrix provides strength, resilience, sorption capacity and processability of the composite obtaining process [8], while the introduced filler particles provide many functional capabilities, such as redox [9], catalytic [10], magnetic [11], optical [12], electrically conductive [13], bactericidal and antifungal activities and properties [14] among others. The combination of the properties of the hydrogel matrix and metal filler of different natures creates unique synergistic effects that are absent in the individual components [15]. In many cases, it is necessary to ensure a change in the behavior of the hydrogel depending on the action of external factors, such as temperature, pressure, pH of the medium, etc. One of the remotely controlled methods of the hydrogel state is to integrate the sensitivity to magnetic fields [16].

Hydrogels that are magnetic and those that are sensitive to the magnetic field are usually obtained by introducing into ferromagnetic fillers (FMFs), such as Fe [17], Ni [18], Co [19], Fe_2O_3 [20], Fe_3O_4, into the hydrophilic polymer matrix [21]. Magnetic properties, in combination with biocompatibility and sorption capacity, provide unique functional capabilities to composite hydrogels, which are used in a wide range of practical applications in the biomedical engineering field [22], including tissue engineering [23], drug delivery and release [24], cancer therapy [25], biosensors [26], etc. FMF-filled hydrogels, due to their unique properties and ability to respond to external stimuli of various natures, are ideal materials for the manufacture of detectors and sensors [16]. On the basis of magnetic hydrogels, polymer-immobilized catalysts are obtained, which possess high selectivity and stability of action, and are able to accelerate the reactions of decomposition, hydrogenation, oxidation, isomerization, etc. [27,28]. Magnetic hydrogel adsorbents are mainly used for heavy metal removal, separation, destruction and adsorption of oil, dyes, toxic organic compounds, several biomolecules and drugs [20]. FMF-filled hydrogels are used to clean the inner surface of oil and gas pipelines, as well as to apply anti-corrosion solutions to the inner surface of pipelines before their operation [29]. The properties of magnetic hydrogels (e.g., magnetic response) rely on several factors, including the type of hydrogel and FMF used, the hydrogel and FMF concentration, and the size and distribution of the FMF within the hydrogels [22].

Methods of obtaining hydrogel magnetic composites are constantly being technologically improved and refined. For example, a promising method is polymerization with simultaneous precipitation of metal, which consists in combining the processes of reduction of metal ions and the synthesis of the polymer matrix [30]. At the same time, one of the main factors influencing the choice of the field of use of FMF-filled hydrogels is the control over the size and shape of the filler particles [31]. Therefore, hydrogels filled with nanoscale ferromagnets, often referred to as nano heterogeneous composite ferrogels or nanocomposite ferrogels, are of particular interest at present [32]. However, methods for producing hydrogels filled with ferromagnet nanoparticles are usually technologically complex, especially on an industrial scale. For example, the filling, during which FMF particles are formed in the polymer matrix by one of the chemical methods, is at least a two-step process [10] and is recommended to be performed when the hydrogel is in the form of a thin film or balls with a diameter of nano- and micro-sizes. During the obtainment of composite hydrogels by the method of polymerization with simultaneous precipitation of the metal, by-products of the reduction reaction are isolated, which affects the increase of the macro-porosity of the polymer matrix [18]. This, in turn, leads to a significant reduction in the strength and resilient characteristics of the composite. At the same time, methods that include the stage of chemical reduction of metal ions (both in the polymer network and during polymerization with simultaneous reduction) require the stage of purification (washing) of hydrogels from by-products of the reduction reaction. A significant disadvantage is the use of toxic reducing agents, which often limits the use of metal-filled hydrogels, especially in the medical field.

For some applications, such as the obtainment of magnetic sorbents or oil and gas pipe cleaners, filling the composite with a nanoscale filler is not always a basic requirement. A disadvantage of such particles may be that the magnetic force is small due to the small volume, so that viscous forces dominate and magnetic separations can take a long time (tens of minutes). This explains the interest in using larger magnetic particles (typically 0.2–5 μm in diameter) for some applications. Such microparticles demonstrate strong response to an external magnetic field. Also, they possess remanent magnetization and coercivity [29]. Therefore, in some cases, it is necessary to use more technologically economical and simpler methods. Currently, there is insufficient information in scientific sources concerning the filling of polymer hydrogels with micro-sized powders of metals. Due to the spatially crosslinked structure of hydrogels, the introduction of prepared FMF particles into their composition is possible by a limited number of methods, namely, during the formation of the hydrogel matrix by polymerization or by precipitation from solution [33,34]. Filling during polymerization is the simplest method from a technological point of view. During the obtainment of hydrogel composite prepared FMF particles are used, which are dispersed in a monomer (or reaction composition) followed by polymerization [33,34]. The technological design of the method does not require expensive equipment. However, the filling during polymerization is accompanied by sedimentation of the filler particles. Sedimentation causes inhomogeneity of the composite in volume and, as a consequence, the appearance of anisotropy of its properties. Therefore, in each case, depending on the nature of the reaction composition and FMF, the dispersity of FMF, it is necessary to make technological decisions to obtain the composite material with the necessary structure and set of properties. The solution to this problem may be the structuring of FMF during the polymerization process of the filled composition in a magnetic field [35]. Also, at the polymerization stage, the filler is introduced at the end of the pot life time of the reaction composition (the time when the composition remains fluid) [36]. During polymerization, high-reactive compositions having a polymerization process of short duration are used for filling; for example, polymer-monomer compositions based on polyvinylpyrrolidone (PVP) with 2-hydroxyethylmethacrylate (HEMA) in the presence of metal ions of variable oxidation state [37].

Unique technological and operational characteristics of copolymers HEMA with PVP (pHEMA-gr-PVP) are provided by PVP [38]. An important characteristic of PVP is its ability to form complexes with low molecular weight substances [39], including monomers [40]. The role of PVP complexes with vinyl monomers is particularly noteworthy [41]. Specifically, this ability has a significant impact on the kinetics of polymerization of HEMA in the presence of PVP (through the formation stage of a complex with charge transfer) and the formation of the polymer matrix structure [42]. At the same time, the use of PVP opens up additional opportunities for modification of various substances [43,44] and improvement of technologies for obtaining new functional materials with a set of unique properties [45,46].

The work [42] presents the study results of the structure and properties of pHEMA-gr-PVP copolymers and their hydrogels obtained by bulk polymerization in the presence of ferrous sulfate (II). The course of grafting polymerization of HEMA on PVP with the formation of a crosslinked copolymer was confirmed by chemical analysis, IR spectroscopy, TGA and DTA. It was established that the directed change of the original composition formulation is an effective way to regulate the structure (copolymer formulation and network density), and, accordingly, the physico-mechanical, transport-diffusion and thermophysical properties of copolymers. As the PVP content in the original composition increases, its grafting efficiency and the crosslinking density of the polymer network decrease. In addition, hardness, heat resistance and sorption capacity of the copolymers in the dry state increase, as well as ion permeability and elasticity in the swollen state, but the tensile strength of the swollen hydrogels decreases. Copolymerization of HEMA with PVP under the action of $FeSO_4$ occurs at high rate in air at room temperature without additional vacuuming, which greatly simplifies the technology of obtaining copolymers and hydrogels based on them, including those filled with FMF particles.

At the same time, the prospects of practical uses of pHEMA-gr-PVP copolymers filled with FMF (FMF/pHEMA-gr-PVP) provide such characteristics of the polymer matrix as large porous structure, swelling ability, sorption capacity of low molecular weight compounds, resistance to aggressive mediums, mechanical strength and elasticity [38,47].

Certainly, the modification of pHEMA-gr-PVP copolymers with ferromagnetic fillers will be accompanied by synergistic effects, the consequence of which is the acquisition of unique properties by composite materials. It is possible to predict that the presence of a metal surface at the synthesis stage of the copolymers will affect the course of the polymerization process, the formation of the polymer matrix structure and, hence, the properties of the composite. Taking into account the processability of HEMA/PVP compositions, their high reactivity in the presence of $FeSO_4$, the unique properties of pHEMA-gr-PVP copolymers, as well as the lack of works related to the study of composites based on FMF/pHEMA-gr-PVP copolymers, research in this direction is relevant. Knowledge of the peculiarities of the effect of FMF on the structure and properties of pHEMA-gr-PVP copolymers will open up prospects of creating new composite materials based on them and expand the fields of their use.

The aim of the work was to investigate the features of the influence of fine powders of ferromagnetic fillers with different natures on the formation of the structure and on the properties of composite pHEMA-gr-PVP copolymers.

2. Materials and Methods

2.1. Materials

The following substances were used: 2-hydroxyethylmethacrylate (Sigma Chemical Co., Saint Louis, MO, USA), which was purified and distilled in a vacuum (residual pressure = 130 N/m^2, TB = 351 K); polyvinylpyrrolidone (AppliChem GmbH, Darmstadt, Germany) of high purity with MM 12,000, which was dried at 338 K in a vacuum for 2–3 h before use; iron (II) sulfate of p.a. grades. For the preparation of the reaction composition, considering its processability, HEMA in the amount of 70–80 mass parts and PVP in the amount of 20–30 mass parts were used. The lower limit of the PVP content was driven by the fact that its lower content significantly increases the hardening time of compositions, especially metal-filled compositions, which were hardened without additional polymerization initiators. The upper limit was due to technological complications. A higher content of PVP increases the duration of its dissolution in HEMA, and increases the viscosity of the composition, which becomes difficult to dose. Fine powders of metals (Ni, Co, Fe) and their alloys (FeCo, $SmCo_5$) with a particle size of 10–50 μm were used as FMF for research.

2.2. *Synthesis Technique of pHEMA-gr-PVP Copolymers and FMF-Filled Composites Based on Them*

Copolymerization of HEMA with PVP was carried out in bulk with the presence of iron sulfate (II) at room temperature [42]. The calculated amount of PVP and $FeSO_4$ were weighed according to the given formulation. HEMA was dosed by the volumetric method. Into 1/3 part of the HEMA amount, required for polymerization, a certain amount of $FeSO_4$ (0.05 wt.%) was dissolved. The required amount of PVP was dissolved in the rest of the HEMA content. Stirring was performed at room temperature until there was complete dissolution of $FeSO_4$ and PVP in HEMA. Solutions of $FeSO_4$ and PVP in HEMA were mixed to obtain a homogeneous mixture, without insoluble agglomerates and mechanical inclusions, which was processed by the casting method into a polymerization mold. The obtained products were rinsed in distilled water until there was complete removal of unreacted HEMA and PVP [42].

Polymerization of HEMA with PVP compositions filled with FMF powders was performed with the presence of iron (II) sulfate and without $FeSO_4$. By mixing the required amounts of HEMA with PVP for polymerization, a polymer-monomer composition (PMC) was obtained, to which the required amount of filler powder was added. If necessary, $FeSO_4$ was previously dissolved in HEMA. The obtained PMC with the filler was stirred occasionally during the time $\tau < \tau_{p.l.}$ ($\tau_{p.l.}$—the pot life time of the composition—the time during which the reaction composition retains fluidity), after which it was dosed into a polymerization mold. The polymerization was carried out at room temperature for 0.3–1 h (depending on the composition formulation, nature and amount of FMF).

2.3. *Measurements and Characterization*

2.3.1. Investigation of Polymerization Kinetics of HEMA/PVP Compositions

The kinetics of polymerization on the initial stages (polymer yield up to 10 wt.%) was investigated by the dilatometric method by changing the volume of the reaction mixture. A 5 cm^3 glass dilatometer with uniform stirring was used for the experiment. The change in volume of the reaction mixture during polymerization was registered optically with a cathetometer [48]. The polymerization reaction rate (V) was calculated with monomer conversion of 5–10% from the following dependence:

$$V = \frac{C_M \cdot A}{\tau}, \quad (1)$$

where: C_M—Monomer concentration, mol/L; A—Monomer conversion; τ—Polymerization time, s.

2.3.2. Standard Methods of Instrumental Research

Attenuated total reflectance Fourier transform infrared (ATR FTIR) spectra were obtained using a PARAGON 1000 FTIR spectrometer equipped with a single-horizontal Golden Gate ATR cell (Perkin-Elmer, Waltham, MA, USA). Samples were used in the form of pellets. The spectra were recorded after 16 scans, at a resolution of 4 cm^{-1}, within the range of 500 to 4000 cm^{-1}. The morphology of samples in the dry state was studied by scanning electron microscopy (SEM). Research was done using a scanning electron microscope-microanalyzer PEMMA-102-02 (JSC "SELMI", Sumy, Ukraine). The range of accelerating voltage change was 0.2–40 keV, the range of magnification change was 10–300,000, and resolution was no more 5.0 nm.

2.3.3. PVP Grafting

The method of determining PVP was based on the formation of its colored complex with iodine. To determine the PVP that entered into the grafting reaction with the monomer, the amount of unbounded PVP in the copolymer was found by studying its water extracts according to the procedure described in [49].

2.3.4. The Molecular Weight between Crosslinks in Polymer Network

The molecular weight between crosslinks in the polymer network (M_C, kg/mol) was determined using the Flory-Rehner method [50].

2.3.5. Physico-Mechanical Characteristics of FMF/pHEMA-gr-PVP Copolymers

Dynamic mechanical characteristics, in terms of the tangent of an angle of mechanical losses $tg\delta$ and the storage modulus G', were studied by the method of dynamic mechanical thermal analysis (DMTA) with the help of a rheometer, Rheometric ARES (Advanced Rheometric Expansion System) (Rheometric Scientific, Inc., Piscataway, NJ, USA), using resilient oscillating deformation with a frequency of 1 Hz in the temperature range 25–200 °C and a heating rate of 3 °C/min. The surface hardness of dry samples (with diameters Ø 12 ± 0.1 mm and height h = 5 ± 0.01 mm) was found according to the conic yield point in a Höppler consistometer (VEB Pruefgeraetewerk, Medingen, Dresden, Germany) at 293 °K by pressing a steel cone with a vertex angle of $58°08'$ into a polymeric specimen under a load of 5.0 kg for 60 s [18]. Boundary water adsorption (W, wt.%) and hydrogel swelling factor (k) were determined according to the procedures described in [37]. Deformation and elastic characteristics of samples in swollen state, such as hardness number (H, MPa), elasticity index (E, %) and plasticity index (P, %), were determined according to the procedures described in [37] and ASTM D2240-15 "Standard Test Method for Rubber Property—Durometer Hardness".

2.3.6. Thermophysical Properties

Determination of heat resistance of copolymers was carried out according to the method ISO 306:2013 "Plastics—Thermoplastic materials—Determination of Vicat softening temperature", which consisted of determining the temperature at which a standard indenter, with a flat bottom surface (Ø 1.128 ± 0.008 mm), was pressed under the action of the load to a depth of 1 mm in the test sample, which was heated at a constant speed. Studies were performed by using a Höppler consistometer [42].

Thermomechanical studies were performed according to ISO 11359-1:1999 "Plastics. Thermomechanical analysis (TMA). Part 1. General principles", the method of which consists in determination of the polymer sample deformation under the action of a permanent load, which is registered as a function of temperature in the form of a thermomechanical curve (TMC). The TMC was obtained on a Höppler consistometer by recording the deformation of the sample (4 mm thick tablets) due to the action on the rod (area 23.7 mm^2) of a load of 5.0 kg. The initial temperature of the study was 293 K. Subsequent deformation measurements were carried out after increasing the temperature every 3 °C. The heating rate was set to ≈1 °C/min. The deformation of the samples (ε) was determined by the formula:

$$\varepsilon = \frac{\Delta l}{h} \cdot 100\% = \frac{l - l_0}{h} \cdot 100\%,, \quad (2)$$

where: l_0—Load-free indicator readings (before lowering the rod), mm; l—Indicator readings after 10 s of holding under load, mm; h—Height of a tablet, mm.

2.3.7. Conductivity

The conductivity characteristics of FMF/pHEMA-gr-PVP composites were estimated by the specific volume resistivity (ρ_v, Ω···m) according to ASTM D4496-21 "Standard Test Method for D-C Resistance or Conductance of Moderately Conductive Materials". Samples of cylindrical shape with a diameter of (12 ± 0.1) mm and a height of (5 ± 0.1) mm were used for ρ_v study. Specific volume resistivity was calculated as [18]:

$$\rho_v = \frac{R_v \cdot S}{h}, \quad (3)$$

where: R_v—Volume electrical resistance, Ω; S—Square of the sample, m^2; h—Sample thickness, m.

Samples of composites obtained by polymerization filling in a magnetic field were investigated in the form of a cube with an edge size of 5 mm. The resistance of such samples was measured in two directions, parallel and perpendicular to the magnetic field lines (FMF particle chain construction).

2.3.8. Magnetic Properties

Magnetic characteristics (specific saturation magnetization, coercivity) of FMF/pHEMA-gr-PVP composites were determined using a magnetometer MPMS-XL5 (Quantum Design, San Diego, CA, USA) with a vibrational sample. For analysis, samples of a cylindrical shape with a diameter of 8 mm and height of 3 mm were used, and a standard carbonyl iron was used [18].

3. Results and Discussion

3.1. Features of FMF Effect on the Formation of the Structure of HEMA-gr-PVP Copolymers

During the filling of HEMA/PVP compositions with FMF powders at the polymerization stage, it was proved that the copolymerization of HEMA with PVP occurred at a higher rate than the polymerization only under the influence of $FeSO_4$ (Figure 1).

At the same time, it was found that although the polymerization of compositions containing FMF occurred at a lower rate, it occurred without additional catalysts, which indicated active participation of the filler surface in initiating of the polymerization (Figure 1) [36]. Therefore, the synthesis of FMF/pHEMA-gr-PVP copolymers could be carried out in two ways, by polymerization with simultaneous filling and by polymerization filling. Polymerization with simultaneous filling consists in mechanical mixing of the original composition with the filler and its subsequent hardening under the action of initiators or catalysts (in this case, in the presence of $FeSO_4$). Polymerization filling differs, because the synthesis of the polymer matrix occurs without using an additional initiator or catalyst. The method consists in polymerization in the presence of metal particles, which, on the one hand, play the role of a filler, and on the other, act as the catalyst of the polymer formation reaction. Polymerization filling is currently a relatively new direction in the synthesis of metal-filled hydrophilic polymers and hydrogel materials based on them [51].

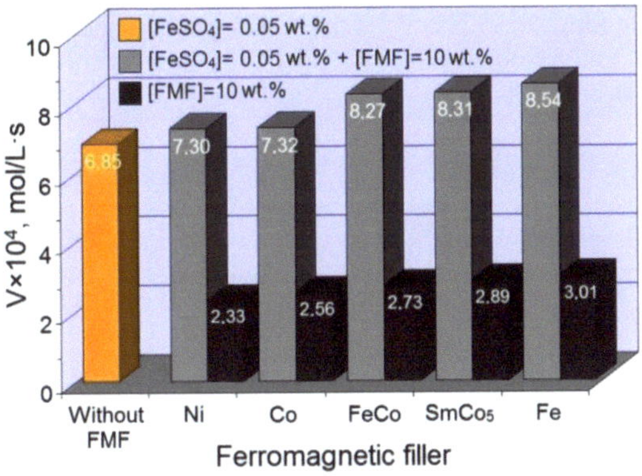

Figure 1. Effect of FMF nature on the initial polymerization rate of HEMA/PVP compositions (HEMA:PVP = 80:20 mass parts).

During the polymerization filling, the introduction of the filler into the polymer was carried out directly at the stage of its synthesis and the metal surface present in the zone of the polymerization reaction affected the whole process of polymer formation. Naturally, the introduction of filler in pHEMA-gr-PVP copolymers at the stage of synthesis affected the formation of their structure, and hence their properties. In order to define the influence of the surface of the metal filler on the formation of the structure of the polymer matrix, the copolymerization of HEMA with PVP was carried out without an additional initiator $FeSO_4$. IR spectra of PVP, HEMA and copolymers of PVP with HEMA obtained in the presence of FMF are presented (using the example of Ni and Co) in Figure 2.

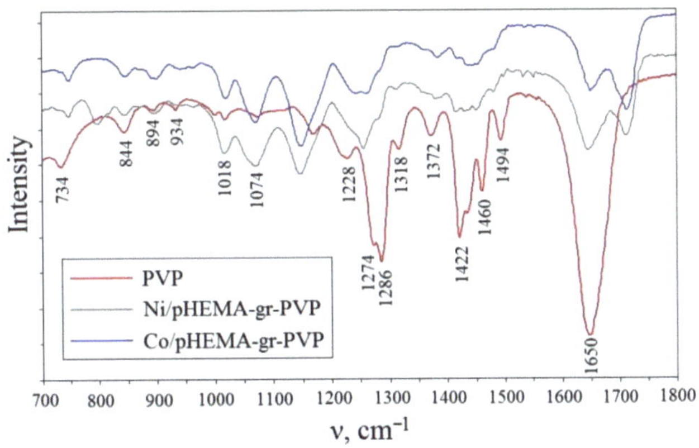

Figure 2. ATR FTIR spectra of PVP and FMF/pHEMA-gr-PVP composites (HEMA:PVP = 80:20 mass parts, FMF content—[FMF] = 10 wt.%).

Analysis of the obtained spectra of copolymers extracted with water until complete removal of unreacted components showed that, in the spectra of filled copolymers, the characteristic bands of PVP were presented; in particular, in the regions of 844 cm^{-1}, 1170 cm^{-1}; 1320 cm^{-1}; 1460 cm^{-1}, 1650 cm^{-1} [52]. This fact indicated the presence of PVP units in the copolymer. The hydrogen abstraction from tertiary carbon during the graft polymerization was indicated by a change in the intensity of adsorption band in the range 1318–1320 cm^{-1} (Figure 2), which decreased significantly in the case of transition from PVP to extracted filled copolymer and was characteristic for deformation vibrations of C-H group of the carbon chain [42,53]. Moreover, the IR spectrum of the metal-filled copolymer showed a significant decrease in the intensity of the characteristic PVP band in the range of 1372 cm^{-1}, which indicated wagging vibrations of the CH_2-CH bond of polyvinylpyrrolidone, proving the PVP tertiary carbon participation in the grafting reaction (Scheme 1) [42].

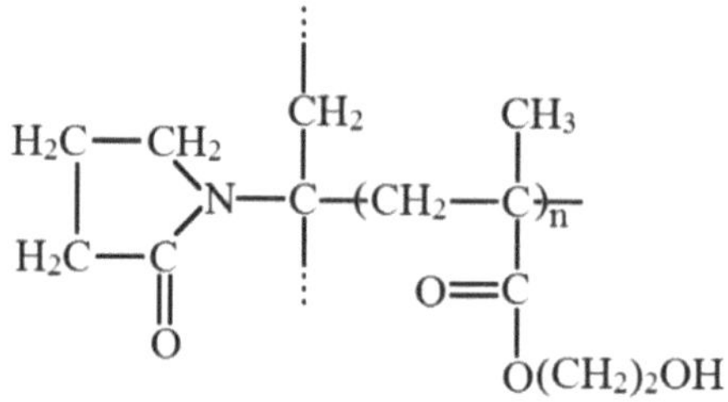

Scheme 1. The structure of pHEMA-gr-PVP copolymer [42].

The group -C-N= of pyrrolidone cycle was characterized by a significant deviation in the direction of vibrating frequencies decrease of its characteristic bands (from 1286 cm^{-1} to 1260 cm^{-1}), which confirmed its direct participation in physical interaction with FMF particles. This change indicated a weakening of the C-H bond, due to the enhancement of the mesomeric effect by the metal (Scheme 2):

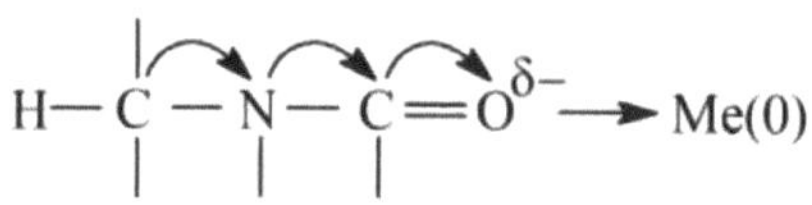

Scheme 2. Weakening of the hydrogen bond with the tertiary carbon atom of the PVP main chain.

Previous studies [38,42] have shown that during the polymerization of HEMA with PVP, grafted spatially crosslinked copolymers are formed. In our case, polymerization system was hetero-phasic, where the solid phase FMF was characterized by a certain activity that affected the polymerization process (Figure 1) and, accordingly, the formation of the polymer matrix structure. Additional confirmation of the formation of the grafted copolymer were the analysis results of the aqueous extract from obtained composite. Due to the fact that not all PVP underwent the grafting reaction, there was a discrepancy between the PVP content in the original composition and in the copolymer composition [38,42]. In order to determine the effect of FMF on the amount of PVP involved in the grafting reaction, the following parameters were calculated: the efficiency of PVP grafting (f_{PVP}, %) and its content in the copolymer (C_{PVP}, wt.%). To characterize the polymer network, the degree of its crosslinking (ν, mol/kg) was determined, which was evaluated by the molecular weight between crosslinks in the polymer network (Mc, kg/mol) (Figure 3).

For comparison, the structural parameters of the polymer network for unfilled copolymer obtained by copolymerization of HEMA with PVP in the presence of Fe^{2+} ions were also presented [42]. The presence of metal particles is the reason for the formation of copolymers with a higher degree of crosslinking, which indicates the formation of additional spatial network, due to the formation of physical nodes, the elements of which can be metal particles. The increase of metal particles content in the original composition led to changes in the efficiency of PVP grafting and its content in the copolymer by extreme dependence (Figure 4a).

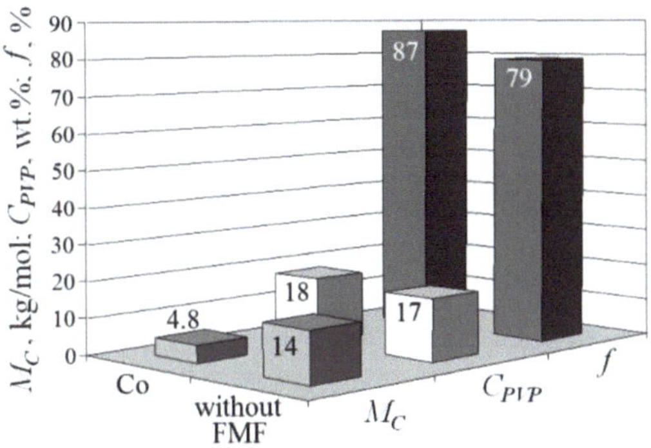

Figure 3. The effect of FMF on structural parameters of pHEMA-gr-PVP copolymers (HEMA:PVP = 80:20 mass parts, [Co] = 10 wt.%): grafting efficiency (f_{PVP}), PVP content (C_{PVP}) and molecular weight between crosslinks in polymer network (M_C).

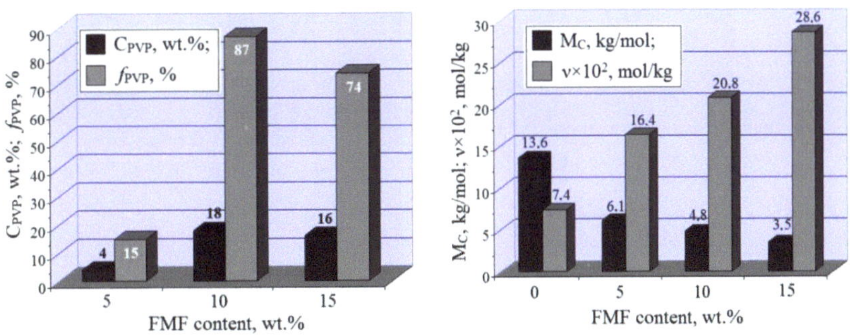

(a) (b)

Figure 4. The effect of cobalt powder content on structural parameters of pHEMA-gr-PVP copolymers (HEMA:PVP = 80:20 mass parts, [FMF] = 10 wt.%): (**a**) grafting efficiency (f_{PVP}) and PVP content (C_{PVP}), (**b**) molecular weight between crosslinks in polymer network (M_C) and the degree of crosslinking (ν).

At the same time, the M_C decreased (Figure 4b), i.e., the degree of crosslinking of the polymer network increased, which could be interpreted as the participation of metal particles in the formation of additional physical fluctuation nodes. The presented results confirm the effect of FMF on the course of formation of the polymer matrix structure. However, this effect can occur only on the surface of metal particles, which is a consequence of the formation of a surface-active layer other than the volume of the reaction mass. Obviously, this is the reason for the difference in the rate and nature of the formation of the structure of the copolymer in the surface layer of the metal particle and in the volume of the composition. This phenomenon can be confirmed by the data of a dynamic mechanical thermal analysis (DMTA). DMTA is especially relevant for polymers filled with metal particles, as it characterizes the effect of the filler on the change in the mobility of macromolecules in the boundary surface layers. At the interface between the phases of polymer and metal particle, the nature of the relaxation processes changes, because, due to adsorption phenomena in the boundary layers of filled polymer systems, changes occur in the molecular mobility of system components. The results of the DMTA are presented in the form of curves of temperature dependence of the tangent of an angle of mechanical loss ($tg\delta$) and the storage modulus (G', Pa) in Figure 5.

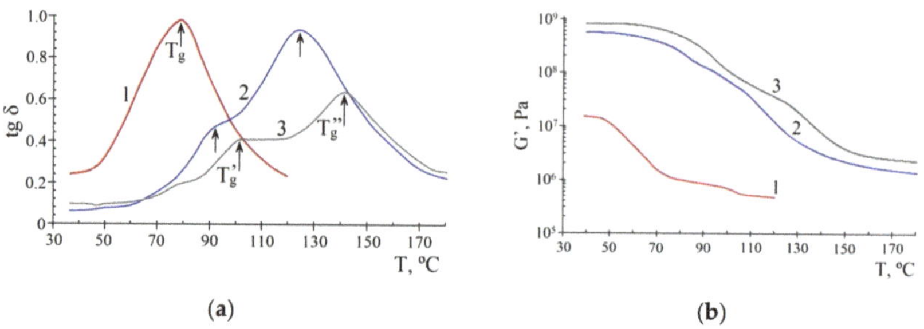

Figure 5. The effect of FMF on the tangent of an angle of mechanical losses $tg\delta$ (**a**) and the storage modulus G' (**b**) of pHEMA-gr-PVP copolymers (HEMA:PVP = 80:20 mass parts; [Me] = 10 wt.%; T_g'—Glass transition temperature in volume, T_g''—Glass transition temperature at the phase boundary): 1—Unfilled copolymer; 2—Co; 3—Ni.

The temperature transitions of filled copolymers (Figure 5, curves 2, 3), compared with unfilled (Figure 5, curve 1), were significantly shifted toward higher temperatures, which was characteristic of both changes in $tg\delta$ and changes in G'. The glass transition temperature (T_g) is the limit of the physical transition of the polymer phase. The value of T_g of the polymer corresponded to the peak on the dependence curves $tg\delta = f(T)$ (Figure 5a). A characteristic feature of the studied materials was the presence of two maxima on the temperature dependence of the tangent of an angle of mechanical losses on temperature, which indicated the existence of two areas, where copolymer with different structure was formed, namely in the interfacial layer on the surface of the metal particle and in the volume (Figure 6).

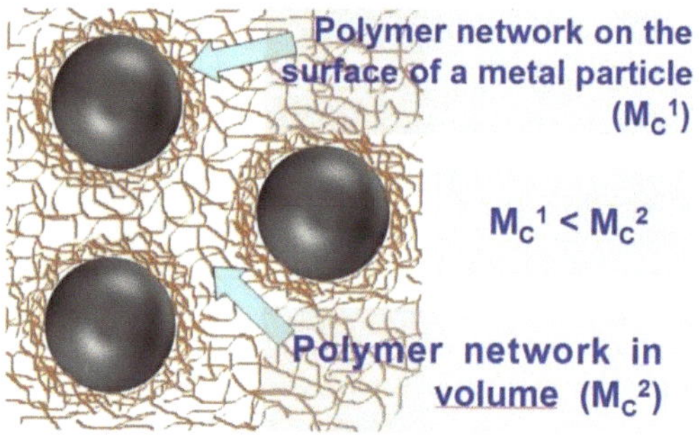

Figure 6. Scheme of polymer matrix structure formation in the presence of a metal surface.

The obtained dependences provided the possibility of comparing the changes in the tangent of an angle of mechanical losses (Figure 5a) and the storage modulus (Figure 5b) of copolymers and to draw a conclusion about the effect of metals on the formation of the polymer matrix. It was found (Figure 1) that the polymerization rate of HEMA/PVP compositions in the presence of Co powder was slightly higher compared to the composition filled with Ni powder. The same regularity was observed for the changes in the intensity of relaxation peaks on the dependences $tg\delta = f(T)$. Therefore, it followed that, with increasing activity of metal in polymerization, molecular mobility and the ability to deform the copolymer increased as well. Meanwhile, a decrease of G' indicated an increase in the resilience of the system.

3.2. Properties of FMF/HEMA-gr-PVP Composites

3.2.1. Sorption Capacity of FMF-Filled pHEMA-gr-PVP Copolymers

One of the important technological characteristics of materials based on pHEMA-gr-PVP copolymers is the ability to adsorb water and other solvents, as well as to dissolve low molecular weight substances in them. The processes of interaction of polymers with low molecular weight liquids are of great importance during the synthesis of polymers, their processing and operation in various liquid media. The study of this interaction is especially important for hydrogels, being materials formed by swelling of hydrophilic copolymers during the sorption of the solvent to equilibrium. PHEMA-gr-PVP copolymers have a spatially crosslinked structure formed by blocks of HEMA grafted on PVP, and contain hydrophilic groups: hydroxyl of methacrylate and carbonyl of pyrrolidone cycle in PVP [42]. Specifically, this structure provides the sorption capacity of pHEMA-gr-PVP copolymers. Sorption capacity was characterized by swelling kinetics curves (Figure 7).

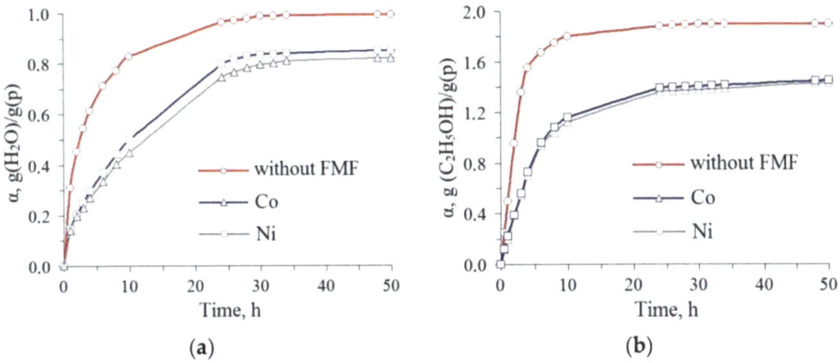

Figure 7. The effect of the presence of FMF in the structure of pHEMA-gr-PVP copolymers on their swelling ability (HEMA:PVP = 80:20 mass parts, [FMF] = 10 wt.%): (**a**) in water; (**b**) in ethanol.

According to the research results, the introduction of FMF particles into the copolymer structure in any case led to a deterioration of the sorption capacity (Figure 7a). The shape of swelling kinetic curves is mainly affected by the structure of the polymer matrix. Since FMF particles act as additional crosslinking nodes, this led to a decrease of M_c (Figure 3), i.e., an increase in the crosslinking density of the polymer network, and, accordingly, the deterioration of the swelling ability. The obtained curves for FMF/pHEMA-gr-PVP composites indicated that the equilibrium degree of swelling in distilled water was achieved for all samples over the same time. At the same time, the difference in the limit values of the degree of swelling indicated the role of the nature of the metal in the polymer matrix structure formation.

To determine the sorption capacity of metal-filled hydrogels relative to low molecular weight substances, the behavior of the obtained materials in the medium of a less polar (relative to water) solvent, ethyl alcohol, was studied (Figure 7b). The sharp increase in the mass of samples in the ethyl alcohol medium could be explained by the intense diffusion of the latter and the higher entropy of its interaction with macro-chains. This was the result of the higher energy of the interaction of alcohol with the matrix, due to which there was a straightening of the chains of the polymer matrix with increasing distance between their individual fragments. In addition, it should be taken into account that the interaction of alcohol molecules occurs with polar groups of the copolymer such as C=O, –OH. The interaction of ethanol molecules with macromolecules of the polymer network was accompanied by blocking of polar groups in the polymer units and, consequently, distancing of network chains among themselves, which, naturally, would favor the increase of the diffusion of ethanol-dissolved low molecular weight compounds in the hydrogel volume with increasing sorption value. Together with solvents, HEMA with PVP copolymers can adsorb dissolved inorganic and organic substances in them [54,55]. In particular, the possibility of sorption of synthetic dyes of organic origin by pHEMA-gr-PVP copolymers was proved in the example of methyl red (Figure 8).

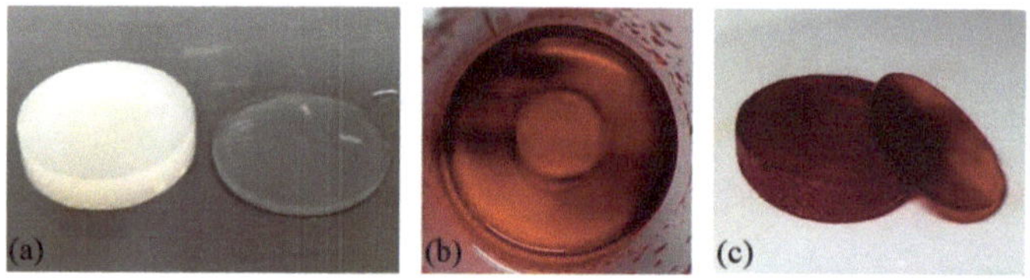

Figure 8. Demonstration of the possibility of sorption of organic dye by pHEMA-gr-PVP copolymer: (**a**) copolymer swollen in water; (**b**) the process of swelling of pHEMA-gr-PVP copolymer in dye solution; (**c**) pHEMA-gr-PVP copolymer with adsorbed dye.

For clarity, unfilled samples of pHEMA-gr-PVP copolymers were used. Thus, the obtained FMF-filled hydrogels could be used as sorbents of inorganic and organic substances.

3.2.2. Influence of Metallic Filler on Thermophysical Properties of pHEMA-gr-PVP Copolymers

Temperature intervals of phase and physical states determine the mechanical properties and, accordingly, the field of practical use of synthesized materials. Heat resistance of metal-filled composites based on pHEMA-gr-PVP copolymers was evaluated by Vicat softening temperature (T_V, °C), phase transitions, using thermomechanical curves (TMCs), which represent the dependence of polymer deformation on temperature (Figure 9).

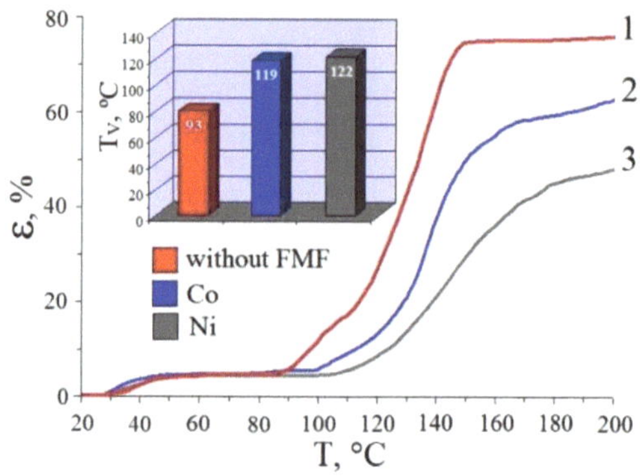

Figure 9. The effect of metal filler on temperature dependence of deformation (ε) and Vicat softening temperature (T_V) of pHEMA-gr-PVP copolymers (HEMA:PVP = 80:20 mass parts, [FMF] = 10 wt.%).

The initial section of the TMCs corresponds to the temperature region of the glassy state, where the energy of thermal motion of macromolecules is insufficient for the manifestation of deformation under load. Accordingly, for polymers only small reversible deformations at low temperatures (up to the glass transition temperature) in the glassy state are observed. The region of the curves characterized by a sharp increase in deformation in a narrow temperature range, corresponds to the structural transition of the system from a glassy state to a rubbery state. The addition of FMF to the copolymer helped to shift the glass transition temperature to the region of higher temperatures (curves 2, 3) compared to the unfilled pHEMA-gr-PVP copolymer (curve 1). During the introduction of metals into the composition, the degree of deformation of the samples decreased. Also, for each test sample there was no transition of the copolymer to a melt state, which was a confirmation of the formation of a spatially crosslinked copolymer. The addition of a metal filler to the copolymer, compared to the unfilled material, was accompanied by an increase in heat resistance (Figure 9).

3.2.3. Magnetic Properties of FMF/pHEMA-gr-PVP Composites

Sorption-capable magnetic composites and hydrogels based on them with samarium-cobalt ($SmCo_5$) ferromagnet powder were obtained by polymerization with simultaneous filling of HEMA/PVP compositions (Figure 10).

Figure 10. Demonstration of the hydrogel composite interaction with a constant magnet: (**a**) swollen sample based on SmCo5/HEMA-gr-PVP copolymer; (**b**) interaction of the sample with a constant magnet; (**c**) demonstration of removal of FMF-filled hydrogel from a liquid medium under the action of a constant magnet.

$SmCo_5$ was chosen as a filler due to the great attention paid to it in modern science and practice [56,57], namely due to its high coercive force, as well as resistance to corrosion and oxidation, which allows the use of composites based on it in humid and aggressive environments. It was found that $SmCo_5$/HEMA-gr-PVP composites possessed magnetic properties after band magnetization using profile inductors. In order to characterize the magnetic properties of the obtained materials, their magnetization curves were studied (Figure 11), from which the value of magnetization (σ, A·m^2·kg^{-1}) and coercive force (H, kA·m^{-1}) were determined.

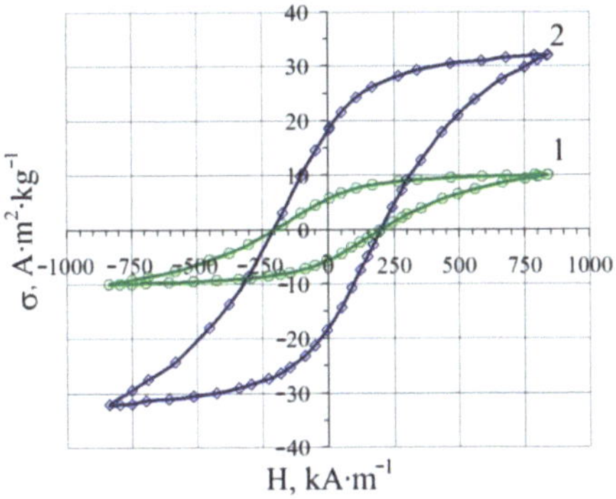

Figure 11. Curves of magnetization of $SmCo_5$/HEMA-gr-PVP composites with different content of $SmCo_5$ (Composition formulation, mass parts: HEMA:PVP = 70:30): 1—20 wt.%.; 2—80 wt.%.

Depending on the content of samarium-cobalt filler, hydrogel magnets were characterized by a coercive force of 200 kA $\times$ m^{-1} and induction of a magnetic field at the poles of 4–5 mT and 10–15 mT, respectively. Apparently, the particles of the ferromagnet after their introduction into the polymer matrix retained their magnetic properties. The obtained composites were characterized by the typical behavior of magnets. Filling hydrogels based on pHEMA-gr-PVP composites with magnetic fillers makes them suitable for obtaining a variety of magnetically sensitive materials with specific characteristics. For example, the combination of the magnetic properties of the filler and the sorption capacity of the hydrogel matrix is a prerequisite for the creation of currently relevant new materials, magnetic sorbents. One of the main problems associated with the usage of these kinds of adsorbents for environmental application and waste water cleaning is the removal of the adsorbents after completion of their tasks [58].

Figure 10c shows the result of the interaction of the filled hydrogel with a constant magnet. The magnetic properties of the composite ensured its unimpeded removal from the liquid medium for its regeneration and reuse.

3.2.4. Electrically Conductive Properties of FMF/pHEMA-gr-PVP Composites

The introduction of fine FMF particles in polymers, along with magnetic properties, provides them with electrical conductivity. Of particular interest to the pHEMA-gr-PVP matrix, as the object of filling, are its hydrophilicity and sorption capacity. The consequence of the synergism of the properties of the hydrophilic polymer matrix and electrically conductive filler is the ability to change the conductive characteristics of the composite depending on external factors, such as the degree of swelling, sorption of low molecular weight substances, temperature, pressure, humidity, and pH medium. The specific volumetric electrical resistance (ρ_V, Ohm·m) of FMF/pHEMA-gr-PVP composites, depending on the nature of FMF particles, was investigated (Figure 12a).

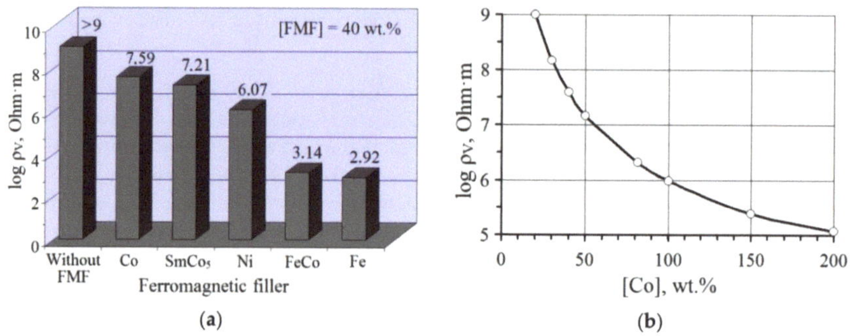

Figure 12. The effect of nature (**a**) and content (**b**) of FMF on electrically conductive characteristics of dry FMF/pHEMA-gr-PVP composites (HEMA:PVP = 70:30 mass parts).

It was established that the electrical resistance of unfilled copolymers obtained by mass polymerization was characteristic for dielectrics. The appearance of electrically conductive properties in Co/pHEMA-gr-PVP composites was observed after the introduction of the filler in the amount of 20–30 wt.% into the structure of the copolymer (Figure 12b). However, as can be seen, ρ_v remained quite high, even with the introduction of Co powder more than 100 wt.%. To significantly increase the electrical conductivity of polymeric FMF/pHEMA-gr-PVP composites filled with a dispersed filler, it was necessary to approach the filler particles together to direct contact or to a distance around several angstroms. In both the first and second cases, the path for the flow of electric current would be in the form of a continuous chain of contacts of sufficiently close filler particles [59]. Therefore, to increase the electrical conductivity, i.e., increase the number of contacts, it was necessary to introduce a larger amount of filler into the composition formulation. However, this method was not always technological, in that, as the weight of the sample increased, its strength characteristics deteriorated. It is also important to note that as the filler content increased, the viscosity of the original composition increased as well. For each filler there is a limit content in the composition, at which the composition still retains fluidity and, accordingly, the ability to form. With a further increase of filler content, the composition becomes non-technological. In this regard, it is of great interest to artificially obtain chain structures of electrically conductive filler in order to increase the conductive properties of polymer composites. In order to increase the electrical conductivity of FMF/pHEMA-gr-PVP composites with relatively small amounts of FMF, it was proposed to structure the filler in the composition by orienting the particles during polymerization in a constant magnetic field (Figure 13a). In a magnetic field, metal particles received an inductive magnetic moment, then, between neighboring magnetized particles, the interaction occurred and due to it they fit into continuous chains (Figure 13b).

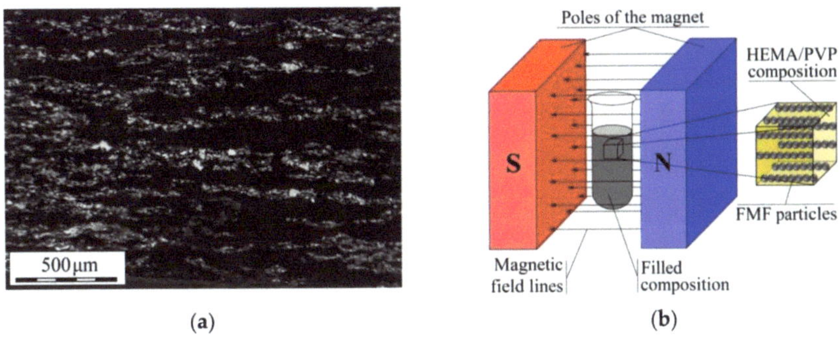

Figure 13. SEM micrographs of Co-filled pHEMA-gr-PVP copolymer with uniaxially ordered particles (**a**) and obtainment process scheme of a composite hydrogel in a magnetic field (**b**).

In Figure 14a, the results of determining the specific resistivity ρ_V for composites filled with FMF of different nature during polymerization in a magnetic field are presented. The electrical resistance of FMF/pHEMA-gr-PVP composites was significantly lower in every case, compared to unfilled pHEMA-gr-PVP copolymer, for which $\rho_V > 10^9$ Ohms.

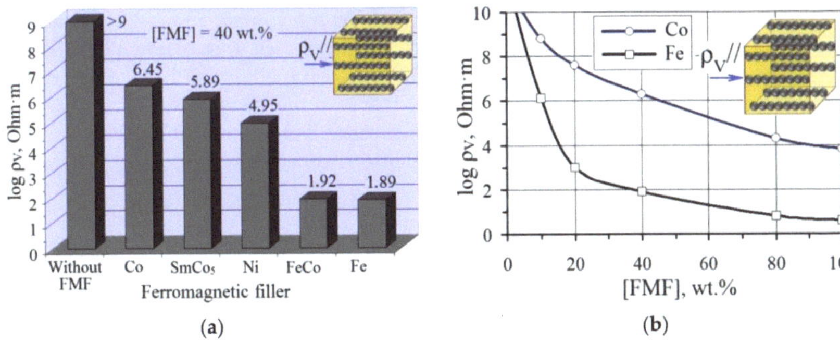

Figure 14. The effect of nature (**a**) and content (**b**) of FMF on volumetric electrical resistivity (ρ_V) of FMF/HEMA-gr-PVP composites obtained in a magnetic field (HEMA:PVP = 70:30 mass parts, H = 23 kA/m).

At the same time, the composites obtained in this way were characterized by anisotropy of electrical conductivity, so that in the direction of the magnetic field lines, $\rho_V//$ was minimal, but in the perpendicular direction, the resistance ($\rho_V\perp$) of composites with a filler content of 40 wt.% was greater than 10^9 Ohms. The specific electrical resistivity of FMF/HEMA-gr-PVP composites obtained without a magnetic field (with statistical distribution of metal particles) was greater by an order (Figure 12) than ρ_V materials with FMF particles oriented in a magnetic field. In Figure 14b, the results of studies characterizing the influence of FMF content on the specific resistivity value of composites, measured in the direction of chains alignment of FMF particles, are presented.

FMF particles under the action of a magnetic field are lined up in chains and interaction between them creates conditions for the passage of electric current. The nature of this interaction will determine the number of contacts between the filler particles, the contact strength and the thickness of the polymer film between the particles. Therefore, it could be predicted that the electrical conductivity of the composite would depend not only on the nature of the metal (magnetic and electrically conductive properties), but also on the magnetic field strength during the polymerization of the original compositions (Figure 15a).

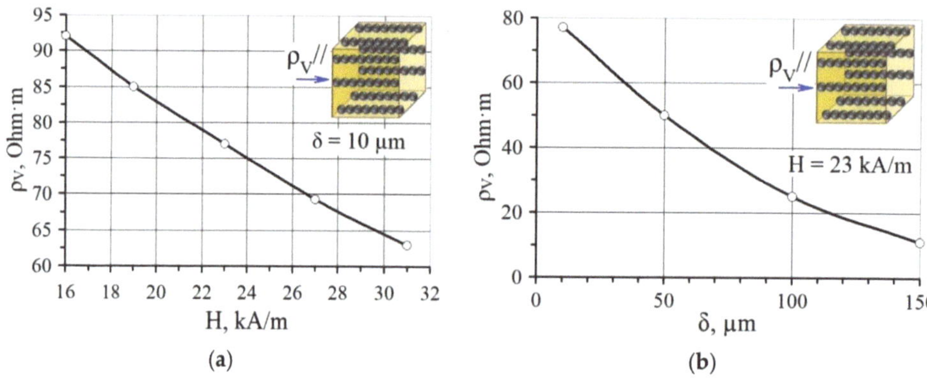

Figure 15. The effect of magnetic field strength (**a**) and FMF particle size (**b**) on resistivity of Fe/HEMA-gr-PVP composites (HEMA:PVP = 70:30 mass parts, [Fe] = 40 wt.%).

If in the case of the distribution of FMF particles in the polymer matrix according to the statistical structure, the electrical conductivity is determined mainly by the number of particles involved in creating contacts, then in the oriented structure the main factor affecting conductivity is the contact resistance between neighboring particles and the number of contacts in a chain [60]. It is a known fact that an increase in particle size causes an increase in electrical conductivity [59]. As the particle size increased, the number of contacts decreased, and, at the same time, the effect on the conductivity of the dielectric polymer matrix decreased and the contact force between the particles increased under the action of the magnetic field, which was a prerequisite for reducing the ρ_V of the composite (Figure 15b).

3.2.5. Physico-Mechanical Properties

On the basis of obtained results from DMTA (Figure 5) and TMC (Figure 9) it was already possible to draw a conclusion about the significant effect of FMF on the viscoelastic and thermophysical properties of pHEMA-gr-PVP copolymers. Since the obtained materials could be operated in two physical states, glassy (hard) and rubbery (soft), so the properties were studied in both dry and hydrated samples. The influence of fine metal filler on the formation of the structure and properties of pHEMA-gr-PVP composites is determined by two factors. The first relates to the physical action on the structure of the polymer. In this regard, the role of the metal particle is, in fact, an element of hetero-phase structure. The second is the result of changes in the polymer matrix, which happen due to interaction at the interface between polymer–metal and the participation of the particle surface in the formation of chemical composition and structure of the polymer, which is manifested by the influence of filler on the grafting efficiency of PVP and M_C (Figures 3 and 4). The overall change in the properties of the filled system, compared with the polymer matrix, takes place due to the simultaneous action of the two factors. On the one hand, during the introduction of FMF, M_C decreased (Figure 3), because the particles played the role of additional nodes of physical crosslinking. On the other hand, the presence of metals affected the efficiency of PVP grafting (Figure 3), the content of which, in the structure of the copolymer, affected its physico-mechanical and thermophysical characteristics [42].

Metal-filled pHEMA-gr-PVP copolymers in the dry state were studied for surface hardness (F, MPa) and swelling ability, which was characterized by water content (W, %) and swelling coefficient (k) (Figure 16a). Copolymers, swollen in water to equilibrium, were studied as elastomers for hardness, resilience and plasticity, which were characterized, respectively, by the hardness number (H, MPa), the elasticity index (E, %) and the plasticity number (P, %) (Figure 16b).

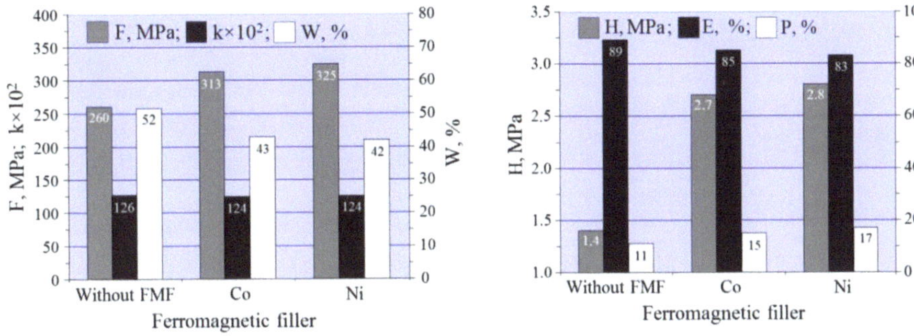

(**a**) (**b**)

Figure 16. The effect of FMF presence on the physico-mechanical properties of pHEMA-gr-PVP copolymers in dry (**a**) and swollen (**b**) states (HEMA:PVP = 80:20 mass parts; [FMF] = 10 wt.%).

The introduction of the metal filler into the original composition formulation, in any case, compared with the unfilled material, promoted the increase of its strength characteristics. As can be seen from the presented results, the surface hardness (*F*, MPa) of dry samples with a FMF content of only 10 wt.% increased by 20–25% (Figure 16a), and the hardness number (*H*, MPa) of the samples in the swollen state doubled (Figure 16b). Obviously, the increase in the strength characteristics of composites occurred due to the increase in the degree of crosslinking and the number of physical nodes in the polymer network (Figures 3 and 4b). The influence of the metal filler particles was associated with the formation of interfacial layers, the structure and properties of which differ from the characteristics of the polymer in volume (Figure 6). At the same time, individual filler particles, due to physical interaction with the components of the original polymer-monomer composition, can play the role of additional physical crosslinking nodes, compensating for the decrease in the concentration of chemical nodes. Naturally, in terms of formation of a more crosslinked polymer network, the sorption capacity of polymers decreased. The water content of composites (*W*, wt.%) decreased by 16–18% (Figure 16a). At the same time, as a result of the filling, the resilience (*E*, %) slightly decreased and the plasticity (*P*, %) increased (Figure 16b). However, the resilience properties of composites remained relatively high (Figure 17).

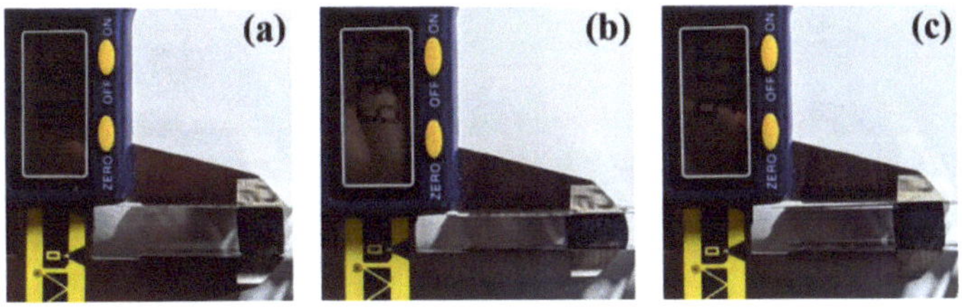

Figure 17. Demonstration of the resilience properties of an equilibrium swollen sample of Co-filled pHEMA-gr-PVP copolymer: (**a**) initial sample; (**b**) sample under load; (**c**) sample after unloading.

During the study of the swelling process of samples of metal-filled copolymers obtained in a magnetic field (with an ordered arrangement of filler particles), the anisotropy of the swelling coefficient was detected (Figure 18).

It was established that the swelling coefficient ($k\perp$ = 1.28) in the perpendicular direction to the filler chains (in the direction of the action lines of magnetic induction forces) was greater than the swelling coefficient in the direction of aligning chains of filler particles ($k//$ = 1.17).

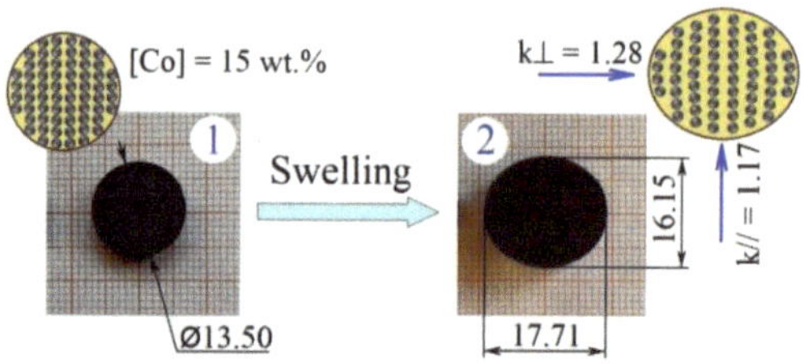

Figure 18. Demonstration of swelling anisotropy of a sample of Co-filled pHEMA-gr-PVP copolymer obtained by polymerization filling in a magnetic field: 1—Dry sample; 2—Equilibrium swollen sample.

4. Conclusions

Composite copolymers of HEMA with PVP, and hydrogels based on them, modified with ferromagnetic fillers (Ni, Co, Fe, FeCo, $SmCo_5$), were obtained by the method of filling during polymerization.

1. It was found that the polymerization of FMF-filled compositions occurred without additional catalysts, which is the evidence of the participation of the FMF surface in initiating the polymerization process of HEMA/PVP compositions. The formation of grafted spatially crosslinked copolymer of pHEMA-gr-PVP was confirmed. The introduction of FMF particles into the original composition formulation caused the formation of copolymers with a greater degree of crosslinking, which indicates the appearance of additional spatial network due to the formation of physical nodes, the elements of which may be metal particles. It was established that in the interfacial layer, on the filler surface and in the volume, a copolymer formed with a different structure.
2. Using the example of water and ethyl alcohol, it was shown that the obtained composites are able to swell in solvents. It was established that the introduction of FMF particles into the structure of pHEMA-gr-PVP copolymers in any case led to a deterioration of the sorption capacity. At the same time, the water content decreased by 16–18% and lay within 42–43% by mass, but remained relatively high. The sorption possibility of pHEMA-gr-PVP copolymers of synthetic dyes of organic origin was shown on the example of methyl red.
3. The filling of pHEMA-gr-PVP copolymers with micro-sized FMF powders increase their heat resistance (Vicat softening temperature is 39–42 °C higher compared to the unfilled material), as well as strength characteristics (in both dry and swollen states). The surface hardness of dry samples with a FMF content of only 10 wt.% increased by 20–25%, and the hardness number of samples in a swollen state is doubled. At the same time, resilience slightly deteriorated, and so plasticity increased.
4. FMF-filled pHEMA-gr-PVP copolymers, after band magnetization with profile inductors, possess magnetic properties. It was established that the obtained composites in dry state are characterized by a coercive force of 200 kA $\times$ m^{-1} and induction of a magnetic field at the poles of 4–5 mT and 10–15 mT, respectively.
5. The introduction of fine FMF particles into pHEMA-gr-PVP copolymers (which are dielectrics in the dry state) provides them with electrical conductivity. Depending on the nature and content of FMF, the resistance of filled materials can be regulated within 10^3–10^6 Ohm·m. The structuring of filler particles into chains using their orientation during polymerization in a constant magnetic field provided a 10-fold increase in the electrical conductivity of composites with relatively small amounts of FMF.
6. Depending on the need and the task in hand, the research results showed, while choosing the nature of the metal-filler, it is possible to obtain hydrogel composite materials with unique predictable properties, having electrically conductive, magnetic, or sorption activities and combinations of the same. The obtained composites were

not contaminated with other substances (by-products from the obtaining reaction of metal-filler), which also allowed the avoidance of an additional longtime stage of purification of the composite, washing. The high reactive capacity of the polymer-monomer composition, based on HEMA and PVP in the presence of $FeSO_4$, provides the ability to obtain hydrogel materials filled with FMF at a high rate using simple technology, at room temperature, in air, without the usage of sophisticated equipment and facilities. Preliminary results indicate the prospects for the use of the obtained materials, for example, as magnetic sorbents for various applications, as well as electrically conductive materials that may respond to changes in conductivity from various factors (e.g., moisture content, differences in pressure, temperature).

Author Contributions: Conceptualization, O.G., L.D., E.S. and B.B.; methodology, O.G., B.B. and L.D.; validation, O.G., L.D., E.S. and B.B.; formal analysis, O.G., L.D. and E.S.; investigation, O.G. and B.B.; resources, O.G., L.D.; data curation, O.G., L.D. and E.S.; writing—original draft preparation, O.G., L.D., E.S. and B.B.; writing—review and editing, O.G., L.D., E.S. and B.B.; visualization, O.G. and L.D.; supervision, O.G., L.D. and E.S.; project administration, O.G. and L.D.; funding acquisition, L.D. and E.S. All authors have read and agreed to the published version of the manuscript.

Funding: This research received no external funding and the APC was funded by grant numbers VEGA 1/0384/20 and 036TUKE-4/2021 granted by The Ministry of Education, Science, Research and Sport of the Slovak Republic.

Institutional Review Board Statement: Not applicable.

Informed Consent Statement: Not applicable.

Data Availability Statement: Not applicable.

Acknowledgments: The emergence of this article was supported by grant numbers VEGA 1/0384/20 and 036TUKE-4/2021.

Conflicts of Interest: The authors declare no conflict of interest.

References

1. Thomas, V.; Namdeo, M.; Murali Mohan, Y.; Bajpai, S.K.; Bajpai, M. Review on polymer, hydrogel and microgel metal nanocomposites: A facile nanotechnological approach. *J. Macromol. Sci.* **2007**, *45*, 107–119. [CrossRef]
2. Kucherenko, A.; Moravskyi, V.; Kuznetsova, M.; Grytsenko, O.; Masyuk, A.; Dulebova, L. Regularities of obtaining metal-filled polymer composites. In *Nanomaterials in Biomedical Application and Biosensors (NAP-2019)*; Pogrebnjak, A., Pogorielov, M., Viter, R., Eds.; Springer Proceedings in Physics; Springer: Singapore, 2020; Volume 244, pp. 59–66. [CrossRef]
3. Hanemann, T.; Szabó, D.V. Polymer-nanoparticle composites: From synthesis to modern applications. *Materials* **2010**, *3*, 3468–3517. [CrossRef]
4. Li, H.; Yang, P.; Pageni, P.; Tang, C. Recent advances in metal-containing polymer hydrogels. *Macromol. Rapid Commun.* **2017**, *38*, 1700109. [CrossRef] [PubMed]
5. Madduma-Bandarage, U.S.K.; Madihally, S.V. Synthetic hydrogels: Synthesis, novel trends, and applications. *J. App. Polym. Sci.* **2021**, *138*, e50376. [CrossRef]
6. Nosova, N.G.; Samaryk, V.J.; Varvarenko, S.M.; Ferens, M.V.; Voronovska, A.V.; Nagornyak, M.I.; Khomyak, S.V.; Nadashkevych, Z.J.; Voronov, S.A. Porous polyacrylamide hydrogels: Preparation and properties. *Vopr. Khimii Khimicheskoi Tekhnologii.* **2016**, *5–6*, 78–86.
7. Jumadilov, T.; Abilov, Z.; Grazulevicius, J.; Zhunusbekova, N.; Kondaurov, R.; Agibayeva, L.; Akimov, A. Mutual activation and sorption ability of rare cross-linked networks in intergel system based on polymethacrylic acid and poly-4-vinylpyridine hydrogels in relation to lanthanum ions. *Chem. Chem. Technol.* **2017**, *11*, 188–194. [CrossRef]
8. Echeverria, C.; Fernandes, S.N.; Godinho, M.H.; Borges, J.P.; Soares, P.I.P. Functional stimuli-responsive gels: Hydrogels and microgels. *Gels* **2018**, *4*, 54. [CrossRef]
9. Tarnavchyk, I.; Voronov, A.; Kohut, A.; Nosova, N.; Varvarenko, S.; Samaryk, V.; Voronov, S. Reactive hydrogel networks for the fabrication of metal-polymer nanocomposites. *Macromol. Rapid Commun.* **2009**, *30*, 1564–1569. [CrossRef]
10. Burkeev, M.Z.; Shibayeva, S.R.; Khamitova, T.O.; Plocek, J.; Tazhbayev, Y.M.; Davrenbekov, S.Z.; Nurmaganbetova, M.T.; Kazhmuratova, A.T.; Zhumagalieva, T.S.; Kezdikbayeva, A.T. Synthesis and catalytic properties of new polymeric monometallic composites based on copolymers of polypropylene glycol maleate phthalate with acrylic acid. *Polymers* **2021**, *13*, 4369. [CrossRef]
11. Zhang, J.; Huang, Q.; Du, J. Recent advances in magnetic hydrogels. *Polym. Int.* **2016**, *65*, 1365–1372. [CrossRef]
12. Shah, L.A.; Ambreen, J.; Bibi, I.; Sayed, M.; Siddiq, M. Silver nanoparticles fabricated hybrid microgels for optical and catalytic study. *J. Chem. Soc. Pak.* **2016**, *38*, 850–858.

13. Abdel-Halim, E.S.; Al-Deyab, S.S. Electrically conducting silver/guar gum/poly(acrylic acid) nanocomposite. *Int. J. Biol. Macromol.* **2014**, *69*, 456–463. [CrossRef]
14. Grytsenko, O.; Pukach, P.; Suberlyak, O.; Shakhovska, N.; Karovič, V. Usage of mathematical modeling and optimization in development of hydrogel medical dressings production. *Electronics* **2021**, *10*, 620. [CrossRef]
15. Chen, T.; Hou, K.; Ren, Q.; Chen, G.; Wei, P.; Zhu, M. Nanoparticle-polymer synergies in nanocomposite hydrogels: From design to application. *Macromol. Rapid Commun.* **2018**, *39*, 1800337. [CrossRef]
16. Filipcsei, G.; Csetneki, I.; Szilágyi, A.; Zrínyi, M. Magnetic field-responsive smart polymer composites. In *Oligomers-Polymer Composites-Molecular Imprinting. Advances in Polymer Science*; Springer: Berlin/Heidelberg, Germany, 2007; Volume 206, pp. 137–189. [CrossRef]
17. Yin, C.S.; Cheung, P.Y.; Kohler, A.C.; Compton, J.T.; Hsu, L.L.; Lau, K.M.; Kim, S.; Lee, B.W.; Lee, F.Y.; Sia, S.K. Additive manufacturing of hydrogel-based materials for next-generation implantable medical devices. *Sci. Robot.* **2017**, *2*, eaah6451. [CrossRef]
18. Grytsenko, O.; Gajdos, I.; Spišák, E.; Krasinskyi, V.; Suberlyak, O. Novel Ni/pHEMA-gr-PVP composites obtained by polymerization with simultaneous metal deposition: Structure and properties. *Materials* **2019**, *12*, 1956. [CrossRef]
19. Ansari, T.M.; Ajmal, M.; Saeed, S.; Naeem, H.; Ahmad, H.B.; Mahmood, K.; Farooqi, Z.H. Synthesis and characterization of magnetic poly(acrylic acid) hydrogel fabricated with cobalt nanoparticles for adsorption and catalytic applications. *J. Iran. Chem. Soc.* **2019**, *16*, 2765–2776. [CrossRef]
20. Kharissova, O.V.; Rasika Dias, H.V.; Kharisov, B.I. Magnetic adsorbents based on micro- and nano-structured materials. *RSC Adv.* **2015**, *5*, 6695–6719. [CrossRef]
21. Zulfiqar; Afzal, S.; Khan, R.; Zeb, T.; Rahman, M.; Burhanullah; Ali, S.; Khan, G.; Rahman, Z.; Hussain, A. Structural, optical, dielectric and magnetic properties of PVP coated magnetite (Fe_3O_4) nanoparticles. *J. Mater. Sci. Mater. Electron.* **2018**, *29*, 20040–20050. [CrossRef]
22. Li, Y.; Huang, G.; Zhang, X.; Li, B.; Chen, Y.; Lu, T.; Lu, T.J.; Xu, F. Magnetic hydrogels and their potential biomedical applications. *Adv. Funct. Mater.* **2013**, *23*, 660–672. [CrossRef]
23. Zasońska, B.; Brož, A.; Šlouf, M.; Hodan, J.; Petrovský, E.; Hlídková, H.; Horák, D. Magnetic superporous poly(2-hydroxyethyl methacrylate) hydrogel scaffolds for bone tissue engineering. *Polymers* **2021**, *13*, 1871. [CrossRef]
24. Senapati, S.; Mahanta, A.K.; Kumar, S.; Maiti, P. Controlled drug delivery vehicles for cancer treatment and their performance. *Signal Transduct. Target. Ther.* **2018**, *3*, 7. [CrossRef]
25. Meenach, S.A.; Hilt, J.Z.; Anderson, K.W. Poly(ethylene glycol)-based magnetic hydrogel nanocomposites for hyperthermia cancer therapy. *Acta Biomater.* **2010**, *6*, 1039–1046. [CrossRef]
26. Mun, S.J.; Ko, D.; Kim, H.U.; Han, Y.; Roh, Y.H.; Kim, B.-G.; Na, H.B.; Bong, K.W. Photopolymerization-based synthesis of uniform magnetic hydrogels and colorimetric glucose detection. *Materials* **2020**, *13*, 4401. [CrossRef]
27. Cai, H.; Lu, P.; Dong, J. Robust nickel–polymer nanocomposite particles for hydrogen generation from sodium borohydride. *Fuel* **2016**, *166*, 297–301. [CrossRef]
28. Ajmal, M.; Aftab, F.; Bibi, I.; Iqbal, M.; Ambreen, J.; Ahmad, H.B.; Akhtar, N.; Haleem, A.; Siddiq, M. Facile synthesis of porous anionic hydrogel embedded with nickel nanoparticles and evaluation of its catalytic performance for the rapid reduction of 4-nitrophenol. *J. Porous Mater.* **2019**, *26*, 281–290. [CrossRef]
29. Philippova, O.; Barabanova, A.; Molchanov, V.; Khokhlov, A. Magnetic polymer beads: Recent trends and developments in synthetic design and applications. *Eur. Polym. J.* **2011**, *47*, 542–559. [CrossRef]
30. Grytsenko, O.M.; Naumenko, O.P.; Suberlyak, O.V.; Dulebova, L.; Berezhnyy, B.V. Optimization of the technological parameters of the graft copolymerization of 2-hydroxyethyl methacrylate with polyvinylpyrrolidone for nickel deposition from salts. *Vopr. Khimii Khimicheskoi Tekhnologii* **2020**, *1*, 25–32. [CrossRef]
31. Yuan, Q.; Williams, R. Large scale manufacture of magnetic polymer particles using membranes and microfluidic devices. *China Particuology* **2007**, *5*, 26–42. [CrossRef]
32. Jean, I.; Baek, J. Nanocomposites derived from polymers and inorganic nanoparticles. *Materials* **2010**, *3*, 3654–3674. [CrossRef]
33. Rao, P.; Lo, I.; Yin, K.; Tang, S. Removal of natural organic matter by cationic hydrogel with magnetic properties. *J. Environ. Manag.* **2011**, *92*, 1690–1695. [CrossRef] [PubMed]
34. Farag, R.K.; Labena, A.; Fakhry, S.H.; Safwat, G.; Diab, A.; Atta, A.M. Antimicrobial activity of hybrids terpolymers based on magnetite hydrogel nanocomposites. *Materials* **2019**, *12*, 3604. [CrossRef] [PubMed]
35. Hu, K.; Sun, J.; Guo, Z.; Wang, P.; Chen, Q.; Ma, M.; Gu, N. A novel magnetic hydrogel with aligned magnetic colloidal assemblies showing controllable enhancement of magnetothermal effect in the presence of alternating magnetic field. *Adv. Mater.* **2015**, *27*, 2507–2514. [CrossRef] [PubMed]
36. Suberlyak, O.; Grytsenko, O.; Hischak, K.; Hnatchuk, N. Research of influence the nature of metal on mechanism of synthesis polyvinilpyrrolidone metal copolymers. *Chem. Chem. Technol.* **2013**, *7*, 289–294. [CrossRef]
37. Suberlyak, O.; Grytsenko, O.; Kochubei, V. The role of $FeSO_4$ in the obtaining of polyvinylpirolidone copolymers. *Chem. Chem. Technol.* **2015**, *9*, 429–434. [CrossRef]
38. Suberlyak, O.; Skorokhoda, V. Hydrogels based on polyvinylpyrrolidone copolymers. In *Hydrogels*; Haider, S., Haider, A., Eds.; IntechOpen: London, UK, 2018; pp. 136–214. [CrossRef]

39. Teodorescu, M.; Bercea, M. Poly(vinylpyrrolidone)—A versatile polymer for biomedical and beyond medical applications. *Polym. Plast. Technol. Eng.* **2015**, *54*, 923–943. [CrossRef]
40. Khromiak, U.; Levytskyi, V.; Stepova, K.; Tarnawsky, A. Synthesis and properties of adhesive polymer-methylmethacrylate materials. *Int. J. Polym. Sci.* **2018**, *2018*, 1–9. [CrossRef]
41. Suberlyak, O.V.; Levitskij, V.E.; Skorokhoda, V.Y.; Godij, A.B. Physical-chemical fenomena on phase boundary vinyl monomer-water solution of polyvinylpyrrolidone. *Ukr. Khimicheskij Zhurnal* **1998**, *64*, 122–125.
42. Grytsenko, O.; Dulebova, L.; Suberlyak, O.; Skorokhoda, V.; Spišák, E.; Gajdos, I. Features of structure and properties of pHEMA-gr-PVP block copolymers, obtained in the presence of Fe^{2+}. *Materials* **2020**, *13*, 4580. [CrossRef]
43. Krasinskyi, V.; Suberlyak, O.; Dulebová, L.; Antoniuk, V. Nanocomposites on the basis of thermoplastics and montmorillonite modified by polyvinylpyrrolidone. *Key Eng. Mater.* **2017**, *756*, 3–10. [CrossRef]
44. Xie, P.; Zhang, L.; Shen, H.; Wu, H.; Zhao, J.; Wang, S.; Hu, L. Biodegradable $MoSe_2$-polyvinylpyrrolidone nanoparticles with multi-enzyme activity for ameliorating acute pancreatitis. *J. Nanobiotechnol.* **2022**, *20*, 113. [CrossRef]
45. Levytskyi, V.Y.; Masyuk, A.S.; Suberlyak, O.V. Preparation and properties of polymer-silicate composites based on hydrophilic polymers. *Vopr. Khimii Khimicheskoi Tekhnologii* **2017**, *6*, 68–74.
46. Krasinskyi, V.; Suberlyak, O.; Zemke, V.; Klym, Y.; Gaidos, I. The role of polyvinylpyrrolidone in the formation of nanocomposites based on acompatible polycaproamide and polypropylene. *Chem. Chem. Technol.* **2019**, *13*, 59–63. [CrossRef]
47. Grytsenko, O.; Pukach, P.; Suberlyak, O.; Moravskyi, V.; Kovalchuk, R.; Berezhnyy, B. The Scheffe's method in the study of mathematical model of the polymeric hydrogels composite structures optimization. *Math. Modeling Comput.* **2019**, *6*, 258–267. [CrossRef]
48. Ciardelli, G.; Cristallini, C.; Barbani, N.; Benedetti, G.; Crociani, A.; Travison, L.; Giusti, P. Bioartificial polymeric materials: -amylase, poly(2-hydroxyethyl methacrylate), poly(N-vinylpyrrolidone) system. *Macromol. Chem. Phys.* **2002**, *203*, 1666–1673. [CrossRef]
49. Muller, K. Identification and determination of polyvinylpyrrolidone (PVP) as well as determination of active substances in PVP-containing drug preparations. *Pharm. Acta Helv.* **1968**, *43*, 107–122.
50. Flory, P.J.; Rehner, J. Statistical mechanics of cross-linked polymer networks. *J. Chem. Phys.* **1943**, *11*, 521–526. [CrossRef]
51. Wahid, F.; Zhong, C.; Wang, H.; Hu, X.; Chu, L. Recent advances in antimicrobial hydrogels containing metal ions and metals/metal oxide nanoparticles. *Polymers* **2017**, *9*, 636. [CrossRef]
52. Bühler, V. *Polyvinylpyrrolidone Excipients for Pharmaceuticals: Povidone, Crospovidone and Copovidone*; Springer: Berlin/Heidelberg, Germany, 2005; pp. 1–254. [CrossRef]
53. Krasinskyi, V.; Suberlyak, O.; Sikora, J.; Zemke, V. Nanocomposites based on polyamide-6 and montmorillonite intercalated with polyvinylpyrrolidone. *Polym.-Plast. Technol. Mater.* **2021**, *60*, 1641–1655. [CrossRef]
54. Suberlyak, O.; Melnyk, J.; Skorokhoda, V. Formation and properties of hydrogel membranes based on cross-linked copolymers of methacrylates and water-soluble polymers. *Eng. Biomater.* **2009**, *12*, 5–8.
55. Xu, J.; Zhao, P.; Zhang, Y. Iodine-responsive poly(HEMA-PVP) hydrogel for self-regulating burst-free extended release. *J. Biomater. Sci. Polym. Ed.* **2017**, *28*, 470–485. [CrossRef] [PubMed]
56. Tang, H.; Mamakhel, M.; Christensen, M. High coercivity $SmCo_5$ synthesized with assistance of colloidal SiO_2. *Sci. Rep.* **2021**, *11*, 4682. [CrossRef] [PubMed]
57. Zhang, D.; Zhu, R.; Yue, M.; Liu, W. Microstructure and magnetic properties of $SmCo_5$ sintered magnets. *Rare Met.* **2020**, *39*, 1295–1299. [CrossRef]
58. Ozay, O.; Ekici, S.; Baran, Y.; Aktas, N.; Sahiner, N. Removal of toxic metal ions with magnetic hydrogels. *Water Res.* **2009**, *43*, 4403–4411. [CrossRef]
59. Yaman, K. Fractal characterization of electrical conductivity and mechanical properties of copper particulate polyester matrix composites using image processing. *Polym. Bull.* **2022**, *79*, 3309–3332. [CrossRef]
60. Patsula, V.; Moskvin, M.; Dutz, S.; Horák, D. Size dependent magnetic properties of iron oxide nanoparticles. *J. Phys. Chem. Solids* **2003**, *257*, 24–40. [CrossRef]

MDPI AG
Grosspeteranlage 5
4052 Basel
Switzerland
Tel.: +41 61 683 77 34

Materials Editorial Office
E-mail: materials@mdpi.com
www.mdpi.com/journal/materials

www.ingramcontent.com/pod-product-compliance
Lightning Source LLC
Chambersburg PA
CBHW041354210926
51850CB00013B/3081